Speciality Inorganic Chemicals

Special Publication No 40

# Speciality Inorganic Chemicals

The Proceedings of a Symposium organised by the Inorganic Chemicals Group, the Fine Chemicals and Medicinals Group, and the N.W. Region of the Industrial Division of the Royal Society of Chemistry, in association with the Dalton Division

University of Salford, September 10th—12th 1980

Edited by
**R. Thompson**
Borax Consolidated Ltd

The Royal Society of Chemistry
Burlington House, London W1V 0BN

**British Library Cataloguing in Publication Data**

Speciality inorganic chemicals.—(Special publication/
Royal Society of Chemistry, ISSN 0260-6291; no. 40)
1. Chemistry, Inorganic—Congresses
I. Thompson, R. II. Royal Society of Chemistry
*Inorganic Chemicals Group*
III. Royal Society of Chemistry.
*Fine Chemicals and Medicinals Group*
IV. Royal Society of Chemistry. *Industrial Division.*
*N.W. Region*
546 QD151.2

ISBN 0-85186-835-5

Cover design by members of the Graphics and Advertising Section of the Salford College of Technology

Printed in Great Britain by
Whitstable Litho Ltd., Whitstable, Kent

# PREFACE

As a sequel to the symposium "The Modern Inorganic Chemicals Industry" held as part of the 1977 Annual Congress in London, the Inorganic Chemicals Group decided to organise a similar meeting to review some of the industrially important inorganic chemicals which it was not possible to include previously. On this occasion it was joined in sponsorship by Fine Chemicals and Medicinals, a parallel subject group of the Society's Industrial Division, and by its North-West Region. South Lancashire and the surrounding area was so much a birthplace of the country's inorganic chemicals industry that the venue for the symposium was a natural choice, and we are indebted to the University of Salford in whose Department of Chemistry and Applied Chemistry the lectures were held.

"The Modern Inorganic Chemicals Industry" was conceived as a book to meet a special need, and the 1977 meeting was built around it. There was no up-to-date textbook on the subject suitable (and suitably priced) for use in schools and universities. Contributors who were acknowledged as authorities in their respective fields were accordingly invited to write chapters for the book, and to deliver corresponding lectures at the symposium. Their typescripts, appropriately illustrated, provided the camera-ready copy for the book which became Special Publication No. 31. The first impression sold out within a year, and was immediately reprinted.

"Speciality Inorganic Chemicals" follows the same formula. A selection of industrial inorganic chemicals was made, and with it authors from Great Britain, Germany and the United States who are known authorities on the chosen topics. The organisers again considered the first essential was for the publication to be, and to be seen to be, *authoritative*. All chapters have been written by contributors who have had considerable experience with the industrial chemistry of the compounds they describe, through employment with companies or organisations recognised as being amongst the world leaders.

A succinct, all-embracing title for the symposium and book was difficult to phrase. Views differ as to what is a "speciality", as the opening chapter indicates. In the overall context of this book, "inorganic" qualifies speciality and it is taken to mean those inorganic chemicals which usually have *special* applications, as distinct from the broad and general uses of the mainly commodity-tonnage materials described in "The Modern Inorganic Chemicals Industry". It does not necessarily imply that they are made in smaller quantities, or that they cost more; the production and selling price of aluminium hydroxide, for example, show this not to be the case. By contrast the compounds of rare earth metals are usually very expensive and many are manufactured in equipment which is closer in scale to the laboratory than to industrial plant. However, most of the other compounds included in the book fall between these extremes.

The treatment of the subject is aimed to be representative across the periodic table and, while it could not in under 500 pages be exhaustive, the compilation complements Special Publication No. 31 in covering most of the inorganic chemicals used in industry today.

*Raymond Thompson*
*Chairman, Inorganic Chemicals Group*

"The term 'speciality chemicals' can mean different things to different people. They can be defined as high cost, low tonnage products, or products having high added value. What is a speciality chemical to one user could easily be a commodity material to another."

— Danny Fagandini, *"Speciality Chemicals"*,
Volume 1, No. 1., February 1981

"When *I* use a word" Humpty Dumpty said in a rather scornful tone "it means just what I choose it to mean, neither more nor less".

— Lewis Carroll, *"Through The Looking Glass,* 1872.

CONTENTS

# Speciality Inorganic Chemicals

By T. Matthews
KLINE SA, RUE FROISSART 89, BOITE 8, 1040 BRUXELLES, BELGIUM

Much confusion exists as to what precisely constitutes a speciality chemical. Very often what are fine chemicals or even low volume commodity chemicals are referred to as specialities. Before looking in detail at speciality inorganic chemicals it is worth while to examine the characteristics which make a chemical product "special", for these are the same whether the product is of inorganic or organic nature. The key to specialities can be put in a single word - DIFFERENTIATION.

Product types

Broadly speaking, chemical products can be divided into two main types:

- Undifferentiated or composition products.
- Differentiated or performance products.

*UNDIFFERENTIATED* products have the following principal characteristics:

- Usually described by chemical formulae or have clear statements of their content.
- Produced by various suppliers in substantially identical form.
- Frequently used in numerous applications.
- Sold to specifications of composition for what they contain.

*DIFFERENTIATED* products have the following characteristics:

- Usually cannot be characterized by chemical formulae or statement of chemical content alone.
- Produced with real differences between the products of different suppliers - or, alternatively, marketed with imputed differences.
- Often used in only a few applications.
- Sold to performance specifications for what they do.

Differentiation of chemicals can be achieved in a number of ways, some of which are shown below:

- Develop products designed to perform specific functions, such as a herbicide or a dyestuff.
- Develop formulated end products, such as compound fragrances or boiler water treatment chemicals.
- Provide extensive technical service on the use of the product.
- Make the chemical an integral part of a complete system, such as the toner of a Xerox copy machine.
- Sell the product under a code name or brand especially where the specification of the material is easily identifiable.
- Patent new synthetic products.
- Innovate continuously, either in the products themselves, in their packaging, dispensers, or systems; or in their distribution or service.

## Classes of chemical products

Going one stage beyond the concept of differentiation and combining this concept, with the normal division of chemical products into high volume and low volume items, we can broadly classify all types of chemical products into four classes as shown in Figure 1:

Figure 1 - CLASSIFICATION OF CHEMICAL TYPES

| Volume | Undifferentiated (composition) | Differentiated (performance) |
|---|---|---|
| High | True Commodities | Pseudo Commodities |
| Low | Fine Chemicals | Speciality Chemicals |

*TRUE COMMODITIES* are organic or inorganic chemicals synthesized in large volume, often from captive raw materials. They are made to clearly defined standards of composition, and products from different manufacturers are essentially identical. Commodities are often, but not always, used for a wide variety of applications and there are many customers. However, a few large customers with huge demands often dominate the market. Moreover, customers are usually very price conscious.

*FINE CHEMICALS* are simply low volume undifferentiated products. Many chemical intermediates, particularly for the pharmaceutical industry, fall into this category. The number of producers of any given fine chemical may be lower than the number producing a given commodity chemical, due to the lower demand. However, fine chemicals from different producers are essentially identical and are often produced to well defined standards such as those of the various national Pharmacopoeias or Food Chemicals Codex.

*PSEUDO-COMMODITIES* are large volume differentiated products, which are often synthesized from captive raw materials, and like commodities often have their sales concentrated in the hands of a relatively small number of customers. Unlike commodities, pseudo-commodities are made to performance specifications rather than to composition specifications. Typical pseudo-commodities are man-made fibres, tonnage plastic resins, and tonnage surfactants.

*SPECIALITY CHEMICALS* the subject of this seminar, are the fourth class of product shown in Figure 1. Speciality chemicals comprise a very diverse group of differentiated products, formulated or synthesized in small volumes, generally from bought-in raw materials. Specialities are usually formulated to solve a particular problem and are often sold to a wide circle of customers. Most importantly, specialities produced by one manufacturer are not identical to those of the competition. Real differences in quality, composition and performance do exist between the products of different producers.

## Types of speciality

Table 1 lists the principal groups of products which can be considered as speciality chemicals.

Products such as cosmetics and toiletries, paint, printing ink and pharmaceuticals are not included in Table 1 as they are considered to be produced by industries which are outside the true chemical industry in the strictest sense of the definition.

Table 1

PRINCIPAL TYPES OF SPECIALITY CHEMICALS

Adhesives -a
Antioxidants
* Biocides
* Catalysts -a

* Chelates
* Corrosion inhibitors -a
* Cosmetic additives
* Diagnostic aids

Dyes
Elastomers -a
* Electronic chemicals
Enzymes

* Flame retardants
Flavours and fragrances
Flotation reagents -a
Food additives

Foundry chemicals
* Industrial and institutional cleaning products
* Laboratory chemicals

* Metal plating and finishing chemicals
* Oil-field chemicals -a
* Paint additives

Paper additives
* Pesticides
Petroleum additives
* Photographic chemicals

* Pigments -a
Plasticizers -a
Plastics and resins -a
* Plastics additives -a

* Printing chemicals
Rubber processing chemicals
Surfactants -a
Textile specialities

* Thickening agents -a
Ultraviolet absorbers
* Water management specialities -a

a- low-volume types

The speciality categories marked with an asterisk in Table 1 all have a sizeable inorganic component in their make-up, and some, such as metal plating and finishing chemicals, are mainly made up of inorganic raw materials. Table 2 lists some of the inorganic chemicals which are used in formulating the specialities of Table 1. The chemical products in Table 2 are <u>not</u> themselves specialities. They are either commodity or fine chemical <u>raw</u> materials used in formulating specialities.

Table 2

EXAMPLES OF INORGANIC RAW MATERIALS FOR SPECIALITY CHEMICALS

**BIOCIDES**

Chlorine
Sodium hypochlorite
Lithium hypochlorite
Salts & oxides of:
  copper, selenium, chromium, zinc, arsenic, mercury
Barium metaborate
Sodium nitrite & nitrate

**CATALYSTS**

Zeolites
Oxides:
  cobalt-molybdenum
  nickel-molybdenum
  nickel-tungsten
Precious metals on supports
Titanium halides
Vanadium compounds

**COSMETIC ADDITIVES**

Calcium pyrophosphate
Sodium metaphosphate
Dicalcium phosphate
Calcium carbonate
Bismuth oxychloride
Titanium dioxide coated mica
Aluminium chlorohydrate
Aluminium-zirconyl hydroxychloride
Aluminium chloride
Stannous fluoride
Sodium monofluorophosphate
Sodium fluoride

**LABORATORY CHEMICALS**

Many inorganics

**PESTICIDES**

Zinc salts
Copper salts
Arsenic compounds

**CHELATES**

Phosphates
Sodium carbonate
Sodium silicate

**CORROSION INHIBITORS**

Sodium bichromate
Sodium/potassium phosphates
Inorganic borates, nitrites, chromates
Zinc compounds

**FLAME RETARDANTS**

Antimony trioxide & pentoxide
Silicates
Zinc borate

**INDUSTRIAL CLEANERS**

many inorganic components

**METAL PLATING & FINISHING**

Chromates, phosphates
Base and precious metal salts & many other chemicals

**OIL FIELD CHEMICALS**

Sodium phosphate
Sodium bichromate
Hydroxy alumina
Zirconium oxychloride
Ammonium bifluoride
Zinc carbonate

**PAINT ADDITIVES**

Cuprous oxide
Fumed silica

**PRINTING CHEMICALS**

Ferric chloride
Mineral acids
Bichromates

Table 2 cont.

EXAMPLES OF INORGANIC RAW MATERIALS FOR SPECIALITY CHEMICALS

PHOTOGRAPHIC CHEMICALS

Borax
Sodium sulphite
Potassium carbonate
Potassium bromide
Ammonium thiosulphate
Potassium ferricyanide

PIGMENTS

Many products

PLASTICS ADDITIVES

Antimony oxides
Aluminium hydroxide

THICKENING AGENTS

Fumed silica
Modified bentonite
Sodium magnesium silicate

WATER MANAGEMENT SPECIALITIES

Hydrazine
Ammonia
Sodium sulphite
& many inorganics

The problem facing the inorganic chemical manufacturer is that his products are so often the basic raw materials used by other companies in making true specialities. Many inorganic chemical companies may see themselves reflected in the hypothetical example shown in Table 3, where the Kline DIFFERENTIATION INDEX is illustrated. No chemical product completely lacks differentiation and very few indeed have total differentiation. The Kline differentiation index offers a semi-quantitative method for assessing the degree of differentiation in a product. On the index, the product is ranked by judgement on each of ten key criteria and its DIFFERENTIATION INDEX is taken as the average of the ten individual ratings.

Basic Inorganics Ltd, as shown in Table 3, is a typical inorganic chemical company and makes, among other things, sodium sulphite. One group of customers are the water management chemical companies, for whom sodium sulphite is a key ingredient in formulating a boiler water de-oxygenating fluid. If one puts the question as to what, in the eyes of the water management companies, differentiates the sodium sulphite supplied by Basic Inorganics from that of any other manufacturer, the short answer provided by Table 3 is not much! Sodium sulphite is bought by the water management companies as a commodity.

Aquapure Ltd, is one of Basic Inorganics customers. It formulates the sodium sulphite into a proprietary speciality fluid, Nilox

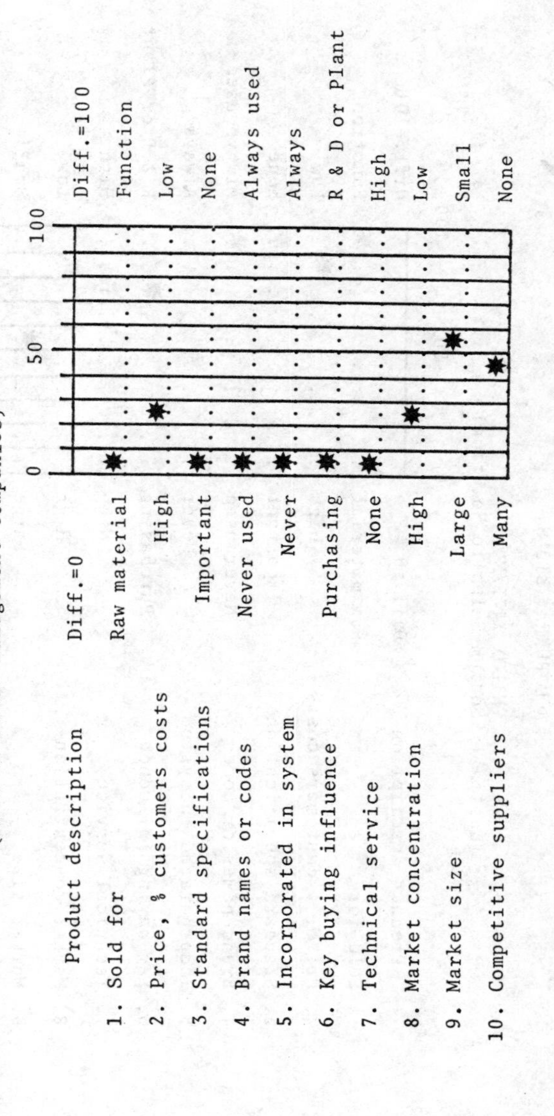
Table 3
THE DIFFERENTIATION INDEX
BASIC INORGANICS LTD'S
SODIUM SULPHITE
(sold to water management companies)
0
50
100
Product description
Diff.=0
Diff.=100
1. Sold for
Raw material
Function
2. Price, % customers costs
High
Low
3. Standard specifications
Important
None
4. Brand names or codes
Never used
Always used
5. Incorporated in system
Never
Always
6. Key buying influence
Purchasing
R & D or Plant
7. Technical service
None
High
8. Market concentration
High
Low
9. Market size
Large
Small
10. Competitive suppliers
Many
None
OVERALL DIFFERENTIATION
0
50
100

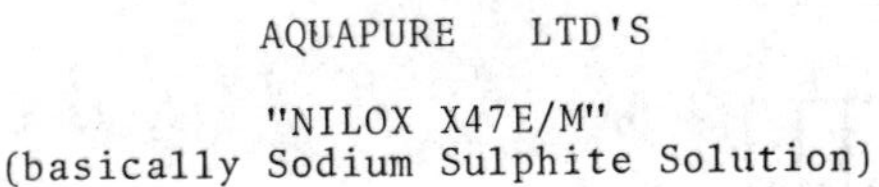

Table 4

THE DIFFERENTIATION INDEX

AQUAPURE LTD'S

"NILOX X47E/M"
(basically Sodium Sulphite Solution)

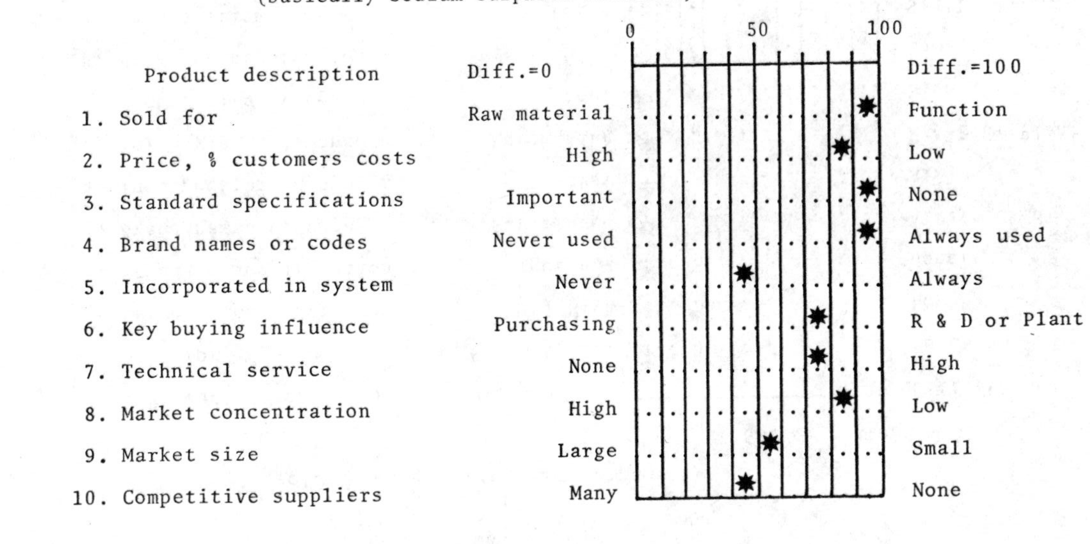

| Product description | Diff.=0 | Diff.=100 |
|---|---|---|
| 1. Sold for | Raw material | Function |
| 2. Price, % customers costs | High | Low |
| 3. Standard specifications | Important | None |
| 4. Brand names or codes | Never used | Always used |
| 5. Incorporated in system | Never | Always |
| 6. Key buying influence | Purchasing | R & D or Plant |
| 7. Technical service | None | High |
| 8. Market concentration | High | Low |
| 9. Market size | Large | Small |
| 10. Competitive suppliers | Many | None |

OVERALL DIFFERENTIATION

X47E/M, which it sells with much technical service as a feed water deoxygenator to operators of medium pressure boilers. Such boiler operators are basically interested only in the well-being of their boiler and are happy to entrust its care to the Aquapure service man. The sodium sulphite solution which is sold as Nilox X47E/M is sold for what it does rather than for what it is. By selling function rather than composition, Aquapure has turned Basic Inorganic's undifferentiated sodium sulphite into a highly differentiated speciality chemical.

Because many inorganic chemical manufacturers are better equipped to produce rather than to indulge in sophisticated technical development and marketing, how can they introduce a degree of differentiation into their basic inorganic chemicals and therebye command a higher price? One method worth considering is that of MARKET SEGMENTATION. This is a particularly useful tool in helping to maximize profits from lower volume products and upgrading the overall return on bulk chemicals.

Table 5 shows diagramatically the idea behind market segmentation. The basic aim is to define those segments of the total market for any given product where you can meet a distinct need. Table 5 is self explanatory.

By way of an example, another inorganic chemical company, Quality Inorganics Ltd., made an effort to identify those segments of the acid and basic inorganic chemicals market where it could offer a special grade or quality at a higher price. It identified the market for electronic chemicals as a growing area, requiring higher than normal purity mineral acids and ferric chloride. Unable to formulate the etching solutions itself, Quality Inorganics took a decision to try to dominate the supply of high quality raw materials to companies formulating etching solutions. By actively promoting its mineral acids and ferric chloride into this market segment, Quality Inorganics Ltd managed to differentiate its products from those of its competitors and thereby sell at higher prices. Table 6 shows the degree of differentiation achieved in attacking this one market segment. Quality Inorganics does not turn its basic products into speciality chemicals, but it is a step in the right direction.

Excellent Etchants Ltd, a speciality chemical company, supplying etching solutions to the electronics industry, then goes on to formulate the ultrapure acids and ferric chloride it buys from

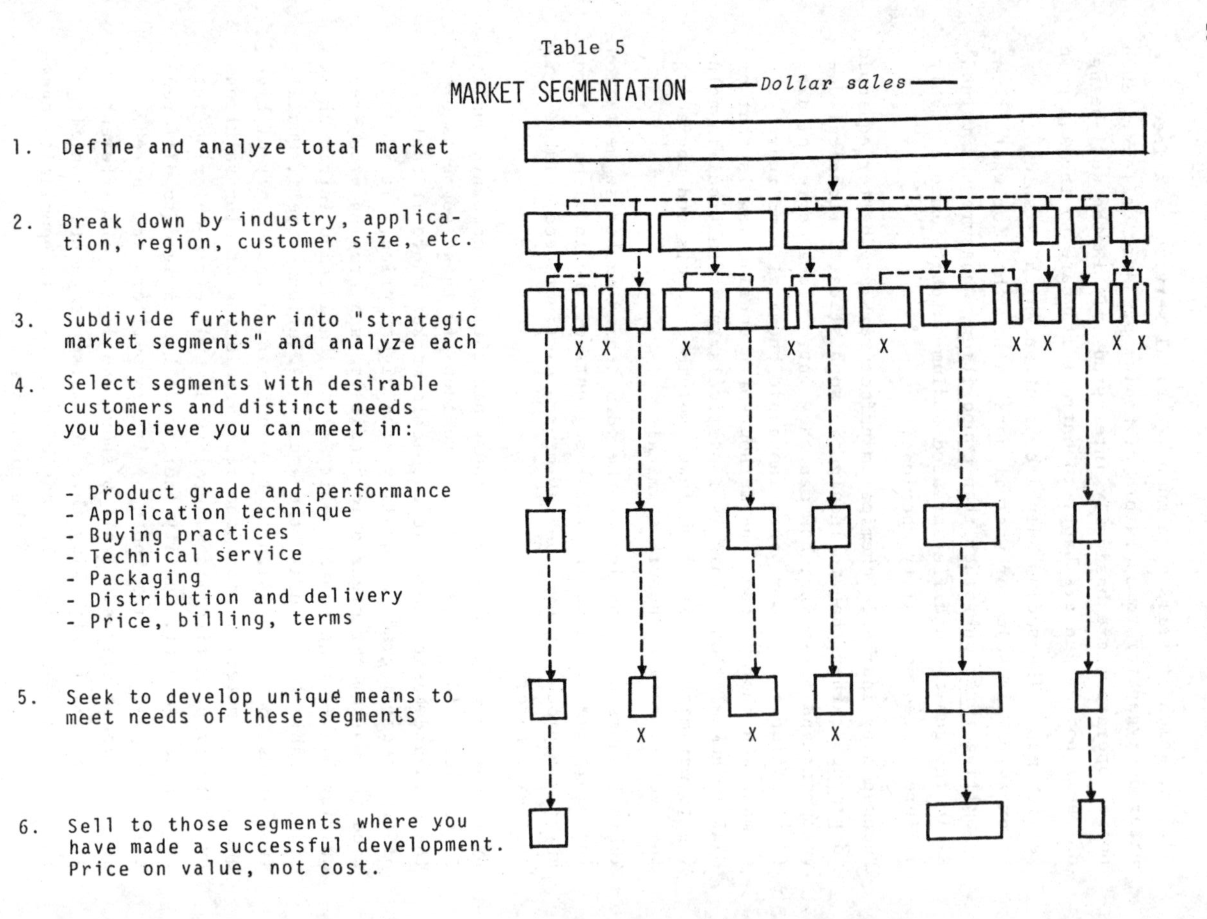
Table 5
MARKET SEGMENTATION
Dollar sales
1. Define and analyze total market
2. Break down by industry, application, region, customer size, etc.
3. Subdivide further into "strategic market segments" and analyze each
4. Select segments with desirable customers and distinct needs you believe you can meet in:
- Product grade and performance
- Application technique
- Buying practices
- Technical service
- Packaging
- Distribution and delivery
- Price, billing, terms
5. Seek to develop unique means to meet needs of these segments
6. Sell to those segments where you have made a successful development. Price on value, not cost.
X

Table 6

THE DIFFERENTIATION INDEX

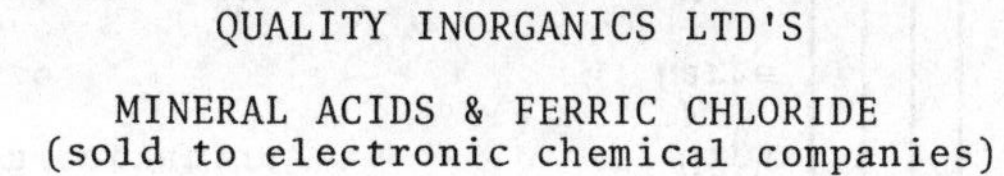

QUALITY INORGANICS LTD'S

MINERAL ACIDS & FERRIC CHLORIDE
(sold to electronic chemical companies)

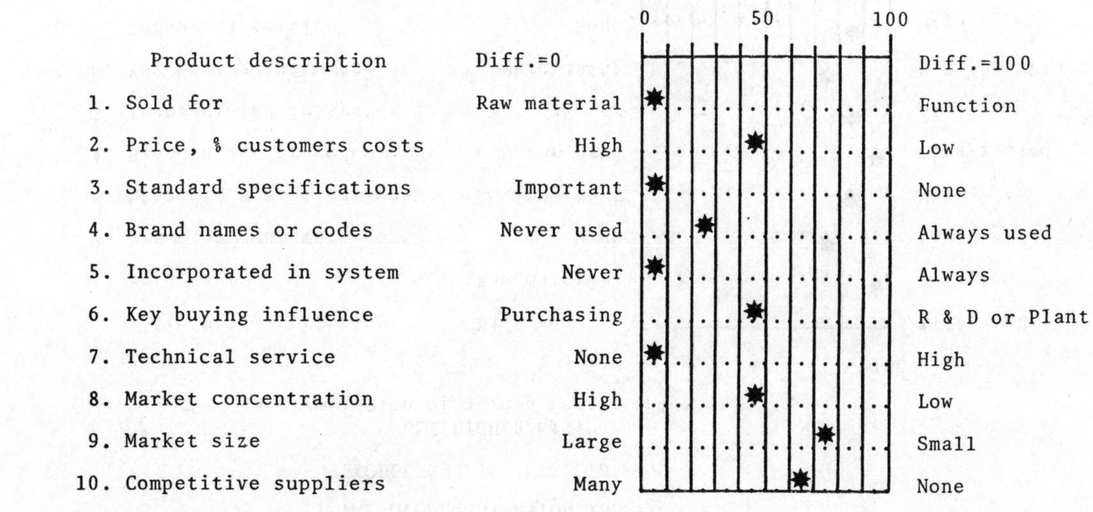


| Product description | Diff.=0 | Diff.=100 |
|---|---|---|
| 1. Sold for | Raw material | Function |
| 2. Price, % customers costs | High | Low |
| 3. Standard specifications | Important | None |
| 4. Brand names or codes | Never used | Always used |
| 5. Incorporated in system | Never | Always |
| 6. Key buying influence | Purchasing | R & D or Plant |
| 7. Technical service | None | High |
| 8. Market concentration | High | Low |
| 9. Market size | Large | Small |
| 10. Competitive suppliers | Many | None |

OVERALL DIFFERENTIATION

Table 7

THE DIFFERENTIATION INDEX

EXCELLENT ETCHANTS LTD'S

"MULTIETCH B/6424"
(a formulation of acid & ferric chloride)

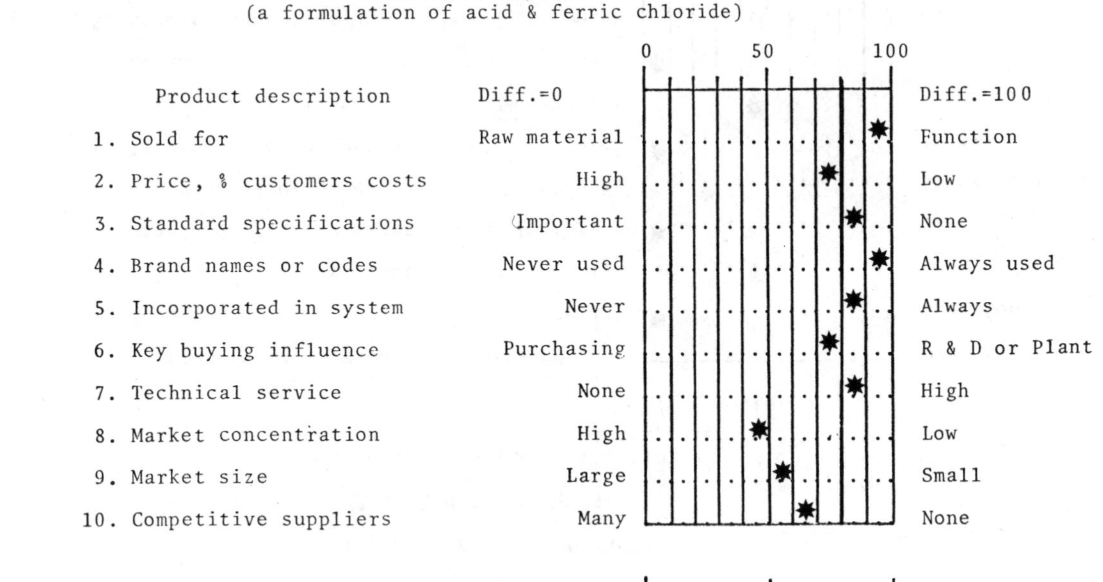

Quality Inorganics into a branded etching solution, Multietch B/6424. By formulating these products and selling them for a function to a well defined market segment, Excellent Etchants achieves a high degree of differentiation of the product, as shown in Table 7.

Market size

In Table 8 are shown some estimates of the relative sizes of the four classes of chemical products for the period 1977 through 1985.

In the USA, the market for speciality chemicals is forecast to reach $25 billion in value in 1980, or roughly 24.0% of the total chemical market. The free world market for specialities is estimated to reach $65 billion in 1980, or 20% of the total chemical market. Overall growth rate for all chemicals 1980-1985 in the free world is forecast to be higher than that in the USA due to the less mature state of third world markets.

Table 8

FORECAST SALES OF INDUSTRIAL CHEMICALS
BY CLASS 1977-1985

U.S.

| Class | Billions of 1977 dollars | | | Increase |
|---|---|---|---|---|
| | 1977 | 1980 | 1985 | %/Year |
| Commodities | $28 | $ 33 | $ 42 | 5.0% |
| Pseudo commodities | 23 | 28 | 41 | 7.3 |
| Speciality chemicals | 19 | 25 | 31 | 6.0 |
| Fine chemicals | 15 | 18 | 26 | 7.2 |
| Total | $85 | $104 | $140 | 6.4% |

Free World

| Class | Billions of 1977 dollars | | | Increase |
|---|---|---|---|---|
| | 1977 | 1980 | 1985 | %/Year |
| Commodities | $ 83 | $107 | $147 | 7.4% |
| Pseudo commodities | 69 | 91 | 143 | 9.3 |
| Speciality chemicals | 51 | 65 | 96 | 8.2 |
| Fine chemicals | 45 | 58 | 92 | 9.2 |
| Total | $248 | $321 | $478 | 8.5% |

## Key business characteristics

Differentiation is the key to speciality chemicals. However, each of the four classes of chemical products are distinct and different kinds of businesses. Success in any or all of the classes is conditioned to a large extent by recognition of this fact, by reaching a clear understanding of the factors influencing success and by shaping the manufacturing, R & D, sales, marketing and technical service functions accordingly. The greatest contrast in "cultures" can be seen by examining commodity and speciality chemicals, for these two classes are diametrically opposite in their characteristics.

*RESEARCH AND DEVELOPMENT* in commodity products is most heavily targeted towards process technology. The main aim is to look for better and more economical processes, for which the main skills required are synthesis and process technology. This type of work typically suits the dedicated chemist and chemical engineer, whose main interest in life centres around the development laboratory and pilot plant.

Because of the importance of raw materials, major emphasis is also placed on these and leading commodity companies are in the forefront in exploiting all advantages to be gained from the ability to utilize alternative forms of raw materials.

In speciality products, the main emphasis is on the customer and his needs. Formulation and knowledge of end uses are important and research and development activity focuses on application and field testing. Such work might seem banal to the dedicated researchers into high technology processes common to commodity production. Staff with an entrepreneurial spirit is what is needed here.

*MANUFACTURING* units for commodities are normally dedicated to producing one chemical, or at best very few. Plants tend to be large and centralized, continuous in operation round the clock, 365 days a year. Plant equipment is complex, automated and usually very inflexible, and production is planned for long periods on rigid schedules.

By contrast, the speciality plant is often small and multipurpose and key operations are blending and formulating. Production is scheduled to meet fluctuating demand of customers. Raw materials are very often bought in. Synthetic patented herbicides and similar agricultural products are an exception here, but a small central synthesis unit, often batchwise in operation, can serve several formulating plants.

*MARKETING* of commodities usually involves making huge sales to a few customers (soda ash to a glass manufacturer for instance). The commodity chemical is a key raw material to the customer who often converts it into another commodity product (the glass itself for instance).

Technical service requirements are virtually non-existent. Market research is used extensively to determine anticipated demand for well-defined products, based on customers' views of the market outlook. It does not define new products or problems.

Typically, speciality chemicals are sold to many small customers, usually as an additive to his main product, or as a service item (flame retardants to the plastics converter or boiler water chemicals to any steam raiser). Volume of sales per salesman is low and sales are often coupled with extensive technical service and continuous liaison with the customer. Markets are difficult to research by standard techniques and test marketing leading to establishment with one client followed by extension to others is standard practice.

Because specialities are priced on performance and are often of an additive nature, they show greater cyclical stability with the rise and fall in the business cycle. Commodities, being identical whatever their source, are much more sensitive to the overall supply and demand balance. Prices are less stable and rise and fall with the business cycle.

*FINANCING* of commodity operations is usually protracted and heavy. Huge investments are needed and typically in the USA $1.25 of fixed capital investment are needed to generate $1.0 of sales.

On the other hand, marketing and selling expenses are low because customers are easily identified and are often tied to some supply contract. Although margin on sales is usually high, return on capital is low due to the capital intensity of the business.

The financing of specialities is completely different. Usually, low capital investment is needed, typically from $0.25-0.50 of fixed capital per $1.00 of sales. This is usually true even where synthetic specialities are involved because the scale of the plant is small. Contrast, for instance, a plant making the active ingredient for a herbicide with the same company's fertilizer operations.

The marketing and selling of specialities is very labour intensive. Because of the close contact with the customer and continuous technical liaison, some categories of product require huge sales forces. Typically these would be the lower grade technological products, such as industrial and institutional cleaning products, where the potential customer circle is enormous.

## Profit from specialities

In the last few years, much has been said about the potential profits to be earned from specialities. Are they more profitable than commodities, pseudo-commodities and fine chemicals? For many years, the Kline Group has systematically collected and analysed profitability data on U.S. chemical companies. The overall conclusion of this ongoing analysis is basically this:

> *"A COMPANY CAN EARN GOOD PROFITS AND SHOW SATISFACTORY GROWTH IN ANY OR ALL CLASSES OF CHEMICAL PRODUCT, PROVIDED THE CORRECT STRATEGY IS ADOPTED"*

However, based on U.S. experience and data, it does seem clear that specialities tend to give higher rates of return on capital and greater cyclical stability than other categories. But, as speciality chemicals tend to be smaller volume items, larger companies will probably have also to be in the larger volume commodity and pseudo-commodity market, as well as specialities, if maximum profits are to be made.

To illustrate the apparent higher profitability of specialities, data on the overall profitability of some U.S. chemical companies is shown in Tables 9 and 10. Chemical companies have been defined as those companies having more than 50% of their sales in chemicals. Thus, such large petro-chemical companies as Exxon and Shell are excluded from the analysis. In the analysis, the sales figures relate to total sales volume of all products, chemical and non-chemical.

Table 9

INVOLVEMENT IN SPECIALITIES AND PROFITABILITY OF 19 U.S. CHEMICAL COMPANIES WITH SALES IN EXCESS OF $ 1 BILLION IN 1979

| Company | Sales $ billion | % Specialities | Profits, % on Sales | Profits, % on Equity |
|---|---|---|---|---|
| 1 Dow Chemical | $ 9.26 | 15% | 8.4% | 20.1% |
| 2 Hercules | 2.35 | 35 | 7.4 | 18.2 |
| 3 Du Pont | 12.57 | 20 | 7.5 | 17.7 |
| 4 Diamond Shamrock | 2.36 | 15 | 7.6 | 17.4 |
| 5 Air Products | 1.23 | 30 | 7.9 | 17.2 |
| 6 Ethyl | 1.66 | 35 | 5.9 | 16.0 |
| 7 Stauffer Chemical | 1.53 | 35 | 8.9 | 16.0 |
| 8 IMC | 1.47 | 5 | 8.2 | 15.3 |
| 9 Rohm & Haas | 1.59 | 50 | 6.0 | 15.0 |
| 10 Union Carbide | 9.18 | 15 | 6.1 | 13.8 |
| 11 Pennwalt | 1.08 | 30 | 4.6 | 13.7 |
| 12 Celanese | 3.15 | 10 | 4.5 | 13.6 |
| 13 American Cyanamid | 3.19 | 35 | 5.3 | 12.5 |
| 14 Monsanto | 6.19 | 20 | 5.3 | 11.9 |
| 15 Akzona | 1.01 | 25 | 2.5 | 7.9 |
| 16 BASF | 1.10 | 40 | 0.9 | 3.7 |
| 17 Allied Chemical | 4.33 | 10 | 0.2 | 0.9 |
| 18 American Hoechst | 1.17 | 40 | 2.7 | N/A |
| 19 Ciba Geigy | 1.32 | 50 | N/A | N/A |

Table 9 summarizes profitability data on the largest U.S. chemical companies, with sales of $1 billion or more in chemicals in 1979. The companies have been ranked in order of return on shareholder's equity. We have attempted, from our knowledge of these companies' activities, to indicate their involvement in speciality chemicals. Few concrete conclusions can be drawn from an examination of the figures. There seems little correlation between sales volume, involvement in specialities and profitability. Indeed, Dow, a leading commodity company heads the list, whilst Rohm and Haas, the largest U.S. company with highest activity in specialities is rank-ranked only ninth. In most cases, the businesses of these largest companies are so diverse and complex that analysis of published data allows no firm conclusions to be drawn about the contribution of specialities to their profitability.

Table 10 shows the profitability of U.S. chemical companies with sales in the range $100-999 million. In 1979, there were 36 companies in this category. We have attempted to estimate the involvement in specialities by these companies. Examination of the data indicates fairly clearly that those companies with a high percentage of their sales in specialities do have a higher profitability on equity - and sales - than those companies with lower speciality involvement. Of the top ten companies, seven have 76-100% of their sales in specialities. Also worth noting is that profitability and total sales volume are unrelated. Some of the smallest companies are most profitable.

Although not presented here, an analysis of the performance of some 25 U.S. chemical companies with sales in the range of $25-$99 million also shows that greater involvement in specialities gives a higher return on equity.

## Growth rates of U.S. chemical companies: 1968 - 1978

Our summary analysis for 1979 is not yet completed, due to the movement of companies from one category of sales to another or - in some cases - complete elimination from the listings. Since 1968, we have measured the increase in sales and profits of

Table 10

INVOLVEMENT IN SPECIALITIES AND PROFITABILITY OF 36 U.S. CHEMICAL COMPANIES WITH SALES IN THE RANGES $100-999 MILLION IN 1979

| Company | Sales $million | % Specialities 0-25 | 26-50 | 51-75 | 76-100 | Profits % on Sales | Equity |
|---|---|---|---|---|---|---|---|
| 1 Beker Industries | 227.8 | x | - | - | - | 16.0% | 39.7% |
| 2 Freeport Minerals | 487.7 | x | - | - | - | 20.8 | 26.7 |
| 3 Loctite | 161.5 | - | - | - | x | 14.4 | 25.2 |
| 4 Great Lakes Chem. | 122.5 | - | - | x | - | 15.2 | 24.6 |
| 5 Lubrizol | 724.7 | - | - | - | x | 12.5 | 24.2 |
| 6 Nalco | 578.9 | - | - | - | x | 10.9 | 24.2 |
| 7 Lea Ronal | 114.4 | - | - | - | x | 3.8 | 22.8 |
| 8 Dow Corning | 610.2 | - | - | - | x | 11.7 | 21.8 |
| 9 Intl.Flav. & Frag. | 409.3 | - | - | - | x | 15.0 | 21.5 |
| 10 Betz Lubricants | 180.2 | - | - | - | x | 11.0 | 21.3 |
| 11 National Chemsearch | 260.6 | - | - | - | x | 8.7 | 18.9 |
| 12 Tremco | 110.6 | - | - | - | x | 7.0 | 18.8 |
| 13 Essex Chemicals | 127.4 | - | x | - | - | 3.9 | 18.1 |
| 14 Mallinckrodt | 392.5 | - | - | - | x | 10.5 | 18.0 |
| 15 Mississippi Chems. | 262.1 | x | - | - | - | 9.1 | 17.2 |
| 16 Chemed | 323.6 | - | - | - | x | 8.0 | 17.2 |
| 17 Economics Labs. | 464.4 | - | - | - | x | 6.5 | 16.9 |
| 18 Mobay | 955.1 | - | - | x | - | 6.1 | 16.8 |
| 19 Petrolite | 186.9 | - | - | x | - | 8.3 | 16.7 |
| 20 A.Schulman | 247.4 | - | x | - | - | 2.5 | 16.1 |
| 21 Texasgulf | 789.3 | x | - | - | - | 17.3 | 15.7 |
| 22 Big Three Ind. | 509.4 | x | - | - | - | 12.3 | 15.6 |
| 23 National Starch | 474.5 | - | x | - | - | 6.7 | 15.1 |
| 24 Morton Norwich | 732.0 | - | x | - | - | 6.3 | 14.5 |
| 25 Philip A.Hunt | 106.0 | - | - | - | x | 7.5 | 13.9 |
| 26 Virginia Chemical | 131.3 | x | - | - | - | 4.2 | 13.3 |
| 27 H.B.Fuller | 258.7 | x | - | - | - | 7.4 | 13.1 |
| 28 Liquid Air | 369.1 | x | - | - | - | 24.7 | 11.8 |
| 29 Stepan Chemical | 145.0 | - | x | - | - | 5.3 | 11.3 |
| 30 Crampton & Knowles | 239.6 | - | - | x | - | 7.1 | 10.8 |
| 31 Reichhold Chems. | 874.9 | - | x | - | - | 12.2 | 7.5 |
| 32 Terra Chemicals | 254.1 | x | - | - | - | 2.7 | 5.8 |
| 33 First Mississippi | 178.4 | x | - | - | - | 4.7 | 4.9 |
| 34 Royster | 227.9 | x | - | - | - | 3.0 | 4.4 |
| 35 Witco | 966.8 | - | x | - | - | 42.8 | 1.9 |
| 36 Degussa | 100.6 | - | x | - | - | (4.0) | N/A |

different sizes and categories of company. The results of this analysis are shown in Table 11.

Table 11

GROWTH RATES OF U.S. CHEMICAL COMPANIES
1968 - 1978

| | Companies | | Average increase %/Year | |
|---|---|---|---|---|
| Sales million | No | Type | Sales | Profits |
| > $1,000 | 1 | Speciality | 11.6% | 10.5% |
| | 13 | Diversified | | |
| 100 - 999 | 9 | Speciality | 15.8% | 18.5% |
| | 19 | Diversified | 14.2 | 10.9 |
| | | Difference | 1.6% | 7.6% |
| 25 - 99 | 8 | Speciality | 14.8% | 15.5% |
| | 8 | Diversified | 12.8 | 11.8 |
| | | | 2.0% | 3.7% |

The results are striking. Speciality companies have not only increased sales value at a greater rate than diversified companies, but have shown even more dramatic increases in their profits.

## Size of speciality markets

In 1979, we estimated the total value of the speciality chemicals included in this paper to be $24.37 billion. Table 12 shows the value of each of the product categories. We have observed that the products can be segregated into three groups, categorized by the type of company producing them. The upper group, headed by pesticides and plastics and resins, tend to be the preserve of large diversified chemical companies. Many of the products in this group tend to have the highest R & D content and are technically the most complex.

A middle group of products, including four types, each with markets valued above $1 billion, is made by both large diversified and smaller specialized companies.

Table 12

SPECIALITY CHEMICALS PRODUCED CHIEFLY BY LARGE DIVERSIFIED COMPANIES

| Product | U.S. Sales 1979 $ million |
|---|---|
| Pesticides | $3,600 |
| Plastics and resins -a | 1,460 |
| Dyes | 830 |
| Pigments, organic | 530 |
| Elastomers -a | 500 |
| Plasticizers -a | 490 |
| Catalysts | 450 |
| Rubber processing chemicals | 440 |
| Paper additives | 350 |
| Antioxidants | 320 |
| Pigments, inorganic | 310 |
| Flame retardants | 250 |
| Flotation reagents | 110 |
| Chelating agents | 90 |
| Ultraviolet absorbers | 45 |
| Total | $9,775 |

a- Speciality types only

SPECIALITY CHEMICALS PRODUCED BOTH BY LARGE DIVERSIFIED COMPANIES AND SMALLER SPECIALIZED COMPANIES

| Product | U.S. Sales 1979 $ million |
|---|---|
| Petroleum additives | $1,470 |
| Plastics additives | 1,240 |
| Food additives | 1,110 |
| Automotive chemicals | 1,000 |
| Textile specialities | 770 |
| Photographic chemicals | 625 |
| Metal plating and finishing chemicals | 520 |
| Biocides | 470 |
| Cosmetic additives | 440 |
| Thickening and sizing agents | 430 |
| Adhesives and sealants -a | 420 |
| Surfactants | 350 |
| Paint additives | 320 |
| Corrosion inhibitors | 170 |
| Electronic chemicals | 140 |
| Enzymes | 135 |
| | $9,610 |

a- Speciality types only (Contd.)

Table 12 (Contd.)

SPECIALIZED CHEMICALS PRODUCED CHIEFLY BY SMALLER SPECIALIZED COMPANIES -a

| Category | U.S. Sales 1979 $ million |
|---|---|
| Industrial and institutional cleaning products | $1,780 |
| Water management chemicals | 1,110 |
| Diagnostic aids | 690 |
| Flavours and fragrances | 470 |
| Oil field chemicals | 300 |
| Laboratory chemicals | 210 |
| Refinery chemicals | 175 |
| Printing chemicals | 130 |
| Foundry additives | 120 |
| | $4,985 |

a- Or recently acquired divisions of larger companies

Table 13

FORECAST GROWTH OF U.S. SPECIALITY CHEMICALS 1979 - 1984

| | |
|---|---|
| ✱ Electronic chemicals | 11.0% |
| Speciality plastics | 11.0 |
| ✱ Diagnostic Aids | 10.0 |
| ✱ Oil field chemicals | 10.0 |
| Speciality adhesives and sealants | 9.0 |
| Speciality elastomers | 8.0 |
| Ultraviolet absorbers | 8.0 |
| Plasticizers | 7.0 |
| Plastics additives | 7.0 |
| ✱ Water management chemicals | 7.0 |
| ✱ Biocides | 6.5 |
| ✱ Flame retardants | 6.5 |
| Formulated flavours & fragrances | 6.5 |
| All speciality chemicals | 6.0 |

The lowest group, headed by industrial and institutional cleaning products and water management chemicals, each worth more than $1 billion in 1979, are usually made by smaller more specialized companies. In many cases they tend to be products high in development content but involving little basic research. Whereas a small company might adequately finance the development of a series of cleaners or water treatment chemicals, it would be hard put to undertake the multimillion dollar synthesis, screening and registration procedures leading up to the marketing of a new synthetic pesticide or polymer.

It is worth noting that the total size of the U.S. market for many specialities is modest. Competition is fierce in all segments and unless a chemical company is highly diversified into many types of speciality, it is often difficult to build up a huge company dedicated entirely to speciality production. Most speciality companies, therefore, tend to have sales of less than $1 billion.

## Best growth prospects

Not all specialities are at the same stage in their life cycle. Some, such as dyestuffs, many ordinary plastics and resins, and rubber chemicals, are already very mature. Nevertheless, certain categories, if not showing high double digit growth, have rates of growth which look super-attractive to today's commodity producer.

In the USA the highest growth areas today are electronic chemicals, speciality plastics, diagnostic aids and oil field chemicals. We estimate that speciality categories overall will grow at an average annual rate of 6.0% from 1979 through 1981.

Table 13 shows the 14 categories of product exhibiting higher than average growth. Those categories marked with an asterisk have high inorganic content.

## Conclusions

From this very brief review, based on large quantities of detailed data on the U.S. chemical industry collected over many years, the key conclusions appear to be:

- Specialities offer the chemical manufacturer greater stability than commodities.
- Specialities are less price sensitive than commodities (or pseudo-commodities or fine chemicals) often because they perform an additive or service role and are not key raw materials.
- Chemical companies with high involvement in specialities do seem to be more profitable than most commodity producers.
- Growth of companies heavily involved in specialities is overall higher than that in other chemical products.
- Because many specialities markets are modest in size even in the U.S. they are likely to be even smaller in each of the national markets of Europe. In many categories, competition in Europe is both severe and fragmented, with a complex supplier profile.

All in all, specialities do offer an attractive area for development and diversification and most are evolving markets. How does a company diversify into - specialities?

Well, that is the subject of another paper.

Note: This paper is based on ideas and concepts developed by the Kline Group, management consultants into the chemical industry.

# Sodium Borohydride and Its Derivatives

By R. C. Wade
VENTRON DIVISION, THIOKOL CORPORATION, CONGRESS STREET, BEVERLY, MASSACHUSETTS 01915, U.S.A.

## 1. Introduction

Sodium borohydride, first synthesized and identified over 35 years ago, is one of the most unique specialty inorganic chemicals being manufactured today. The diversity of applications which have been developed for it are truly amazing. So much so that one of its discoverers, Dr. Herbert C. Brown, shared the 1979 Nobel prize in chemistry for his work with this compound and its organoborane derivatives.

In 1950 Metal Hydrides Incorporated (now THIOKOL/Ventron Division) obtained the first commercial license for the manufacture and use of sodium borohydride under issued patents and patent applications from the inventors, H. I. Schlesinger and H. C. Brown. The first commercial sales occurred in the same year, sodium borohydride selling for about $100/lb. A small scale commercial plant was put into operation in 1954 and the price of sodium borohydride dropped to approximately $35/lb and potassium borohydride to approximately $24/lb.

In the fall of 1957 a large scale manufacturing plant was designed, built and operated by Metal Hydrides, Inc. for the U. S. Navy's high energy fuels program. After the contract was terminated Metal Hydrides purchased this plant in 1960 and operated it while developing new uses for sodium borohydride. The price for sodium borohydride further decreased at that time. A second plant to produce stabilized aqueous borohydride solutions was built in 1975 by Ventron Corp. in Elma, Washington, to supply the pulp industry on the West Coast of the U. S. and Canada and the Asian market.

A small facility to produce sodium borohydride powder was built by Bayer, A. G. in Germany during the early 1960's and continues to operate at present--mainly to supply the internal needs of this company.

In late 1979, the Italian company, ANIC, announced their intention to start supplying sodium borohydride powder in Europe in 1980.

While gaining large scale industrial acceptance worldwide, sodium borohydride still remains an inorganic specialty chemical.

## 2. Method of Manufacture

Sodium borohydride is manufactured commercially by reacting sodium hydride and trimethylborate in a high boiling mineral oil medium at about 275°C.

$$4NaH + (CH_3O)_3B \xrightarrow{oil} NaBH_4 + 3NaOCH_3$$

A line drawing for the Ventron process is shown in Fig. 1.

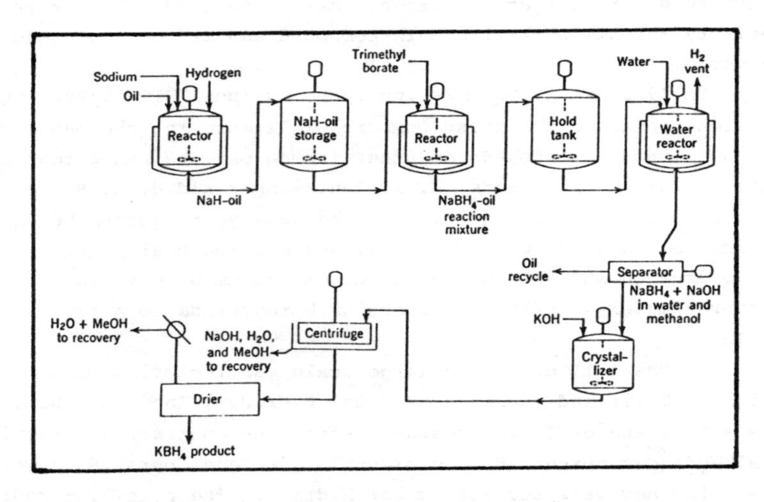


Fig. 1 Sodium and potassium borohydride production plant.

When the above reaction slurry is added to water, the oil separated from the aqueous layer and the methanol stripped from the solution, there results a very important commercial product--a solution containing 12% sodium borohydride in aqueous caustic soda called SWS™. Proprietary blends of this solution account for the major uses of sodium borohydride in many different applications. Sodium borohydride is then extracted from SWS, purified, crystallized and dried to a 97+% pure powdered product.

The Bayer process for manufacturing sodium borohydride is a high temperature dry-way reaction of sodium borosilicate, sodium metal and hydrogen. The overall reaction is:

$$Na_2B_4O_7 + 16Na + 8H_2 + 7SiO_2 \longrightarrow 4NaBH_4 + 7Na_2SiO_3$$

Sodium borohydride is recovered from the reaction mass by extraction with aqueous ammonia. The borohydride-ammonia solution is fed to a dryer which drives off the ammonia which is recovered and recycled. The sodium borohydride crystals are further dried and packaged. SWS cannot be produced directly in this process.

## 3. Properties

### 3.1. Physical and Thermodynamic

The important physical and thermodynamic properties of pure sodium borohydride powder are listed in Tables I and II[3].

TABLE I

Selected Physical Properties of Sodium Borohydride

| Property | |
|---|---|
| Formula | $NaBH_4$ |
| Molecular Wt. | 37.84 |
| Purity | 97+% |
| Color | White |
| Crystalline Form (anhydrous) | Face centered cubic a = 6.15 A.U. |
| (dihydrate) | Exists below 36.4°C. |
| Melting Point | 505°C. (10 atm. $H_2$) |
| | Decomposes above 400°C. in vacuum |
| Thermal Stability | Will not ignite at 300°C. on hot plate |
| | Ignites from free flame in air, burning quietly |
| Density | 1.074 g/cc. |
| Apparent Bulk Density | 5 lbs./gal. |

TABLE II
Thermodynamic Properties of Sodium Borohydride

| | Function | Value | Ref. |
|---|---|---|---|
| Free Energy of Formation | $\Delta F^{\circ}_{298}$ | −30.1 kcal/mole | (2) |
| Heat of Formation | $\Delta H^{\circ}_{298}$ | −45.53 kcal/mole | (1) |
| Entropy | $S^{\circ}$ | +24.26 cal/°mole | (4) |
| Heat Capacity | $C_p^{\circ}$ | +20.67 cal/°mole | (2) |
| Free Energy of Ionization $NaBH_4(s)=Na^{+}+BH_4^{-}$ | $\Delta F^{\circ}_{298}$ | −5660 cal/mole | (3) |
| Borohydride Ion, $BH_4^{-}$ (aq.) Free Energy of Formation | $\Delta F^{\circ}_{298}$ | +28.6 kcal/mole | (3) |
| Heat of Formation | $\Delta H^{\circ}_{298}$ | +12.4 kcal/mole | (3) |
| Entropy | $S^{\circ}_{298}$ | +25.5 cal/°mole | (3) |
| Hydrolysis $BH_4^{-}+H^{+}+3H_2O$ (liq.)= $H_3BO_3+4H_2(g)$ | $\Delta F^{\circ}_{298}$ | −88.8 kcal/mole | (3) |
| Oxidation $BH_4^{-}+8\ OH^{-}=B\ (OH)_4^{-}+4\ H_2O+8e^{-}$ | $\Delta F^{\circ}_{298}$ | −228.9 kcal/mole | (3) |
| | $E^{\circ}_{298}$ | +1.24 volts | (3) |

3.2. Solubility

3.2.1. Water. The solubility of sodium borohydride in water, the most commonly used solvent, has been accurately measured at different temperatures. The data presented in Fig. 2 show the equilibrium temperature of the two crystal forms $NaBH_4$ and $NaBH_4 \cdot 2H_2O$. The curve below 36.4°C represents the solubility of the dihydrate, and above 36.4°C the solubility of anhydrous $NaBH_4$.

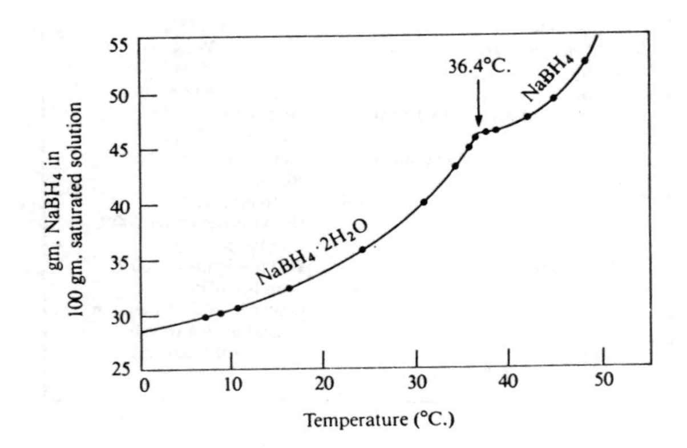


Fig. 2 The solubility of sodium borohydride in water at different temperatures.

3.2.2. Nonaqueous Solvents. In general sodium borohydride is soluble in polar solvents containing hydroxyl or amine groups. See Table III.

TABLE III
$NaBH_4$ Solubility in Various Solvents (g./100 g. solvent)

| Solvent | Temp. (°C.) | Solubility |
|---|---|---|
| water | 0 | 25.0 |
| | 25 | 55.0 |
| | 60 | 88.5 |
| liquid ammonia | 25 | 104.0 |
| methylamine | −20.0 | 27.6 |
| ethylamine | 17 | 20.9 |
| n-propylamine | 28 | 9.7 |
| isopropylamine | 28 | 6.0 |
| n-butylamine | 28 | 4.9 |
| cyclohexylamine | 28 | 1.8 |
| morpholine | 25 | 1.4 |
| | 75 | 2.5 |
| aniline | 75 | 0.6 |
| pyridine | 25 | 3.1 |
| | 75 | 3.4 |
| monoethanolamine | 25 | 7.7 |
| ethylenediamine | 75 | 22.0 |
| methanol | 20 | 16.4 (reacts) |
| ethanol | 20 | 4.0 (reacts slowly) |
| isopropanol | 25 | 0.37 |
| | 60 | 0.88 |
| t-butanol | 25 | 0.11 |
| | 60 | 0.18 |
| 2-ethylhexanol | 25 | 0.01 |
| tetrahydrofurfuryl alcohol | 20 | 14.0 (reacts slowly) |
| ethylene glycol dimethyl ether | 0 | 2.6 |
| | 20 | 0.8 |
| diethylene glycol dimethyl ether | 0 | 1.7 |
| | 25 | 5.5 |
| | 45 | 8.0 |
| | 75 | 0.0 |
| triethylene glycol dimethyl ether | 0 | 8.4 |
| | 25 | 8.7 |
| | 50 | 8.5 |
| | 100 | 6.7 |
| tetraethylene glycol dimethyl ether | 0 | 8.7 |
| | 25 | 9.1 |
| | 50 | 8.4 |
| | 75 | 8.5 |
| | 100 | 4.2 |
| dimethyl acetamide | 20 | 14.0 |
| dimethyl sulfoxide | 25 | 5.8 |
| acetonitrile | 28 | 2.0 |
| tetrahydrofuran | 20 | 0.1 |

3.3. Stability

Sodium borohydride is very stable thermally. It decomposes slowly at temperatures above 400°C in vacuum or under a hydrogen atmosphere. Sodium borohydride absorbs water rapidly from moist air to form a dihydrate which decomposes slowly forming hydrogen and sodium metaborate. Decomposition in air is therefore a function of both temperature and humidity.

3.3.1. Aqueous Solutions. The stability of sodium borohydride in water is dependent upon the temperature and the pH. The hydrolysis reaction is accelerated by increasing the temperature and lowering the pH.

$$NaBH_4 + 2H_2O \longrightarrow NaBO_2 + 4H_2$$

As the borohydrides are alkaline, the higher the concentration, the more stable the solution. See Table IV and Section 3.4.2.3

TABLE IV
pH of Solutions of $NaBH_4$ at 24°C.

| Concentration of $NaBH_4$ | pH |
|---|---|
| 1.000 M | 10.48 ± .02 |
| 0.100 M | 10.05 ± .02 |
| 0.010 M | 9.56 ± .02 |

The hydrolysis of sodium borohydride in water causes a rise in pH value, and the rate of decomposition therefore decreases. For example, a 0.01 M solution of $NaBH_4$ has an initial pH of 9.6 which changes during hydrolysis to 9.9.

3.3.1.1. Effect of pH. It is obvious, therefore, that the addition of sodium hydroxide will stabilize aqueous sodium borohydride solutions. (Fig. 3) At higher pH values there is essentially no decomposition during storage.

3.3.1.2. Effect of Temperature. If the temperature is increased, the stability decreases, as shown in Fig. 4. This can be compensated for, however, by adding more caustic or increasing the sodium borohydride concentration.

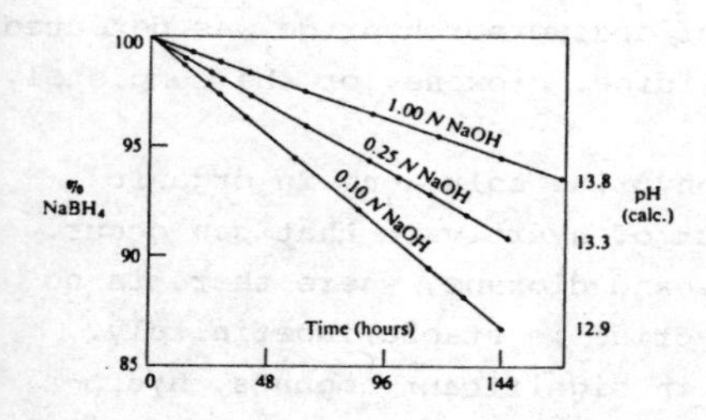


Fig. 3. Effect of pH on stability of $NaBH_4$ solutions.

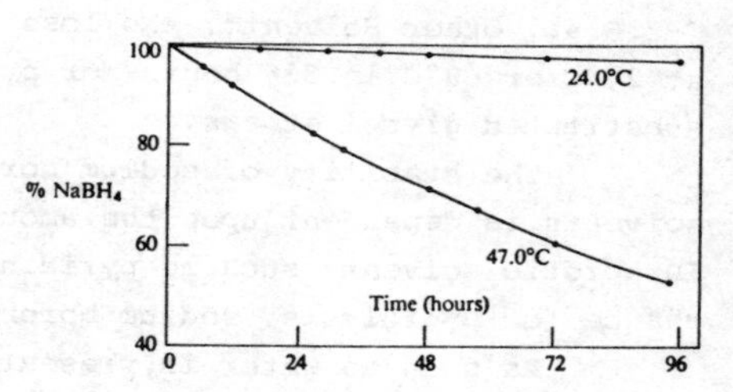


Fig. 4. Stability of an alkaline solution (1.00 N NaOH) of sodium borohydride (0.10 M $NaBH_4$) at 24°C and 47°C.

3.3.1.3. Effect of Catalysts. In addition to the above mentioned variables many finely divided transition metals (which may result from the borohydride reduction of the metal ions) can catalyze the hydrolysis reaction even in strongly basic solutions. These metals in order of decreasing catalytic activity are:

Ru = Rh > Pt > Co > Ni > Os > Ir > Fe.

3.3.2. SWS™ is a stabilized water solution of 12% sodium borohydride in caustic soda. 12% solutions of sodium borohydride in 41% sodium hydroxide decompose at a rate of only 0.008% per day at 100°C and at 0.002% per day at 54°C.

3.3.3. Alcohol Solutions. $NaBH_4$ is unstable in both methanol and ethanol due to solvolysis, but is indefinitely stable in higher alcohols such as isopropanol, t-butanol and 2-ethylhexanol, even at elevated temperatures. There is no appreciable decomposition of sodium borohydride in isopropanol or t-butanol at temperatures as high as 60°C. The higher the molecular weight of the aliphatic alcohol, the higher the temperature at which sodium borohydride can be used.

The instability in lower alcohols can be overcome by the addition of base, as in aqueous solutions. For example, in ethanol only 5.7% of the sodium borohydride is lost in 144 hours at 24°C in the presence of 2N sodium hydroxide.

3.3.4. Other Solvents. No loss of sodium borohydride was detected at 24°C or 60°C in 336 hours for pyridine, dioxane, or the completely substituted glycol ethers.

The stability of sodium borohydride solutions in organic solvents is dependent upon the amount of hydrolysis that can occur. In aprotic solvents such as pyridine and dioxane, where there is no chance for hydrolysis, sodium borohydride is stable indefinitely.

As soon as water is present in significant amounts, hydrolysis can occur and affect the stability.

Concentrated solutions of $NaBH_4$ in dimethylformamide (DMF) (>2.0 M) in the absence of a reducible substrate are unsafe at elevated temperatures, and violent exothermic reactions can occur, generating flammable gases ($H_2$, $HN(CH_3)_2$ and $N(CH_3)_3$). Dissolving $NaBH_4$ in DMF, in large quantity, without efficient cooling of the vessel is hazardous, due to the high heat of solution (estimated to be >1.5 kcal/mole). To avoid the potential violent reaction, it is recommended that the use of dimethylacetamide be evaluated instead, especially at higher temperatures.

3.4. Chemical Properties

3.4.1. Organic Reductions. Historically, the first commercial uses for sodium borohydride were for the reduction of organic compounds containing carbonyl groups. Classical techniques for accomplishing these reductions have been developed.

Sodium borohydride attacks the carbon atom which has the largest positive charge.

$$4\left[\begin{matrix} H \\ | \\ -\overset{}{C}=O \\ \delta+ \end{matrix}\right] + BH_4^- \longrightarrow 4\left[\begin{matrix} H \\ | \\ -C-OH \\ | \\ H \end{matrix}\right]$$

Because of this, any substituent which increases the fractional positive charge on the carbonyl carbon atom will increase the rate of reduction. If the fractional positive charge is decreased by substituents, then the rate is slowed down.

Sodium borohydride is an attractive reducing agent for organic substrates because of its convenience as well as its selectivity and efficiency. The general techniques of its use are by now well known to the practitioner of organic synthesis, who also knows that modifications are sometimes dictated by the properties (solubility, thermal stability, pH sensitivity) of the material being reduced.

The reduction of a large number of organofunctional groups has been discussed in detail in a number of recent reviews[1-3]. Excellent references to the original literature are available in these reviews. A summary of functional groups reduced by various borohydrides and borohydride derivatives is shown in Table V. Industrial applications are covered in Section 4.1.

TABLE V

Organic Functional Group - Reducing Agents

| | Aldehyde | Ketone | Acid Chloride | Ester | Carboxylic Acid | Carboxylic Salt | Amide | Imide | Epoxide | Lactone | Carbinol | Imine | Nitrile | Nitro | Unsat. Quat. | Halide | Tosylhydrazone | Olefin |
|---|---|---|---|---|---|---|---|---|---|---|---|---|---|---|---|---|---|---|
| $NaBH_4$ | + | + | + | ±[a] | − | − | − | − | ± | ± | | + | − | − | + | − | | − |
| $LiBH_4$ | + | + | + | + | − | − | − | | + | + | | | − | − | | − | | − |
| $Zn(BH_4)_2$ | + | + | + | − | | | | | − | − | | | | | | | | − |
| $NaBH_4/AlCl_3$ | + | + | + | + | + | − | + | | + | + | | | + | − | | | | + |
| $NaBH_3CN$ | +[b] | +[b] | | − | | | − | + | | | | + | − | − | + | + | + | + |
| $NaBH_2S_3$ | + | + | − | − | | | + | | | + | | | + | + | | | | − |
| $LiBH(R)_3$ | + | + | + | + | | | | | + | + | | | | | | | | |
| $NaBH(OAc)_3$ | + | ±[c] | | | | | | | | | | | | | + | | + | +[d] |
| $NaBH_3(OAc)$ | | | | − | | | + | | | | | | + | − | | | | |
| $NaBH(OCOCF_3)_3$ | | | | | | | | | | | + | | | | | | | +[d] |
| $NaBH_3(OCOCF_3)$ | | | | − | | | + | | | | | | + | − | | | | |
| $NaBH(OR)_3$ | + | + | + | +[e] | | | | | | | | | +[e] | | | | | |
| $NaBH_3(OH)$ | | + | | + | | | | | | | | | + | + | | − | | |
| $NaBH_3$(anilide) | + | + | + | + | | | − | | | | | | − | + | | | | |
| $NaBH_2$(ethanedithiolate) | | | | | | | + | + | | | | | | + | | | | |
| $THF \cdot BH_3$ | + | + | − | ±[a] | + | − | + | | + | + | | | + | − | | | | + |
| $(3\text{-Me-2-Bu})_2BH$ | + | + | − | − | − | | | | − | +[f] | | | + | + | | − | | + |
| $(CF_3COO)_2BH$ | | | | | | | | | | | | | | | | | | +[g] |

*a* - slow reaction
*b* - pH 3 - 4
c - <10% in refluxing benzene
*d* - activated double bonds
*e* - elevated temp.
*f* - reaction in 1:1 ratio
g - indole to indoline
*h* - reverse addition gives aldehyde

3.4.2. Inorganic Reactions. Sodium borohydride is a tremendously versatile reducing agent and ligand for inorganic reactions, as shown by the wealth of literature which has appeared in the last ten years. Some excellent reviews have appeared in the literature and are recommended[4 - 7].

3.4.2.1. Metal Cation Reductions. A substantial number of metal cations are reduced by borohydride in protic or aprotic solvents. Reduction can be classified according to the product obtained. The products may be a lower valence compound, the free element, a volatile hydride or a metal "boride". Some of these reductions are summarized in Table VI.

Table VI

Cations Reduced by $NaBH_4$

| Reactant | Lower Valence | Element | Boride | Volatile Hydride |
|---|---|---|---|---|
| $Fe^{3+}$ | $Fe^{2+}$ | Fe° | | |
| $Ru^{3+}$ | | Ru° | | |
| $Pd^{2+}$ | | Pd° | | |
| $Cu^{2+}$ | | Cu° | Cu(B) | |
| $Ag^{1+}$ | | Ag° | | |
| $Hg^{2+}$ | | Hg° | | |
| $Pb^{2+}$ | | Pb° | | |
| $Ce^{4+}$ | $Ce^{3+}$ | | | |
| $Cr_2O_7^{2-}$ | $Cr^{3+}$ | | | |
| $MnO_4^{2-}$ | $Mn^{3+}$ | | | |
| $Ni^{2+}$ | | | $Ni_2B$ | |
| $Ge^{4+}$ | | | | $GeH_4$ |
| $As^{5+}$ | | | | $AsH_3$ |

A large number of industrial uses have been developed based on the borohydride reduction of metal cations which are treated under Section 4.4.1. - 4.4.6.

3.4.2.2. Anion Reductions. An important industrial use of sodium borohydride is the reduction of the bisulfite anion to produce the dithionite (hydrosulfite) anion[8,9]:

$$BH_4^- + 8HSO_3^- \longrightarrow 4S_2O_4^{2-} + BO_2^- + 6H_2O$$

Hydrosulfite $(S_2O_4)^{2-}$ generated by $NaBH_4$ reduction of tetravalent sulfur species is widely applied industrially in bleaching mechanical pulps[10], vat dye reductions[11], and clay leaching[12]. (See Section 4.5.1. - 4.5.3. for further details.)

Some of the more important anions which are easily reduced by borohydride are shown in Table VII.

Table VII

Anion Reductions by $NaBH_4$

| Reactant | Product | Remarks |
|---|---|---|
| $NO_3^-$ | $NH_3$ | (Ni, Co catalyst) |
| $OCl^-$ | $Cl^-$ | |
| $Fe(CN)_6^{3-}$ | $Fe(CN)_6^{4-}$ | |
| $HSO_3^{1-}$ | $S_2O_4^{2-}$ | pH = 5 - 6 |
| $SO_3^{2-}$ | No reaction | pH = 9 |

3.4.2.3. Reaction with Water. The reaction of sodium borohydride with water $NaBH_4 + 2H_2O \longrightarrow 4H_2 + NaBO_2$ is of enormous practical importance. If this hydrolysis reaction occurs in competition with the reduction of an organic or inorganic compound, borohydride obviously is wasted and its utilization effiency is lowered. On the other hand, sodium borohydride is a remarkably concentrated source of hydrogen. One gram, dissolved in water, will release 2.37 liters of molecular hydrogen. Important industrial use is made of this property.

Factors which control the rate of hydrolysis include concentration of $BH_4^-$, concentration of NaOH or base (pH) and temperature. The effect of these variables on the rate of hydrolysis has been extensively studied. A computer program has been developed to predict these hydrolysis rates[13]. A comparison of reported and predicted values for % borohydride decomposition in water is shown in Table VIII.

Table VIII

Comparison of Reported and Predicted Values for % Decomposition of Borohydride

| $[BH_4^-]$ | [NaOH] | % NaOH by wt | Temp °C | % Decomp. Reported | % Decomp. Predicted | Time in Days |
|---|---|---|---|---|---|---|
| 0.10M | 0.1 N | ---- | 24 | 9.21 | 6.98 | 4 |
| 0.10M | 0.25N | ---- | 24 | 6.52 | 4.67 | 4 |
| 0.10M | 1.00N | ---- | 24 | 3.75 | 1.87 | 4 |
| 0.10M | 1.00N | ---- | 47 | 50.06 | 22.15 | 4 |
| 0.27% | ---- | 1.0% | 21 | 0.545 | 0.82 | 1 |
| 1.34% | ---- | 5.0% | 21 | 0.233 | 0.253 | 1 |
| 2.68% | ---- | 10.0% | 21 | 0.0783 | 0.105 | 1 |
| 10.00% | ---- | 10.0% | 21 | 0.0538 | 0.105 | 1 |
| 5.35% | ---- | 20.0% | 21 | 0.00538 | 0.0243 | 1 |
| 8.02% | ---- | 30.0% | 21 | 0.0020 | 0.00428 | 1 |
| 9.35% | ---- | 35.0% | 21 | 0.000038 | 0.00117 | 1 |
| 11.00% | ---- | 40.0% | 21 | 0.000005 | 0.000116 | 1 |
| 12.90% | ---- | 46.9% | 21 | 0.0000003 | ---- | 1 |
| 5.00% | ---- | 35.0% | 54 | 0.00215 | 0.0496 | 1 |
| 9.35% | ---- | 35.0% | 54 | 0.0021 | 0.0496 | 1 |
| 11.00% | ---- | 40.0% | 54 | 0.00045 | 0.0055 | 1 |
| 12.90% | ---- | 46.9% | 54 | 0.0002 | ---- | 1 |

The times for complete hydrolysis of sodium borohydride solutions in water at 25°C which have been buffered at various pH's are shown in Table IX.

Table IX

$NaBH_4$ Hydrolysis Time vs pH

| pH | $NaBH_4$ Completely Hydrolyzed In |
|---|---|
| 4.0 | 0.02 sec |
| 5.0 | 0.22 sec |
| 6.0 | 2.2 sec |
| 6.25 | 3.9 sec |
| 6.5 | 7.0 sec |
| 6.75 | 12.4 sec |
| 7.0 | 22.1 sec |
| 8.0 | 3.7 min |
| 9.0 | 36.8 min |
| 10.0 | 6 hr 8 min |

3.4.2.4. Organometallic Reactions. The rapid growth of organometallic chemistry in recent years has given rise to numerous applications of sodium borohydride's reducing capabilities. From a survey of literature citations using $NaBH_4$ in this specialized field, it quickly becomes apparent that, in general four major types of reactions are involved: initial formation of organometallic compounds and complexes, reduction to lower-valent metal compounds, demetallation or cleavage of organometallics to the metal and organic species, and conversion of organometallic halides to the corresponding hydride or hydride halide.

These reactions can be generalized as follows, where L = ligand, M = metal, X = anion, and n + m = valence of M.

$$L_nMX_m \xrightarrow{NaBH_4} L_n + mM^\circ \quad \text{(metal precipitate)}$$

$$L_nMX_m \xrightarrow[\text{excess L}]{NaBH_4} L_{n+1}MX_{m-1} \quad \text{(lower-valent metal complex)}$$

$$L_nMX_m \xrightarrow[\text{excess L}]{NaBH_4} L_nMX_{m-1}H \longrightarrow L_nMH_m \quad \text{(complexed metal hydride)}$$

$$L_nMX_m \xrightarrow[\text{excess L}]{NaBH_4} L_nMX_{n-1}BH_4$$

$$L_nM\underbrace{HBH_4}_{m} \quad \text{(complex metal hydride -- borohydride)}$$

$$L_nM(BH_4)_m$$

Where borohydride is ligand, coordination with the metal can be monodentate, bidentate or tridentate.

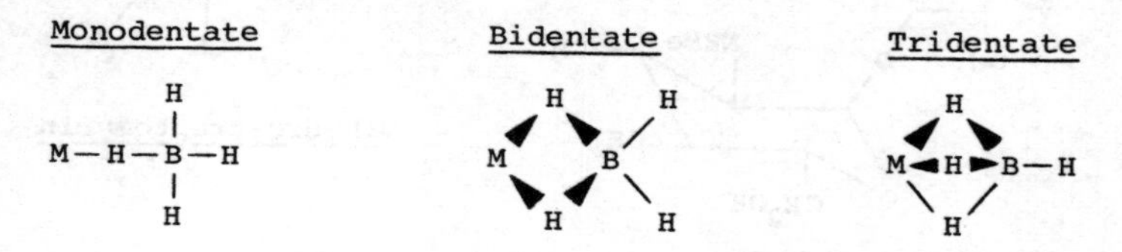


Several important industrial processes have been developed based on organometallic catalysts prepared from sodium borohydride. These are discussed in Section 4.4.3.

## 4. Industrial Applications

### 4.1. Pharmaceutical-Fine Chemical

Reduction of organic compounds by $NaBH_4$ is a major use area and continues to grow. Early in its history, $NaBH_4$ was found to be a nucleophile and to reduce aldehydes rapidly and ketones more slowly to the corresponding alcohol. Its ability to reduce these functional groups quantitatively in the presence of other functional groups, i.e. its selectivity, led to its broad acceptance and use in the synthesis of pharmaceuticals and fine organic chemicals.

Examples of some of the products made on an industrial scale by borohydride reduction are listed below:

#### 4.1.1. Vitamin A (Retinol)[14].

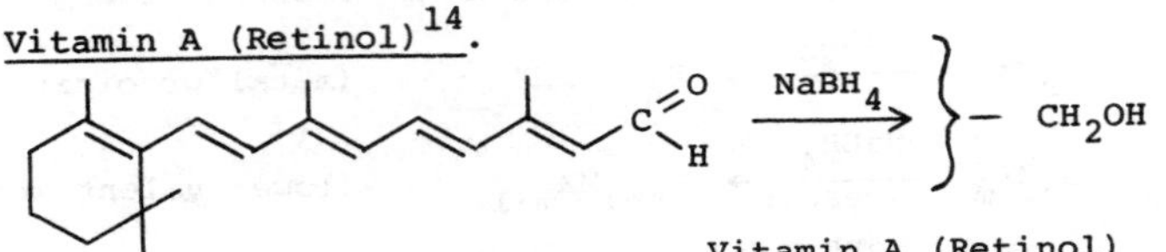

C(=O)H —$NaBH_4$→ — $CH_2OH$

Vitamin A (Retinol)

#### 4.1.2. Dihydrostreptomycin[15].

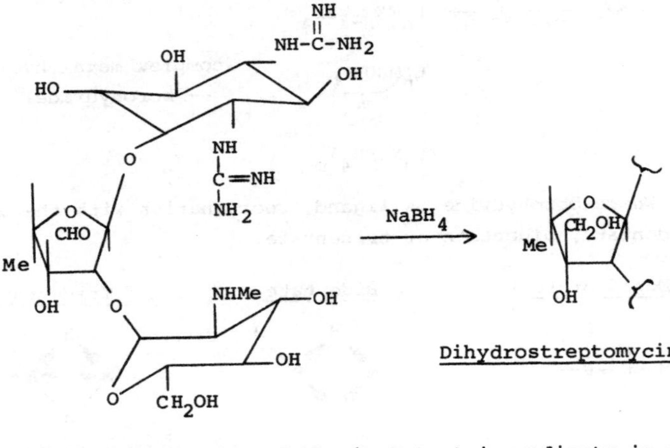

NH, NH-C-NH$_2$, OH, HO, OH, O, NH, C=NH, NH$_2$, O, CHO, Me, OH, O, NHMe, OH, OH, O, $CH_2OH$

$NaBH_4$ →

O, $CH_2OH$, Me, OH

Dihydrostreptomycin

4.1.3. Isohumulone. One of the important ingredients in the brewing of beer are the hop extracts. An active ingredient is the compound isohumulone, a photosensitive material. When the unmodified hop extract containing isohumulone is added to beer, the beer must be bottled in colored glass bottles, or else the photo reaction of isohumulone will generate a "skunky" odor and flavor in the beer. Reduction of isohumulone with sodium borohydride eliminates the photosensitive carbonyl group and beers containing this extract can be bottled in clear glass containers[16].

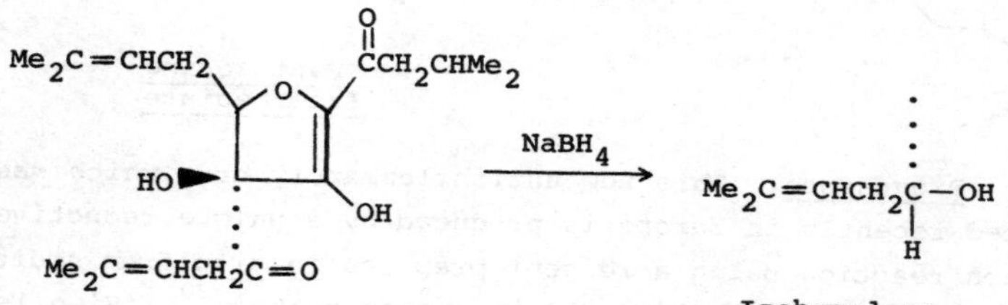

$$Me_2C{=}CHCH_2 \ldots CCH_2CHMe_2,\ HO,\ OH,\ Me_2C{=}CHCH_2C{=}O \xrightarrow{NaBH_4} Me_2C{=}CHCH_2\overset{}{C}(H){-}OH$$

Isohumulone

4.1.4. [2-(3 phenoxy phenyl) propanoic acid]. This compound is an important antiinflammatory used in treating arthritis. One of the intermediate steps in its manufacture involves the reduction of a substituted acetophenone to the alcohol with sodium borohydride[17].

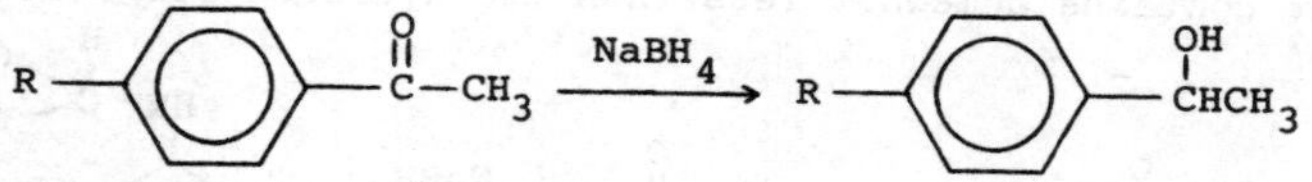

$$R{-}C_6H_4{-}C(O){-}CH_3 \xrightarrow{NaBH_4} R{-}C_6H_4{-}CH(OH)CH_3$$

Similar reductions are used in the preparation of perfume intermediates.

4.1.5. Chloramphenicol. A major application for sodium borohydride in Europe is for the production of Chloramphenicol, a broad spectrum antibiotic. The acylazide group of the precursor is reduced to the carbinol[18].

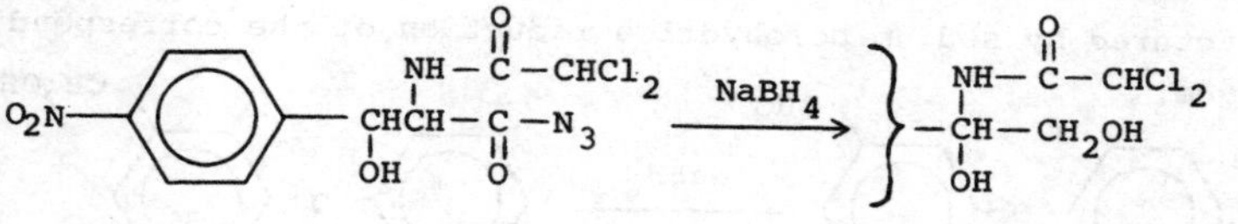

$$O_2N{-}C_6H_4{-}CH(OH){-}CH(NH{-}C(O){-}CHCl_2){-}C(O){-}N_3 \xrightarrow{NaBH_4} O_2N{-}C_6H_4{-}CH(OH){-}CH(NH{-}C(O){-}CHCl_2){-}CH_2OH$$

Chloramphenicol

4.1.6. Pentazocine. An analgesic which is widely used as an aspirin substitute uses interesting carbon-carbon double bond reduction with sodium borohydride in its synthesis[19].

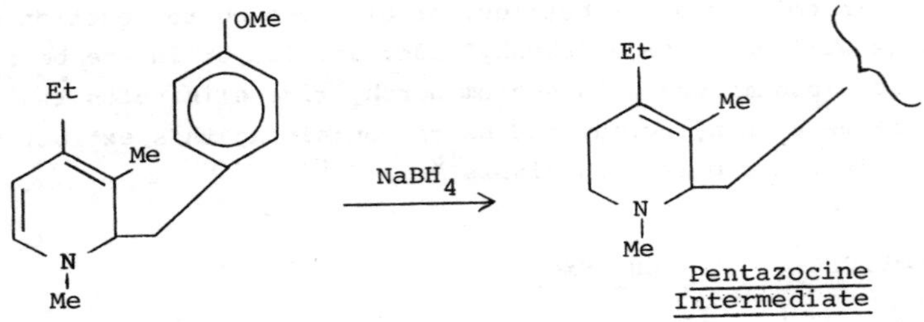

OMe

Et

Me

N

Me

NaBH$_4$

Et

Me

N

Me

Pentazocine Intermediate

4.1.7. Proquazone. This new antiinflammatory drug which was introduced recently in Europe is produced by a unique reductive alkylation reaction using a reagent prepared in situ from sodium borohydride and terephthalic acid in excess acetone[20]. Simplistically, the reagent appears to be

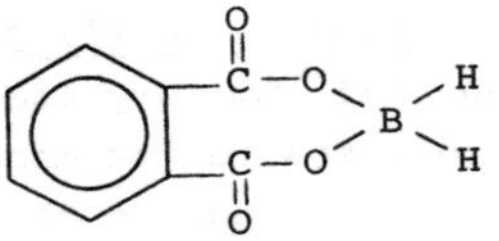

O, C—O, B, H, C—O, H, O

but in fact the reagent contains somewhat less than two hydrides per boron,

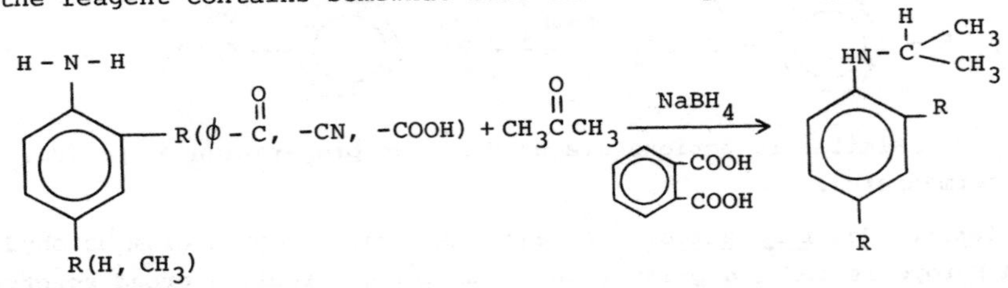

H - N - H

R(φ - C(=O), -CN, -COOH) + $CH_3C(=O)CH_3$ —NaBH$_4$ / COOH, COOH→

HN - C(H)($CH_3$)($CH_3$)

R

R

R(H, $CH_3$)

4.1.8. m-Phenoxybenzylacohol. This alcohol is an intermediate in the production of certain synthetic pyrethroid insecticides. It is manufactured by sodium borohydride reduction of the corresponding aldehyde.

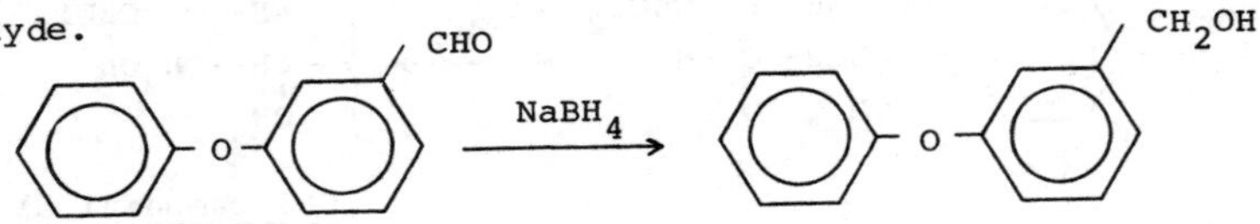

CHO

O

NaBH$_4$

O

$CH_2OH$

Many other pharmaceutical and fine chemical products use small amounts of sodium borohydride in their manufacture and several more are in development which will be of commercial importance.

4.2. <u>Process Stream Purification</u>

4.2.1. <u>Background</u>. About 20 years ago, several manufacturers of Oxo alcohols developed processes using sodium borohydride to reduce the aldehyde and other impurities in these alcohols[21]. The presence of these impurities caused the plasticizer esters made from the alcohols to have an undesirable color or odor. The borohydride purification technique proved to be technically and economically superior to alternative methods such as catalytic hydrogenation, distillation, absorbtion or oxidation. The purification of many high volume organic products which contain reducible carbonyl, peroxy or metal ion impurities has grown rapidly in the ensuing years and now is a major market for sodium borohydride.

4.2.2. <u>Present Uses</u>. Well over 1 billion pounds per year of the following products are treated with sodium borohydride according to these recommendations: (It should be emphasized that many of these applications are protected by patents currently in force.)

Table X

| Product/Process | Recommended $NaBH_4$ Use Levels (Form) | Applications | Comments |
|---|---|---|---|
| Amines | 5-1000 ppm (powder) | Color stabilization, decolorization, prevention of color development in subsequent product. | Higher levels required for decolorization. |
| Aromatic Amines | 5-500 ppm (powder, possibly SWS wash) | Color stabilization, decolorization, prevention of color development in subsequent product. | Higher levels required for decolorization. |
| Alcohols | 5-150 ppm (powder, pellet, or SWS) | Aldehyde removal, improvement in acid color, metal cation removal, odor removal, prevention of color in subsequent products. | Higher levels required on stripped alcohol. |
| Amides | 500-2000 ppm (powder or SWS wash) | Decolorization, color stabilization. | |
| Alkoxylation of Alcohols, Amines or Amides | 100-1000 ppm (powder, possibly SWS) | Prevention of color. | Add SBH either to alcohol or after alkaline catalyst. |
| Quaternization | 100-1000 ppm (powder) | Prevention of color. | Disperse or dissolve SBH in amine prior to reaction. |
| Esterification | 500-2000 ppm (SWS) | Decolorization of crude ester (see also alcohol treatment). | Added to caustic wash. |
| Neutralization/ Saponification | 500-2000 ppm (SWS) | Prevention of color. | Add SWS to caustic prior to reaction. |

4.2.3. <u>Techniques for Treatment</u>. Sodium borohydride is commercially available in three product forms: powder, SWS™ (a stabilized water solution of 12% $NaBH_4$ in sodium hydroxide) and 1 inch pellets. Techniques to use all three forms have been developed by THIOKOL/Ventron Division in cooperation with many customers.

- Sodium Borohydride Powder
  The powder may be added directly, as a slurry, or as a solution to the raw material storage, product storage or to tank cars. Some method of mixing is desirable.
- Sodium Borohydride Pellets
  Pellets are normally used in packed columns where the product is pumped through just before entering the storage tank. Alternatively the material in storage is recycled through the column.
- Sodium Borohydride - SWS™ (TM - THIOKOL/Ventron Division)
  SWS solution is used where caustic is not detrimental to the product stream. SWS is conveniently metered into the product stream, storage tanks, or tank cars. The small amount of added caustic and water may be left in the treated material or removed in subsequent finishing steps.

Another purification procedure involves metering the SWS solution directly into the final product distillation and adding SWS below the product take-off point permits efficient utilization of borohydride and prevents carry-over of caustic into the purified product.

4.2.4. Cost Considerations. The average use level of sodium borohydride for PSP applications is between 50 and 500 ppm. Assuming a cost of $15.00/lb of $NaBH_4$, the cost of $NaBH_4$ to treat 1 pound of product will be 0.075 to 0.75 cents. Obviously, the cost-benefit ratio is high and so this market for sodium borohydride is growing rapidly.

### 4.3. Polymer Processes

4.3.1. Background. The performance and useful lifetime of polymeric materials, including elastomers, plastics, fibers and coatings can be dramatically influenced by the presence of impurities which are frequently introduced during polymerization or polymer processing

For example, catalyst residues, metal ion contaminates and monomer impurities can adversely affect the polymerization process and physical properties of the final product, resulting in discoloration, odor and embrittlement. Using the same rationale described in the preceding section, sodium borohydride can effectively improve the performance of polymeric materials by chemically reducing the trace oxidized and metallic impurities to inert forms. This has led to the development of many applications in the polymer processing industry. Some of these are summarized below.

4.3.2. Monomer and Solvent Purification. Monomers such as styrene, vinyl chloride, butadiene, etc., must be free of oxidized or metallic impurities in order to meet rigid quality standards. Aldehyde impurities, e.g. acetaldehyde in vinyl chloride, can act as chain transfer agents and give a product with poor physical properties. The following illustrates the effect of borohydride purification on the carbonyl (divinylbenzene) or peroxide (styrene) level in two typical monomers.

| Monomer | Initial | After $NaBH_4$ Purification |
|---|---|---|
| Styrene | 0.02% | 0.0012% |
| Divinylbenzene | 33 ppm | N.D. |

In this example, the level of carbonyl and peroxide was reduced to acceptable concentrations with a typical sodium borohydride level of 100 - 200 ppm.

Olefin oxides are used to produce polymeric adducts with many substrates, mostly used in surfactants. The impurities present in olefin oxides, as in many other monomers, are primarily aldehydes (up to 500 ppm).

Because sodium borohydride does not react with olefin oxides, it can be effectively used for the purification of ethylene oxide, propylene oxide, and higher molecular weight compounds[22].

Phenol, a monomer in thermoset polymers, is produced mainly by the cleavage of cumene hydroperoxide into phenol and acetone. Impurities in this phenol develop undesirable color when it is chlorinated or reacted with formaldehyde. Aqueous solutions of the phenol can be purified by treating with as little as 0.3% $NaBH_4$ in 1 - 2 hours[23].

4.3.3. Monomer Inhibition and Stabilization. Sodium borohydride was examined at 100 - 200 ppm as a polymerization inhibitor for methyl methacrylate and found to be more effective than hydroquinone-type inhibitors in preventing polymerization during monomer storage under nitrogen at elevated temperature. $NaBH_4$ appears to function by reducing peroxide and hydroperoxide impurities that form during monomer storage. Unlike conventional free radical inhibitors, $NaBH_4$ does not interfere with azo initiator systems. In peroxide-initiated systems, the inhibitory effect of $NaBH_4$ can be overcome by simple acidification[24].

4.3.4. Polymerization Initiators. Borohydrides have been investigated for a variety of purposes in initiator systems for polymerization.

- To prepare reduced metal, metal complex or "Ziegler-Natta type" catalysts for oligomerization and polymerization.
- As a reductant in redox initiator systems[25].
- As an initiator in anionic polymerization[26].
- As an inhibitor or short stop in free radical polymerizations[27].

The major focus of attention has been on borohydride-reduced metal and metal complex catalysts for oligomerization and polymerization. Currently, borohydride-reduced nickel complexes are used in a commercial process for the manufacture of alfa-olefins.

The use of sodium borohydride as the reductant in redox initiator systems offers another attractive potential for improved polymers. Initial studies indicate that sodium borohydride may be employed as the primary reducing agent to promote the decomposition of persulfate initiators at low temperature to polymerize reactive monomers such as styrene and the acrylates.

For less reactive monomers, such as vinyl acetate, sodium borohydride may be employed as a secondary reductant to activate common redox systems of the persulfate/bisulfite or persulfate/metal ion types and achieve polymerization at higher rates or lower temperature.

For example, a polymerization rate of 14%/hour was observed using the common redox recipe $K_2S_2O_8/NaHSO_3$ with vinyl acetate at 20°C. The addition of sodium borohydride to this polymerization system increased the rate 20-fold, i.e. to 5%/minute (300%/hour)[28].

4.3.5. Stabilization of Polymers. Comprehensive studies on oxidized cellulose have shown that borohydride reduction of the carbonyl groups in this polymer greatly improves the stability of the cellulose to heat, light, and moisture.

Borohydrides have been found to be effective in the stabilization of polyvinyl alcohol, PVC, polyolefins, polyacrylates, polyamides,phenolics, polyethers, as well as cellulosics[29].

Borohydride concentrations of 0.01 to 1.0% by weight are generally effective. Sodium borohydride or its derivatives (quaternary ammonium borohydrides) may be added to the polymer in solution, during melt processings, or in some cases employed as a quench bath surface treatment.

4.4. Metal Reductions

4.4.1. Pollution Control. In aqueous solution, sodium borohydride is a very effective reducing agent for conversion of soluble metal cations, weakly associated complexes, and many organometallic compounds to the insoluble elemental metal:

$$NaBH_4 + 8OH^- \longrightarrow NaBO_2 + 6H_2O + 8e^-$$
$$8M^+ + 8e^- \longrightarrow 8M^\circ$$
$$NaBH_4 + 8M^+ + 8OH^- \longrightarrow NaBO_2 + 8M^\circ + 6H_2O$$

Since one mole of sodium borohydride can reduce eight equivalents of metal ion, low-level borohydride usage results in substantial reduction of the dissolved metal. To meet the discharge limits now set for many toxic heavy metals, processes using sodium borohydride have been developed and are in use by several industries.

The use of sodium borohydride for effluent treatment offers several advantages over conventional technologies. The procedure is irreversible, extremely rapid, and results in virtual complete removal of the dissolved metals even at low initial metal concentrations. The use of borohydride converts the soluble, oxidized metal species to an easily recoverable, compact precipitate and affords the user a practical, economical method of meeting existing and future discharge limits.

4.4.1.1. Lead. The removal of trace amounts of lead from the effluent waters of tetra alkyl lead manufacturers by treatment with sodium borohydride has been patented and is practiced industrially[30]. A secondary treatment with borohydride reduces the plant discharge to < 0.1 mg/L of lead. An excess of borohydride is required to completely reduce the organolead complex discharged. The reactions involved are summarized.

$$6R_3Pb + NaBH_4 + 6OH^- \longrightarrow 2R_6Pb_2 + NaBO_2 + H_2 + 4H_2O + 6X^-$$

$$2R_6Pb_2 \longrightarrow 3R_4Pb + Pb$$

Lead settles out in a settling pond and the effluent water is polish filtered and discharged.

Due to the excess sodium borohydride required, the chemical cost of treatment is $1 - 4 per pound ($2.20 - 8.80 per kilogram) of lead recovered.

4.4.1.2. Mercury[31]. In the electrolytic production of chlorine/caustic by mercury cells, mercury enters the effluent stream from several sources. These sources containing 10 - 50 mg/L of mercury are collected into a single pit into which sodium borohydride SWS™ is continuously metered. Reduction occurs as follows:

$$Hg^{+2} + NaBH_4 + 8OH^- \longrightarrow 4Hg^\circ\downarrow + NaBO_2 + 6H_2O$$

An excess of borohydride is used to maintain a reducing environment to prevent reoxidation of the mercury. From the collection pit the effluent is pumped to an agitated reaction tank to ensure good mixing and permit a 15 - 30 minute holding time. Then the effluent is filtered through a precoated pressure leaf filter which lowers the mercury concentration to 0.1 - 0.8 mg/L.

The chemical cost of the sodium borohydride treatment is $2 - 4 per pound ($4.40 - 8.80 per kilogram) of mercury recovered.

4.4.1.3. Silver (cadmium) from Photographic Film Manufacturers[32]. Sodium borohydride is used to recover silver from effluent stream not only for pollution control, but also for the economic value of the silver.

$$8Ag^+ + BH_4^- + 2H_2O \longrightarrow 8Ag + BO_2^- + 8H^+$$

Main sources of these effluents are from large scale photographic film manufacturers. The effluent may contain 10 - 120 mg/L of $Ag^+$ and in some cases 5 - 60 mg/L of $Cd^{+2}$. After borohydride treatment, the discharge shows an average of 0.09 mg/L Ag and 0.09 mg/L Cd. The chemical cost averages about $0.50 per troy oz for a metal which now sells at several times this cost.

A large number of smaller film processors use borohydride to recover silver from spent fixer solutions by primary borohydride precipitation or as a secondary treatment after electrolysis or iron wool precipitation.

Electroplaters recover silver (and gold) from electroplating baths by first oxidizing the cyanide present in the bath followed by borohydride reduction of the metal salts present. Borohydride reduction is very efficient in these processes (99.9%) and the silver averages 96 - 98% pure.

4.4.2. Precious Metal Recovery. Not only are silver and gold recovered by borohydride reduction processes as reviewed above, but many other precious metals, especially those present in spent catalysts, are also recovered. These processes generally involve dissolution of the metals from support materials followed by pH adjustment and borohydride precipitation. Although the value of the metals recovered is large, the amount of $NaBH_4$ used is small. In many cases, $NaBH_4$ has replaced the use of zinc for precious metal recovery with a consequent reduction in zinc ion discharge. New processes are being developed to recover copper, nickel and cobalt by borohydride reduction.

4.4.3. Catalyst Preparation. In the past 25 years over 300 publications have appeared in the international technical literature describing the reduction of metallic compounds with borohydride to produce catalysts, and the use of these catalysts in various reactions. Perhaps the nickel/cobalt reductions have received the most attention[33] although precious metals are not far behind. The reaction to produce nickel (and cobalt) boride takes place as follows.

$$2NaBH_4 + NiCl_2 + 4.5\,H_2O \longrightarrow 0.5\,Ni_2B + 1.5\,H_3BO_3 + 6.25\,H_2\uparrow + 2NaCl$$

A side reaction occurs which requires excess sodium borohydride to be used in these preparations.

$$NaBH_4 + 2\,H_2O \xrightarrow{M_2B} NaBO_2 + 4\,H_2\uparrow$$

Catalysts can be prepared as very finely divided slurries in water or alcohol, on inert supports, as colloidal complexes in dimethylformamide, as colloids trapped in polymeric gels, as organometallic complexes, and as polymer bound organometallic complexes.

Catalysts prepared by borohydride reduction in general have shown superior activity and life over catalysts prepared by other methods. Better sulfur tolerance has been noted in many instances.

Industrial processes which have been developed using borohydride reduced catalysts are summarized in Table XI.

Table XI

| Process | Catalyst (System) | Ref |
|---|---|---|
| Aniline → cyclohexylamine<br>phenol → cyclohexanol<br>benzene → cyclohexane | $CoCl_2 \cdot 6H_2O/NaBH_4/H_2O$ | (34) |
| nitrobenzene → aniline | $Ni_2B$/alcohol | (35) |
| nitriles → primary amines | $Ni_2B$, $Co_2B$ | |
| nitriles → secondary amines | Rh/B | (36) |
| nitriles → tertiary amines | Pt/B | |
| succinonitrile → 2-pyrrolidone | $Ni_2B$ | (37) |
| methanol + CO + $H_2$ → ethanol | $Co_2B$ (Rh) | (38) |
| olefin + CO + $H_2$ → aldehyde | Polymer bound Rh or Ir complex borohydride | (39) |
| CO or $CO_2$ + $H_2$ → methane | Ni/B, Co/B + promoters | (40) |
| acrylonitrile + $H_2O$ → acrylamide | Cu/B | (41) |
| auto emmission control | Pd on $Al_2O_3$ (Japan) | (42) |

Several borohydride reduced organometallic catalyst systems have been developed in recent years that are now in full scale use or under active development.

The Shell Higher Olefin Process (SHOP) oligomerizes ethylene to higher olefins using the catalyst system $Ni^{+2}/R_2PCH_2CH_2COONa/C_2\text{-}C_4$ diol/$NaBH_4$. It is critically important to have ethylene present at all times during the preparation of the catalyst[43].

A process for the dimerization of 1-butene to n-octene is being developed which uses the catalyst $NiCl_2$/cyclo-octadiene/$CF_3-C(=O)-CH_2-C(=O)-CF_3/NaBH_4$[44].

A process for the addition of HCN to n-octene-1 to produce pelargononitrile using the catalyst system $Ni^{+2}$/organophosphites/$NaBH_4$ is also under development[45].

The production of 2-substituted pyridines from nitriles and alkynes using a catalyst prepared in situ from $CoCl_2 \cdot 6H_2O/NaBH_4$ is expected to be commercialized shortly in Germany.

4.4.4. Magnetic Materials for Recording Tapes. Several companies in the U.S. and Japan have developed processes for the borohydride reduction of iron salts (together with minor amounts of other magnetic metal salts) to produce very finely divided magnetic iron (oxide) particles for use in high fidelity magnetic recording tapes[46]. Some of the Japanese produced materials are on the market now.

4.4.5. Electroless Plating of Metals. Sodium borohydride is a highly effective reducing agent for the chemical plating of transition metals and the precious metals on a wide variety of substrate materials. By carefully controlling conditions it is possible to plate out either the metal boride or the elemental metal onto both metals and non-metals such as glass, ceramics and plastics. This method provides a means of chemically depositing adherent films of nickel boride or cobalt boride on steel. It may also be applied to the electroless plating of copper and gold onto metals or non-metals such as printed circuit boards[47].

In actual practice, dimethylamine borane, which is derived from sodium borohydride, is used to a greater extent than sodium borohydride because of its better hydrolytic stability in the plating baths.

4.4.6. Metal Coating of Window Glass. For many years one of the major producers of architectural building glass and windows in the U.S. has used a patented process for coating the glass with extremely thin continuous coatings of nickel[48]. This is reduced directly on the glass surface with sodium borohydride. The use is growing with the growing interest in solar windows for energy conservation in large buildings.

4.5. Anion Reductions

4.5.1. Pulp Brightening. The largest single use of sodium borohydride is for the reduction of the bisulfite anion to produce sodium dithionite or sodium hydrosulfite[49].

$$8HSO_3^- + BH_4^- \longrightarrow 4S_2O_4^{-2} + BO_2^- + 5H_2O$$

Industrially, an aqueous solution of sodium borohydride called BOROL™* is reacted with additional sodium hydroxide and sulfur dioxide in a fully automated, pH controlled, all liquid process.

---

* BOROL™ - Trademark owned by THIOKOL/Ventron Division.

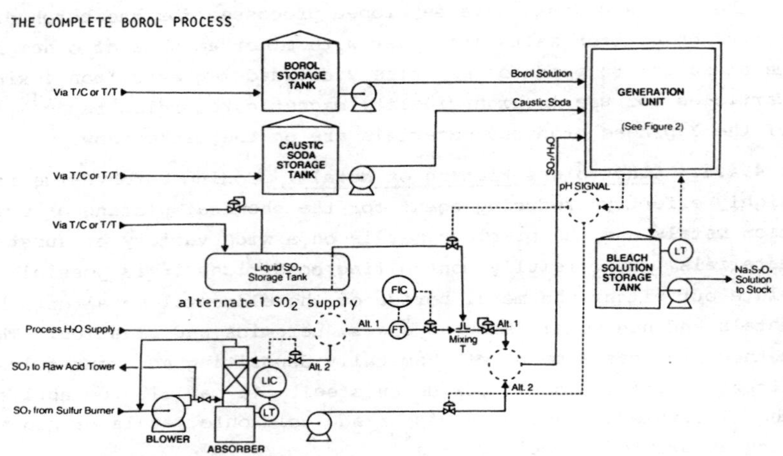


The complete Borol process for generating 100% active sodium hydrosulfite solution. $SO_2/H_2O$ solution can be prepared from either liquid $SO_2$ or burner gas.

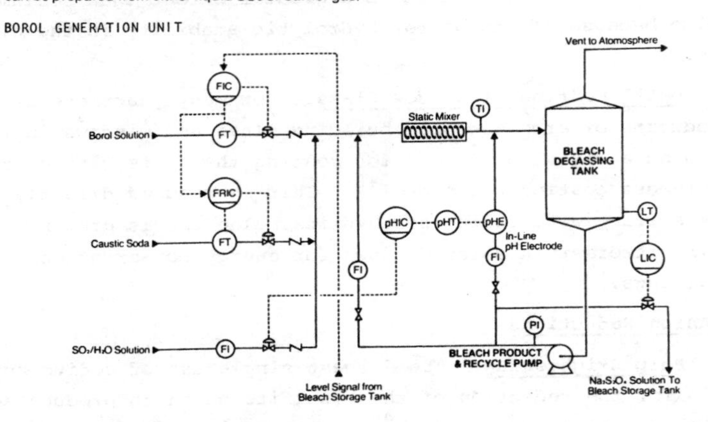


The generation unit provides full automation of all raw materials and generates sodium hydrosulfite solution on demand.

FIGURE 5.

The overall reactions are:

$$\underset{\text{BOROL}}{(1.0\ NaBH_4 + 3.4\ NaOH)} + 5.6\ NaOH + 9\ SO_2 \longrightarrow 4.0\ Na_2S_2O_4 + NaBO_2 + NaHSO_3 + 6\ H_2O$$

The major use for the hydrosulfite produced by this process is for the brightening of groundwood pulp. Over 40 mills in the U.S., Canada, Japan, Sweden, Finland and Italy presently use the process and many others are preparing to convert to it.

Completely built generating units and operating instructions are provided by Ventron, or are designed by Ventron for the individual pulp mills who assemble the units themselves. Solutions of hydrosulfite generated from BOROL, which are prepared only on demand, have demonstrated superior economy, convenience, flexibility and reliability over other sources of hydrosulfite--hence the rapid and continuing growth of this application. Schematic diagrams for the process and generating unit are shown in Figure 5.

4.5.2. Clay Leaching. Kaolin, as mined, contains minor amounts of ferric iron which causes the material to have a slight reddish-yellow color. For many applications, such as paper coatings and chinaware, this must be removed to obtain a product with the highest degree of whiteness possible. This is accomplished by a reductive leaching process using sodium hydrosulfite to reduce ferric to ferrous iron which then is leached from the clay at a pH of 2 - 3.5. Several large clay processors in the U.S. and England now use a sodium borohydride in caustic soda solution called Borite™ to generate hydrosulfite directly in a clay slurry[50]. Sulfur dioxide is first added and dissolved in the clay slurry. The Borite solution is then added with good mixing to form hydrosulfite.

$$NaBH_4 + 8\ SO_2 \longrightarrow 4\ S_2O_4^{-2} + 8H^+ + NaBO_2$$

Part of the acidity ($H^+$) is neutralized by the caustic in the Borite solution, but not all of it. The pH of the bleaching solution is 2.0 - 3.5. About a 90% yield of titratable hydrosulfite is obtained. After the reduction reaction is complete, additional sulfuric acid (and alum) may be added to completely flocculate the clay which is subsequently filtered and washed. Brightness increases of 1 - 8 points are achieved depending on the chemical dosage and the individual characteristics of the clay. Important savings are expected to be realized by clay producers over other forms of sodium hydrosulfite. A flow diagram of the process is shown in Fig. 6.

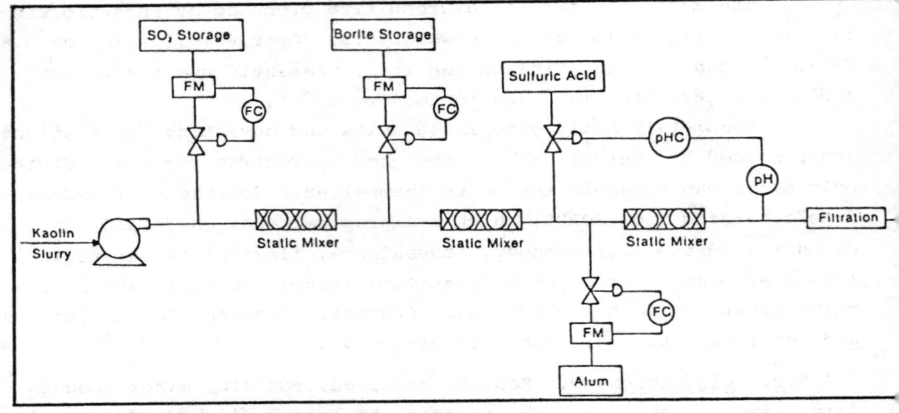


Fig. 6 Borite <u>In Situ</u> Process

4.5.3. <u>Textile Applications - Vat Dye Reductions</u>. Highly alkalin solutions of sodium hydrosulfite have long been used to reduce insol ble anthraquinone type vat dyes to the soluble leuco form. This the penetrates the textile fibres. Reoxidation of the reduced hydroquinone form to the insoluble quinone form precipitates the dye in fibr Vat dyed fabrics are described as being extremely fast--that is, resistant to loss of dye during use of the fabric. Many other types o dyestuffs have functional groups which also are reducible by sodium hydrosulfite such as sulfur dyes which contain the reducible disulfide group.

$$(\text{insol})\ -S-S- \xrightarrow{\text{red.}} -S-H \xrightarrow{\text{NaOH}} -S-Na\ (\text{sol})$$

Several systems have been developed for reducing vat dyes using sodium borohydride[51]. The first of these involves the generation of sodium hydrosulfite solution in a manner analogous to that described above for pulp brightening. The only significant differen is that the hydrosulfite generation reaction is run at higher soluti concentrations to produce a solution which contains about 15% sodium hydrosulfite. Where this solution is to be shipped, it is stabilized by raising the pH from about 6.5 (as generated) to about 12 with added NaOH. Some chelating agent is also added to prevent precipitation of iron hydroxide.

In Japan, a process has been developed for reacting a sodium borohydride-caustic solution with a sodium bisulfite solution in a continuous fashion to produce a reducing solution which undoubtedly contains sodium hydrosulfite. This solution, called "VENHIT", is used in several textile mills for the reductive clearing of dyed polyester-cotton fabrics.

Another system for reducing dyes uses a bath consisting of sodium hydroxide, sodium hydrosulfite and sodium borohydride. Part of the large excesses (up to 10X theoretical) of sodium hydrosulfite normally used in the dye bath can be replaced at considerable savings with small amounts of sodium borohydride with concurrent improvements in the overall quality of the dyeing[52]. This system is mainly used in high temperature (>220°F) dyeing systems. Borohydride appears to function as a stabilizer for the reduced (leuco) dye thus eliminating the need for large excesses of hydrosulfite.

A complex dye reducing system presently used in the U.S. and Europe consists of two components which are mixed in water prior to use to form the reducing solution. The first component consists of a solid mixture of sodium tetracyanonickelate, sodium formaldehyde sulfoxylate and sodium hydroxide. The second component is a solution of sodium hydroxide and sodium borohydride. The mixed reducing solution has superior air stability than sodium hydrosulfite solutions hence lower overall chemical costs are realized[53].

Alkaline sodium borohydride solutions are also used to reduce free formaldehyde on fabrics finished with resins containing this material. The sodium borohydride-sodium hydroxide solution is generally added to a wash box in the final finishing sequence and the formaldehyde is reduced to the less noxious methanol. Costs to remove 0.2% formaldehyde range from 0.1 - 0.5 cents per yard of fabric[54].

4.6. Hydrogen Generation. The hydrolysis of sodium borohydride to generate hydrogen has some unique industrial applications.

$$NaBH_4 + 2H_2O \xrightarrow[\text{metal catalyst}]{\text{acid or}} 4H_2\uparrow + NaBO_2$$

Since 1 g of $NaBH_4$ can release 2.37 liters of $H_2$(STP) and its solubility in water at 25°C is 55 g/100 g solution, then 100 g of solution can produce 130 liters of hydrogen. Sodium borohydride can be regarded as water soluble hydrogen.

4.6.1. Foamed Plastics. This reaction has been used to foam plastic materials. A recent publication describes the foaming of PVC plastisols and epoxy resins with sodium borohydride[55]. Since it will produce 10 - 20 times as much gas as an equal weight of commercial nitrogen blowing agents, substantial cost savings can be realized. Additional advantages are:

- Foaming can take place at room temperature.
- Hydrogen is an excellent heat transfer gas, thus curing takes place rapidly and evenly.
- In PVC systems, heat and light stabilizers can be eliminated from the recipe because sodium borohydride "cleans up" the plastisol system. (See Section 2.3.1.)

Other materials which can be foamed include: sodium silicate, polyurethanes, polyamides, unsaturated polyesters, polysulfides polyolefins, and explosives[56].

4.6.2. HYDRIPILLS™ (Trademark of THIOKOL/Ventron Division). HYDRIPILLS are tablets consisting of 92.5% sodium borohydride and 7.5% anhydrous cobalt chloride. The 3/4 inch (diam) tablet weighs about 1.7 g and will release 3.8 liters of hydrogen per tablet when added to water. A smaller sized tablet, 10/32 inch (diam) weighs about 0.3 g and produces 0.670 liters per tablet. HYDRIPILLS are used on a large scale for producing a reducing atmosphere in anaerobe jars in biology laboratories[57].

4.6.3. Miscellaneous. Other applications which have been develope for borohydride generated hydrogen are in fuel cells and pressurizing oil wells.

4.7. Preparation of Diborane and Derivatives

4.7.1. The preparation of diborane from sodium borohydride is the only practical method for obtaining this material and its derivatives on the industrial scale used today.

$$3\,NaBH_4 + 4\,BF_3 \xrightarrow{THF} 2\,B_2H_6 + 3\,NaBF_4$$

The chemistry of diborane, borane-THF ($BH_3$:THF) and the organoboranes prepared by the hydroboration reaction constitutes a huge body of chemistry which has been pioneered by H. C. Brown and others over the past forty years--but is beyond the scope of this chapter[58]. Small but growing industrial quantities of borane:THF, certain organoboranes, and alkylborohydrides are presently used. Substantial quantities of diborane are converted to carboranes for military applications.

4.7.2. Several amine boranes ($R_3N:BH_3$) derived from sodium borohydride via diborane and primary, secondary or tertiary amines have important industrial uses as developers in color photographic processes, reductants in electroless plating and chemoselective reducing agents for organofunctional groups. The covalently bonded amine boranes are generally more resistant to hydrolysis than sodium borohydride and some of them can be used in aqueous solution at pH of 7 or less without objectionable hydrolysis. Since they are also stable in air, they are a convenient source of borane for hydroboration reactions[59].

5. Conclusion. The extraordinary diversity of techniques to use and modify the remarkable chemical and physical properties of sodium borohydride which have been developed over the past 35 years has led to its industrial applications described in this chapter. There is no end in sight yet of potential new applications. Because of its high energy content, sodium borohydride will always be a relatively high priced reagent. Thus it will likely remain a specialty chemical and not a bulk commodity. Its ability to deliver its energy more efficiently than other materials or processes will determine where it will be used in the future. The accomplishments to date with sodium borohydride certainly are worthy of the Nobel Prize.

## 6. Literature References


1. E. Shenker, "Use of Complex Borohydrides and of Diborane in Organic Chemistry" in "Newer Methods of Preparative Organic Chemistry" Vol. IV, W. Foerst, ed., Academic Press, New York (1968).

2. A. Hajos, "Complex Hydrides and Related Reducing Agents in Organic Synthesis", Elesevier, Amsterdam (1979).

3. "Sodium Borohydride", THIOKOL/Ventron Division, Danvers, MA, 01923 USA (1979).

4. B. D. James and M. G. H. Wallbridge, Progr. Inorg. Chem., 11, 99 (1970).

5. T. J. Marks and J. R. Kolb, Chem. Rev., 77, 263 (1977).

6. V. Kadlec, H. Kadlecova and J. Masek, Chem. Listy, 100, 673 (197

7. T. F. Jula, "Inorganic Reductions with Sodium Borohydride. . .", Ventron Corp. (1974).

8. G. S. Panson and C. E. Weill, J. Inorg. Nucl. Chem., 15, 184 (19

9. P. Luner, R. LaPlaine and R. Wade, Pulp & Paper Mag. Canada, 65, T-101 (1964).

10. J. W. Gerrie, ibid., 75, T-251 (1974).

11. Ventron Corp., "Liquiven®".

12. U. S. 3,937,632 (1976).

13. M. M. Kreevoy and R. W. Jacobson, Ventron Alembic, 15, 2 (1979)

14. British Patent 778,753 (1957); U. S. 2,839,585 (1958).

15. a. U. S. 2,790,792 (1957);
b. U. S. 3,397,197 (1968).

16. U. S. 3,044,879 (1962).

17. U. S. 3,600,437 (1971).

18. G. Ehrhart, W. Südel, and H. Nahm, Chem. Ber., 90, 2088 (1957).

19. U. S. 3,250,678 (1966).

20. Ger. Offen. 2,701,888 (1977).

21. U. S. 2,867,651 (1959); 2,957,023 (1960); Brit. Pat. 981,965 (1

22. Ger. Offen. 1,114,704 (1963); U. S. 3,213,113 (1965).

23. U. S. 3,150,191 (1964).

24. S. L. Snyder, J. Appl. Poly. Sci., 24, 2237 (1979).

25. Brit. Pat. 910,933 (1963); U. S. 3,021,320 (1962); U. S. 3,597, (1971).

26. U. S. 3,021,315 (1962); Brit. Pat. 1,025,437 (1966); 1,031,087 (1966); U. S. Pat. Off Def. Pub. 875,006 (1970).

27. U. S. 3,396,154 (1968); 3,769,268 (1973); 4,085,267 (1978).

28. THIOKOL/Ventron Division, Technical Bulletin, "Redox Initiator Systems Based on Sodium Borohydride", (1979).

29. U. S. 3,086,963 (1963); 3,679,646 (1972); Brit. Pat. 808,108 (1959); 862,867 (1961); Jap. Pat. 75 34,339 (1975); Swed. Pat. 350,975 (1972) U. S. 3,372,140 (1968); 3,413,260 (1968); Jap. Pat. 73 15,473 (1973); 72 41,102 (1972); 71 06,869 (1971); Belg. Pat. 622,701 (1963); Neth. Appl. 6,704,682 (1967); 6,601,470 (1966); U. S. 3,374,275 (1968); 3,091,554 (1963).

30. U. S. 3,770,423 (1973).

31. U. S. 4,098,697 (1978); 3,847,598 (1974).

32. a. U. S. 4,131,455 (1978);
 b. M. M. Cook and J. A. Lander, Environ. Sci. & Tech., PAT Report (1980).

33. R. C. Wade, D. G. Holah, A. N. Hughes and B. C. Hui, Cat. Rev. - Sci. Eng., 14, 211 (1976).

34. Brit. Pat. 1,361,279 (1974).

35. A. D. Smith, "Hydrogenation of Nitrobenzene over P-2 Nickel Boride Catalyst", Thesis, Univ. of Louisville (Ky, USA), Dec. (1979).

36. C. Barnett, Ind. & Eng. Chem., 8, 145 (1960).

37. U. S. 4,036,836 (1977).

38. U. S. 4,171,461 (1979).

39. U. S. 4,178,312 (1979).

40. a. H. Kurita and Y. Tsutsumi, Nihon Kagaku Zaahi, 82, 1461 (1961).
 b. H. Hammer and I. Hakim, Chem. - Eng. Tech., 50, 622 (1978).
 c. R. W. Mitchell, L. J. Pandolfi and P. C. Maybury, J. Chem. Soc., Chem. Commun., 1976, 172.
 d. C. Bartholemew, Gordon Conference on Catalysis, Summer (1979).

41. Jap. Pat. 73 23,717 (1973); 73 36,118 (1973); 73 57,911 (1973); U. S. 3,962,333 (1976); 3,886,213 (1975); 3,944,609 (1976); 4,040,980 (1977).

42. Jap. Pat. 77 104,492 (1977).

43. U. S. 3,737,475 (1973); 3,686,351 (1972); 3,676,523 (1972); 3,825,615 (1974).

44. W. Keim, et al, J. Mol. Catalysis, 6, 79 (1979).

45. W. Keim, Private Communication.

46. U. S. 3,206,338 (1965); 3,535,104 (1970); 3,567,525 (1971); 3,661,556 (1972); 3,669,643 (1972); 3,672,867 (1972); 3,726,664 (1973); 3,837,912 (1974); 3,865,627 (1975); 3,966,510 (1976); 4,009,111 (1977); 4,020,236 (1977); 4,063,000 (1977); 4,069,073 (1978); 4,096,316 (1978); 4,097,313 (1978); 4,101,311 (1978); 4,125,474 (1978).

47. THIOKOL/Ventron Division, "Electroless Plating with Boron Hydrid Reductants", abstracts of over 100 references.

48. U. S. 3,671,291 (1972); 3,672,939 (1972).

49. a. P. Luner, R. LaPlaine and R. C. Wade, Pulp and Paper Mag. of Canada, 65, T-101 (1964).
   b. C. A. Richardson, D. C. Johnson and T. G. Goodart, TAPPI, 53 2275 (1970).
   c. R. G. Guess, Pulp & Paper, June, 1979, p. 74.

50. U. S. 3,937,632 (1976).

51. U. S. 2,991,152 (1961); 3,124,411 (1964).

52. a. U. S. 3,127,231 (1964);
   b. D. Vivilecchia, Am. Dyestuff. Reporter, 55, (13) 68 (1966).

53. M. M. Cook, ibid, March, 1979, 41.

54. D. G. Vivilecchia, ibid, September, 1972.

55. R. C. Wade and C. Letendre, J. Cell. Plastics, 16, 32 (1980).

56. U. S. 2,930,771 (1960); 3,355,398 (1967); 3,331,790 (1967); 2,758,980 (1956); 2,909,493 (1959); 3,084,127 (1963); 3,711,345 (1973);3,823,098 (1974); 3,114,724 (1963); Brit. Pat. 1,147,707 (1969).

57. J. H. Brewer, A. A. Heer, C. B. McLaughlin, Applied Microbiology 3, 136 (1955).

58. H. C. Brown, "Hydroboration", W. A. Benjamin, Inc., New York, NY (1962).

59. A. Pelter and K. Smith, Chapter 14.2 "Boron-Hydrogen Compounds" Comprehensive Organic Chemistry, ed. D. Barton and W. D. Ollis, Pergamon Press, Oxford, England (1979).

# Production and Use of Dithionites

By L. C. Bostian
VIRGINIA CHEMICALS INC., 3340 WEST NORFOLK ROAD, PORTSMOUTH, VIRGINIA 23703, U.S.A.

## Introduction

Dithionites are interesting chemical compounds which have a number of varied industrial uses. By far, the most important compound is sodium dithionite, $Na_2S_2O_4$, which is also known commercially as sodium hydrosulfite and sometimes as sodium hyposulfite. In this paper will be discussed the general properties of dithionites, the methods of manufacture, the structure of the industry and the commercial uses of the compounds.

## Historical

In 1718, G. E. Stahl found that iron dissolves in sulfurous acid forming a reddish yellow liquid. In 1789, C. L. Berthollet observed that no gas was evolved when the iron dissolved. It was soon found that other metals such as tin and zinc gave similar results. Electrolysis of sulfurous acid gave solutions which also had similar reducing powers. The empirical formula for dithionous acid was first found by Shützenberger to be $(HSO_2)_n$, and the correct formula $H_2S_2O_4$ was first determined by Bernthsen.(1)

Interest in dithionites was greatly accelerated by publishing of the book Das Hydrosulfite by K. Jellinek in 1912.(2)

The first commercial use of dithionites was in vat dyeing of cotton. In the early 1930's, dithionites were introduced as brightening agents for newsprint.

For many years zinc dithionite was the chief article of commerce for pulp bleaching, but recent regulations limiting discharges of zinc have resulted in a decrease in use of zinc dithionite and a dramatic increase in production and use of sodium dithionite.

## General Properties of Dithionites

Dithionous acid is unstable, but normal salts of the acid are well characterized. The salts are stable when maintained anhydrous, but solutions decompose slowly. Heating solid dithionites results in the reaction:

$$2\ S_2O_4^{=} \rightarrow S_2O_3^{=} + SO_3^{=} + SO_2$$

which occurs rapidly above 150°C with the sodium salt.[3] Other studies have indicated that sodium sulfate and elemental sulfur are also products of thermal decomposition:

$$S_2O_4^{=} \rightarrow SO_4^{=} + S$$

If the solid is heated to 150°C, rapid decomposition causes the temperature to rise further, eventually resulting in ignition. Anaerobic decomposition of solutions is also complicated[4] and occurs generally according to the equation:

$$2\ S_2O_4^{=} + H_2O \rightarrow S_2O_3^{=} + 2\ HSO_3^{-}$$

Under acid conditions, the reaction is second order and is quite rapid. Since the bisulfite ion is acidic, a neutral solution becomes acidic as hydrolysis occurs with the result that decomposition accelerates with time. To prevent this, it is customary to add some alkali to the solution so as to keep the pH above 9.5. It is also customary to keep the reaction temperature below ambient, to decrease the rate of hydrolysis.

In the presence of excess alkali, sulfide is formed:[5]

$$3\ Na_2S_2O_4 + 6\ NaOH \rightarrow 5\ Na_2SO_3 + Na_2S + 3\ H_2O$$

Consequently, too much excess alkali can also hasten decomposition.

Dithionite solutions are powerful reducing agents under alkaline conditions:

$$4\ OH^- + S_2O_4^= \ = \ 2\ SO_2^= + 2\ e^-$$

$$E_o \ = \ -1.12 \text{ volts}$$

In acid solutions, the reducing power is much less:

$$2\ H_2O + HS_2O_4^- = 2\ H_2SO_3 + H^+ + 2\ e^-$$

$$E_o \ = \ -0.08 \text{ volts}$$

Dithionite solutions are readily oxidized by air, producing bisulfite:

$$Na_2S_2O_4 + H_2O + O_2 \ \rightarrow \ NaHSO_3 + NaHSO_4$$

and it is essential to avoid contact with air in commercial applications. In the presence of 2-anthraquinone sulfonate as catalyst, aqueous $Na_2S_2O_4$ solution is an excellent scavenger of oxygen from inert gases. Studies of the oxidation of solutions have shown that the reaction is half-order with respect to $S_2O_4^=$. Recent electron spin resonance studies have shown the presence of the radical ion $\cdot SO_2^-$ to the extent of about 1 part in 3000.[6] It is probable that the dithionite ion splits as follows:

$$S_2O_4^= \rightleftharpoons 2\ \cdot SO_2^-$$

X-ray crystallographic studies[7] on $Na_2S_2O_4 \cdot 2H_2O$ have shown the dithionite ion to have the peculiar structure shown below:

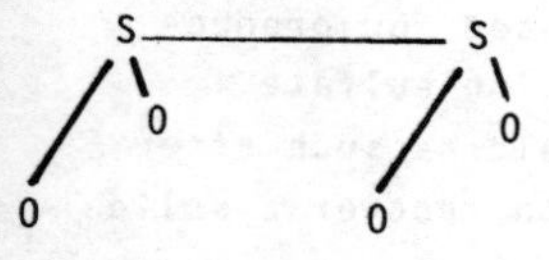

The two $SO_2$ planes are almost parallel, with the angle between the $SO_2$ planes and the S-S bond being 100°. The S-S bond distance is 2.39 Å, which is longer than the usual S-S bonds in polysulfides. The S-O bond distance is 1.15 Å. The crystals are monoclinic of space group P 2/c and have the approximate symmetry $C_{2V}$.

## Industrial Manufacture

Zinc dithionite is prepared by reaction of sulfur dioxide with an aqueous suspension of zinc dust:

$$2\ SO_2 + Zn \rightarrow ZnS_2O_4$$

After filtration to remove unreacted zinc, the resultant solution of zinc dithionite may be used as such, or can be drum dried to produce a powder.

With restrictions on the amount of zinc allowed in plant effluents, sodium dithionite has largely displaced the zinc compound in commercial usage. Sodium dithionite is prepared by four methods: (1) from zinc dithionite, (2) from sodium amalgam, (3) by the formate process, and (4) from sodium borohydride. I will discuss the first three methods, and leave the borohydride process to be discussed by Dr. Wade.

Zinc-Based Sodium Dithionite:[8] A solution of zinc dithionite can be readily converted to a solution of sodium dithionite by reaction with sodium carbonate solution to produce insoluble zinc carbonate:

$$ZnS_2O_4 + Na_2CO_3 \rightarrow Na_2S_2O_4 + ZnCO_3$$

One can also employ caustic soda to produce zinc hydroxide. A flow diagram of the process is shown in Figure 1. The zinc carbonate is filtered and washed, and can be used for production of zinc compounds such as zinc oxide or zinc sulfate. The sodium dithionite solution remaining is sold as such after proper addition of alkali, or can be treated to recover a solid product. To produce the solid, crystalline sodium chloride and alcohol are added to salt out the sodium dithionite, which precipitates as the dihydrate. After filtering the dihydrate, it is slurried with ethanol, heated to convert the dihydrate to the anhydrous form, which is then filtered and dried. An appreciable quantity of dithionite remains in the salt solution and is therefore not recoverable as a solid product.

Figure 1

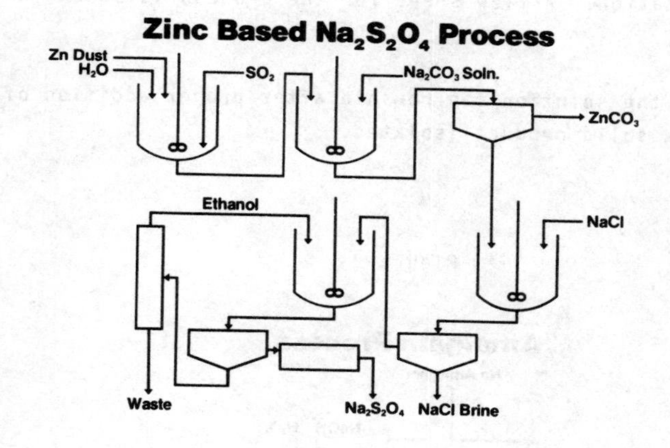


Amalgam Process:[9] When the mercury cell process is used to produce chlorine from sodium chloride, the sodium amalgam normally is reacted with water to produce sodium hydroxide. An alternative is to react the amalgam with sulfur dioxide to produce sodium dithionite. The equations can be written:

$$2\ NaCl + 2\ (Hg) \rightarrow 2(Hg)Na + Cl_2$$

$$2(Hg)Na + 2SO_2 \rightarrow Na_2S_2O_4 + 2(Hg)$$

The sodium amalgam is dropped through a special reactor containing sodium bisulfite solution maintained at a pH of about 5 and a temperature of about 23°C.

The sodium amalgam produced in the normal chlorine cell must be diluted with "used" amalgam, before being introduced to the dithionite reactor. In practice, two reactors in series are employed, and it is necessary to cool the reaction to maintain the temperature at 23°C. After completion of the reaction, the solution is filtered to remove traces of mercury, and then cooled to 0°C, diluted to the desired strength and alkali added for stabilization. A flow sheet for the process is shown in Figure 2.

If desired, the solution can be sold after proper addition of alkali, or a solid product isolated.

Figure 2

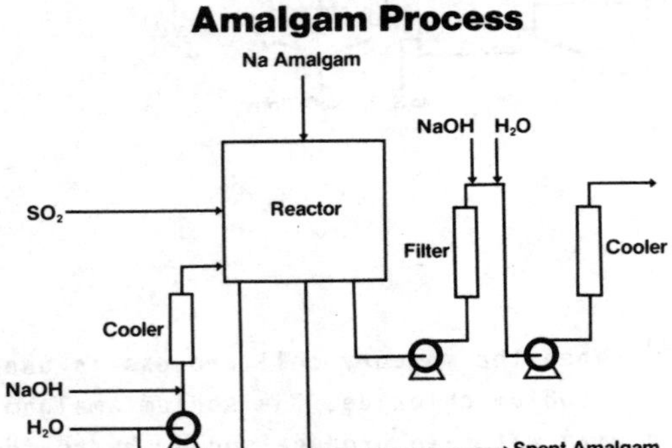

<u>Formate Process:</u>[10] The overall chemical equation for the production of sodium dithionite by the formate process is:

$$2\ SO_2 + NaOH + HCOONa \rightarrow Na_2S_2O_4 + CO_2 + H_2O$$

It is probable that the process occurs in several steps, such as:

$$SO_2 + NaOH \rightarrow NaHSO_3$$

$$SO_2 + HCOONa + H_2O \rightarrow NaHSO_3 + HCOOH$$

$$HCOOH + 2NaHSO_3 \rightarrow Na_2S_2O_4 + CO_2 + 2H_2O$$

The reaction is carried out in an aqueous methanol solution, in which the reactants are soluble, but the product is insoluble. The product precipitates as the anhydrous salt, and is filtered and dried. The filtrate is distilled to recover solvent, and the column bottoms are either sold to Kraft mills as a source of sodium and sulfur, or treated in a biological pond. A flow sheet for the process is shown in Figure 3.

Figure 3

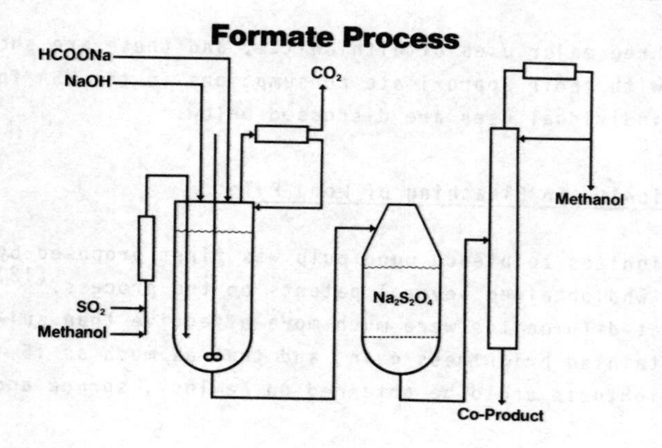

A particularly important piece of equipment is the filter-dryer, which is an invention of Mitsubishi Gas Chemical Company.[11] This vessel is equipped with a filter medium on which the dithionite crystals are retained when the slurry is fed to the vessel under pressure. The crystals are then washed with methanol, and dried by rotating the vessel under vacuum. Heat is supplied by hot water circulating through coils in the vessel, as well as through a jacket surrounding the vessel. After the material is dry, additives can be blended with the dithionite.

## Structure of the Industry

Commercial plants exist using each of the three manufacturing processes discussed above. In addition, there are several plants which produce sodium borohydride, a portion of which is employed to produce sodium dithionite. The dithionite derived from sodium borohydride is not included in the following discussion.

Table 1 summarizes the estimated capacities of the plants by process type. It is interesting that the formate process now has the greatest share of capacity (42%) even though the first such plant was completed only 12 years ago.

## Dithionite Uses

There are three major uses of dithionites, and these are shown in Table 2 with their approximate consumptions in the USA for 1979. The individual uses are discussed below.

## Sodium Dithionite in Bleaching of Wood Pulp

Use of dithionites to bleach wood pulp was first proposed by Hirschkind, who obtained several patents on the process.[12] He found that dithionites were much more effective than sulfites in obtaining brightness gain, and that as much as 15 units of brightness could be obtained on hemlock, spruce and balsam.

Table 1

Worldwide Dithionite Capacity by Process

| | | |
|---|---|---|
| Formate Process | 95,700 metric tons | 42% |
| U.K., Italy, Japan, USA, India | | |
| Amalgam Process | 33,000 metric tons | 15% |
| W. Germany, USA | | |
| Zinc Process | 97,600 metric tons | 43% |
| U.K., Belgium, France, Spain, Switzerland, Czechoslovakia, USA, China, Argentina, Columbia, Mexico, Venezuela, Brazil, Taiwan, Korea | | |
| Estimated World Capacity | 226,300 metric tons | |

Table 2

Estimated Usage of Dithionites in USA (1979)

| | Metric Tons | Percent |
|---|---|---|
| Textile | 26,000 | 40 |
| Pulp and Paper | 23,100 | 36 |
| Clay | 10,400 | 16 |
| Other | 5,300 | 8 |
| | 64,800 | |

At about the same time, Andrews[13] studied the application of dithionites to groundwood bleaching, with the result that zinc dithionite was employed to brighten groundwood newsprint on a mill scale in 1932. During the 1930's, use of dithionites became widespread on the West Coast of North America where a high percentage of hemlock is used in groundwood manufacture. In those early days, zinc dithionite was generally produced at the mills. At the present time, with the advent of strict limitations of the amount of zinc allowed in plant effluents, there has been an almost complete shift to use of sodium dithionite.

Dithionites are particularly good for bleaching groundwood because they bleach without dissolving a significant portion of the lignin present. For the largest brightness gains, a two-step bleaching process is used, employing hydrogen peroxide and then dithionite. In this manner, brightness values can be raised from 50 to 80 with a decrease of less than 10% in the lignin content of the pulps.[14]

It is believed that the main chemical action of dithionite is to reduce colored quinoid structures present in the pulp to the colorless phenolic structure.

Because oxygen destroys dithionite on contact, it is important to keep the reaction system as free of oxygen as possible. It is advisable to deaerate stocks before adding the dithionite solution, as well as to avoid exposing both the dithionite solution and pulp slurry to air.

The amount of dithionite used varies between 10 and 20 pounds per ton of dried pulp. Above 10 pounds per ton, the increase in brightness is proportionately less and therefore more costly. Increased temperature increases the rate of reaction. A temperature of about 140$^{o}$F (60$^{o}$C) is generally used. At a lower temperature of 100$^{o}$F (43.3$^{o}$C), one cannot achieve the same brightness gain with the same quantity of dithionite, because it would be impractical to extend the reaction time to that required. Retention times are generally in the range of one to two hours.

The desired pH for bleaching is best determined by lab studies using the given pulp. Generally a pH of 4-6 is recommended for zinc dithionite and 5-6.5 for sodium dithionite. The optimum pH will be a compromise between that which causes excessive loss of reducing agent and that which results in the most effective bleaching of pulp.[15]

## Use of Dithionites in Vat Dyeing

Vat dyes are water insoluble substances which possess the property of being reduced in alkaline solution to a water soluble form. The reducing solution used for this purpose is a mixture of sodium dithionite and sodium hydroxide. The reduced dye ions diffuse into the fiber (usually cotton) and become bound to the cellulose by hydrogen bonds and van der Waals forces. When the reduced dye is oxidized, it becomes water insoluble and is now mechanically trapped within the fiber.

Figure 4 shows the chemical change which occurs on treatment of indigo with alkaline dithionite solution.

Figure 4

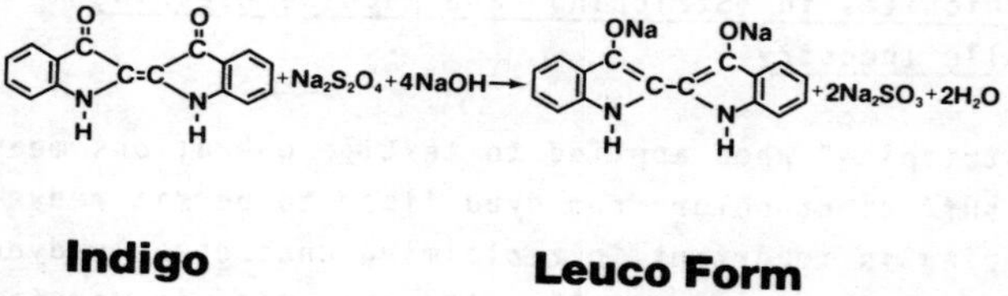

$$\text{Indigo} + Na_2S_2O_4 + 4NaOH \longrightarrow \text{Leuco Form} + 2Na_2SO_3 + 2H_2O$$

**Indigo** **Leuco Form**

For printing with vat dyes, it is usual to use sodium formaldehyde sulfoxylate as the reducing agent:

$$HOCH_2SO_2Na$$

This material will exert its reducing power only when the temperature is elevated (180 - 200°F).

This compound may be prepared by adding formaldehyde and sodium hydroxide to a solution of dithionite:

$$Na_2S_2O_4 + CH_2O + NaOH \rightarrow HOCH_2SO_2Na + Na_2SO_3$$

Alternatively, one may add formaldehyde to zinc dithionite to produce zinc formaldehyde sulfoxylate:

$$2\ ZnS_2O_4 + 4\ CH_2O + 2\ H_2O \rightarrow Zn(HSO_2 \cdot CH_2O)_2 + Zn(HSO_3 \cdot CH_2O)_2$$

Addition of more zinc dust and heating results in conversion of all of the sulfur to the basic zinc formaldehyde sulfoxylate:

$$Zn(HSO_2 \cdot CH_2O)_2 + Zn(HSO_3 \cdot CH_2O)_2 + 2\ Zn + 2\ H_2O \rightarrow 4\ Zn(OH)\ (HSO_2 \cdot CH_2O)$$

Finally, a sodiation of the hot slurry with NaOH results in a solution of the sodium salt:

$$Zn(OH)\ (HSO_2 \cdot CH_2O) + NaOH \rightarrow Zn(OH)_2 + NaHSO_2 \cdot CH_2O$$

Uses of Dithionites in "Stripping" and Machine Cleaning in the Textile Industry

The term "stripping" when applied to textile operations means removal of sufficient color from dyed fiber to permit redyeing. Stripping is important in reclaiming unacceptably dyed cotton and rayon goods. Generally, the operation is carried out in three steps: a preboil with a suitable alkali, treatment with a reducing agent, and washing to remove the solubilized dye.

The preboil removes extraneous matter and loosens the color, allowing the reducing agent to act exclusively on the remaining color. The stock is rinsed to remove dirt and dislodged color.

The reduction is generally carried out at temperatures of 140-160°F at a pH of 7.5 to 11. Three percent dithionite based on the dry weight of the goods is the recommended dosage. As in other applications of dithionite, contact with air should be avoided.

## Bleaching Clay with Dithionites

Both coating and filler grade clays can be bleached to achieve brightness gains of up to 5 points using dithionites. It is probable that the chief chemical reaction occurring is the reduction of ferric iron to the less colored and more soluble ferrous form.

The variables which are important in determining the brightness gains are: dithionite amount, temperature, time and bleaching pH.

Dithionite is quite effective in bleaching when used in amounts up to four pounds per ton of clay. Additional smaller increments of brightness are achieved with use of 4-8 pounds per ton. Other usual conditions are 140°F, retention time of 20 minutes and a bleaching pH of from 2 to 4. Two difficulties are sometimes encountered: a gray cast can appear in the clay when high concentrations of dithionite are employed, and at high temperature a yellow cast may appear. The gray cast can be prevented by use of a low reaction pH.

## Some Miscellaneous Uses for Dithionites

Sodium dithionite is the only generally available chemical which will reduce hexavalent chromium to the trivalent state in alkaline solution. This property makes sodium dithionite uniquely valuable for the purification of waste waters containing hexavalent chromium, for example, from plating solutions. The chemical reaction is:

$$2\ Na_2CrO_4 + 3\ Na_2S_2O_4 + 4\ H_2O \rightarrow 2\ Cr(OH)_3 + 4\ Na_2SO_3 + NaHSO_3$$

If the pH at the completion of reaction is in the range of 8 to 9, essentially all of the chromium will precipitate.

Sodium dithionite is also used to lower costs and improve product quality in the production of cane sugar. Dithionite reacts with some of the impurities such as waxes and gums and thereby improves filterability and results in crystals that are lighter in color.

## The Future Outlook

It is probable that the trend away from zinc-based sodium dithionite will continue, with new plants being constructed with the formate or amalgam processes. The production of dithionite from sodium borohydride also offers potential.

Research will continue to seek less expensive reducing agents for $SO_2$, such as hydrogen or carbon monoxide. It is interesting that Mitsubishi Gas Chemical[16] has recently patented a process using hydroquinone as the reducing agent for bisulfites. Since hydroquinone is produced by hydrogenation of quinone, hydrogen is the ultimate reducing agent in this new process.

References

(1) J. W. Mellor, "Modern Inorganic Chemistry," Longmans Green & Co., Ltd., London, 1927, p. 510.

(2) K. Jellinek, "Das Hydrosulfite," Stuttgart, 1912.

(3) E. Schulek and L. Moros, Magy. Kem. Folyoirat, 1957, 63, 41.

(4) L. Burlamacchi, G. Guarine and E. Tiezzi, Trans. Faraday Soc., 1969, 65, 496.
W. J. Lem and M. Wayman, Canad. J. Chem., 1970, 48, 2778.

(5) E. M. Marshak, Khim. Naukai Prom., 1957, 2, 524.

(6) S. Lynn, R. E. Rinker and W. H. Concoran, J. Phys. Chem., 1964, 68, 2363.

(7) J. D. Dunitz, Acta Cryst., 1956, 9, 579.

(8) W. A. M. Edwards and J. H. Clayton, "Sodium Hydrosulfite and Related Compounds," B109 Reports 422 (PB34027) and 271 (PB22409).

(9) S. Z. Avedekian, U.S. P. 2,938,771 (1960).

(10) Mitsubishi Gas Chemical Co., U.S. P. 3,411,875 (1968).
Virginia Chemicals Inc., U.S. P. 3,576,598 (1971).

(11) Mitsubishi Gas Chemical Co., U.S. P. 3,664,035 (1972).

(12) W. Hirshkind, Paper Trade Journal, 1932, 60, No. 18, 31; U.S. P. 1,873,924 (1932); U.S. P. 2,071,304-9 (1939).

(13) I. H. Andrews, Pulp & Paper Mag. Canada, 1945, 46, 679; Tappi, 1949, 32, 206; Can. P. 347,561 (1935).

(14) J. Polcin, Zellst. Pap., 1973, 22, No. 8, 226.

(15) M. A. Kise, "The Bleaching of Pulp," Third Ed., Tappi Press, 1979, p. 255.

(16) Mitsubishi Gas Chemical Co., U.S. P. 4,177,247 (1979).

# Lower Volume Industrial Inorganic Compounds from Elemental Phosphorus

By J. C. McCoubrey
ALBRIGHT & WILSON LTD., P.O. BOX 80, TRINITY STREET, OLDBURY, WARLEY, WEST MIDLANDS B69 4LN, U.K.

## 1. Introduction

This symposium aims to be complementary with that given in 1977 on the Modern Inorganic Chemical Industry in which Childs[1] summarised the origins, production and use of phosphorus for phosphoric acid and phosphate salts.

To that end major products are reviewed here sufficient to define the remainder of the industrial constellation out of the many phosphorus chemicals known. It has therefore been necessary to include quite large scale inorganic products and materials which range from near commodities to near performance chemicals in this account.

The compounds dealt with vary from oxidation states -3 (phosphine $PH_3$) to +5 (phosphoric oxide $P_4O_{10}$) and include the well known halides and sulphides which play an important role particularly in a number of organic chemical industries.

The chemical variety and problems of the industry are illustrated and help to demonstrate why this group of chemicals may qualify for the classification of speciality.

## 2. Elemental Phosphorus

Below 800°C the vapour of phosphorus consists of tetrahedral $P_4$ molecules[2]. So too does the liquid which has a boiling point of 280.5°C whether formed from vapour or from any of the allotropes[3]. The common solid allotrope is cubic white phosphorus which has a large unit cell of 56 molecules of $P_4$, melts at 44.1°C, is a wax-like non-conductor and ignites in air[3,4] with a self ignition temperature of 34°C.

White phosphorus is insoluble in water and though water dissolves in it to the extent of 3.6 mg $H_2O$/g $P_4$ in the temperature range 25-45°C both solid and liquid phosphorus are normally transported under water. Despite its natural tendency to fire and its poisonous character (TLV-TWA is

0.1 mg /$m^3$), hundreds of thousands of tons of liquid white phosphorus have been moved by sea, rail and road and millions of tons pumped around factories. Plant and transportation to handle this material is designed with great care and precautions to deal with spillages are given considerable attention.

## 2.1 Other allotropes

Of the various polymeric allotropes the black phosphorus varieties are the most stable but are commercially unimportant. The only polymeric allotrope of commercial significance is amorphous red phosphorus which is formed from $P_4$ exothermically[5] to the extent of 26kJ/mole of $P_4$ and used to make safety matchbox strikers and pyrotechnics.

The red phosphorus of commerce is traditionally made by gentle refluxing (280°C) of a tank of white phosphorus for around 48 hours probably with a terminating period at a higher temperature. It can be made over a shorter time cycle by heating to 400°C but this requires more elaborate control procedures. The hard product is broken out of the reactors, wet milled, boiled with alkali and after washing, dressed in whatever way is appropriate to its end use. World production is in the low thousands of tons per annum and while the material commands some premium over white phosphorus in Europe it is sold very competitively in the U.S.A.

A recent increase in interest in amorphous phosphorus arises from its use when mixed with resins as a flame retardant.

## 3. Metal Phosphides[6]

There are a number of direct relevancies of the less reactive metal phosphides to industrial processes. Additions of phosphorus are often made to iron and steel in processes using "ferrophosphorus", a crude mixture of iron phosphides formed in the phosphorus making furnaces in tens of thousands of tons and sold at by-product prices. Metallic copper phosphide $Cu_3P$ is made industrially from its molten elements for its considerable use in bronzes and iridium phosphide $Ir_2P$ is familiar as a hard inert tip for fountain pens. Small quantities of P atoms (0.001%) substituted for Si atoms in the silicon lattice enhance its semi-conductor properties in certain applications.

## 3.1 More reactive Phosphides

A different kind of application is found in the production of aluminium phosphide AlP which is used for the controlled production of phosphine by liquid or atmospheric hydrolysis. Special tablets of aluminium phosphide are sold industrially for slow release fumigation of grain.

In a similar way calcium phosphide $Ca_3P_2$ and magnesium phosphide $Mg_3P_2$ which mainly contain $P^{3-}$ ions in their structure are used in sea flares but are able to produce spontaneously inflammable phosphine probably because of the presence of small amounts of CaP mono-phosphide with $P_2^{4-}$ ion which hydrolyses to $P_2H_4$.

Reactive phosphides are usually made by heating metals with red phosphorus for safe control and a phosphide such as $Zn_3P_2$ zinc phosphide, used as a common rat poison, is produced in this way. Some materials such as "commercial" calcium phosphide can be made by the more direct method of reacting quick lime with phosphorus vapour.

All these products are made on relatively sensitive small scale plant in quantities of tens to hundreds of tons/year. They are all energy and labour intensive and a typical phosphide $Zn_3P_2$ sells at about £2000/ton in the U.K.

4. Phosphorus trihalides

The exothermic reaction of halogens with white (liquid) phosphorus leads to the formation of $P^{III}$ phosphorus halides. These are very reactive and poisonous materials[7] which range from the gaseous tri-fluoride $PF_3$ (b.p. - 101°C) to the tri-iodide, a solid with melting point 61°C.

Although phosphorus trifluoride has been available commercially in the U.S.A. (Ozark Mahoning) for some years it appears to have only a small use mainly for the production of catalyst complexes such as Ni $(PF_3)_4$ or Fe $(PF_3)_5$. Phosphorus tribromide $PBr_3$ (b.p. 106°C) is made in quantities of the order of hundreds of tons per year and is used for a limited range of industrial brominations.

Far the most important commercial trihalide is $PCl_3$ for which world manufacturing capacity can be estimated to be more than 200,000 tons/year. The material is made by the controlled addition of chlorine to white phosphorus in a heel of phosphorus trichloride. The product so made is distilled continuously (b.p. 75.2°C), the removal of heat from the reaction being a primary concern. The purity of commercial $PCl_3$ is well above 99%, the main impurity being $POCl_3$. Major producers in the U.S.A. are Monsanto, Stauffer, FMC, Occidental and Mobil and in Europe Albright and Wilson and Hoechst.

One of the major problems in handling phosphorus trichloride is its great reactivity with water; it fumes when exposed to moist air, reacts violently with a small amount of water and with increasing quantities produces in turn diphosphorous acid and phosphorous acid. The threshold limit value - TWA is 0.5 ppm in air[8]. Phosphorous trichloride can be handled prudently in mild steel equipment but for better safety and protection against hydrolysis products is handled in lead lined, glass or nickel vessels. Transportation in nickel tankers is preferred in the U.K. but stainless steel tankers can be used with care. A resin lined mild steel drum is used for supply in this country.

The general chemistry of the material is summarised[4] in Fig.1.

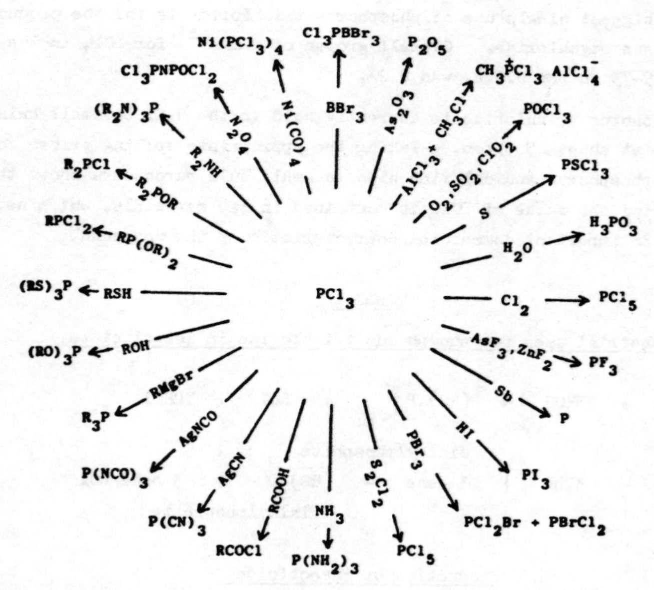


Figure 1 Reactions of phosphorus trichloride
Reproduced with permission from p.68 of ref.4

## 4.1 Applications of $PCl_3$

Phosphorus trichloride plays a very important role in large scale industrial organo-phosphorus chemistry[9]. In Table 1 below are shown some of the typical materials used in the insecticide industry where the alkyl phosphites find considerable outlets in domestic and agricultural pesticides[10]. Basic phosphorus and organo-phosphorus producers do not always manufacture these phosphite based insecticides, this being a further specialised business. A very large producer of this family of insecticides in the U.S.A. is Shell.

Applications of phosphite esters in the rubber and plastic industries for anti-oxidancy, stabilising and flame retardance represent another substantial tonnage demand. A variety of compounds containing phosphorus-carbon bonds are also made from $PCl_3$. These include not only the important nitrogen based materials used as sequestrants but also phosphonates used for making insecticides, and diphosphonates used as complexers. This family of chemicals can also include certain types of nerve gas but there are no satisfactory statistics available for these.

The biggest single use of phosphorus trichloride is for the production of phosphorus oxychloride. Overall growth of demand[11] for $PCl_3$ in the period 1969-79 in the U.S.A. was 8.2%.

Phosphorus trichloride is currently sold in the U.K. to small bulk purchasers at about £500/ton. Taking the appropriate selling prices for elemental phosphorus and chlorine also to small bulk purchasers shows that around 60% of the value of $PCl_3$ is contained in raw materials, which serves to define an important commercial characteristic of the compound.

TABLE 1

Industrial uses for phosphorus trichloride in insecticides.

$$PCl_3 + 3ROH \rightarrow (RO)_2P(=O)H + RCl + 2HCl$$

dialkylphosphite

$$PCl_3 + 3ROH + 3\text{ base} \rightarrow (RO)_3P + 3\text{ base-HCl}$$

trialkylphosphite

Production of insecticides

$$(MeO)_2P(=O)H + OHC.CCl_3 \rightarrow (MeO)_2P(=O)CH(OH)\text{-}CCl_3$$

dimethyl phosphite — chloral — Trichlorfon

$$(MeO)_3P + OHC.CCl_3 \rightarrow (MeO)_2P(=O)O\text{-}CH{=}CCl_2 + CH_3Cl$$

trimethyl phosphite — chloral — Dichlorvos

TABLE 2

Other industrial applications

$PCl_3 + 3Ph.OH \rightarrow P(OPh)_3 + 3HCl$
triphenylphosphite

$P(OPh)_3 + 3R.C_6H_4.OH \rightarrow P(O.C_6H_4.R)_3 + 3Ph.OH$
tri(alkylaryl)phosphite stabiliser/antioxidant
(e.g. tri(isodecylphenyl)phosphite)

$3PCl_3 + NH_3 + 3CH_2O + 6H_2O \rightarrow N(CH_2-P(=O)(OH)(OH))_3 + 9HCl$
general purpose sequestrant

$PCl_3 + 3R.COOH \rightarrow 3R.COCl + H_3PO_3$
acid chloride

$PCl_3 + CH_4 \rightarrow CH_3PCl_2 + HCl$
methyl dichloro phosphine
versatile reactant for flameproofing.

## 5. Phosphorus oxyhalides

Although both phosphoryl fluoride $POF_3$ (b.p. - 39°C) and phosphoryl bromide $POBr_3$ (m.p. 100°C with decomposition) are commercially available, the former in the U.S.A. and the latter in Europe, neither material has substantial commercial significance. By contrast phosphorus oxychloride is produced in many thousands of tons in Europe and the U.S.A.

There are three very interesting routes to the manufacture of $POCl_3$ as detailed below. Unfortunately the third of these, despite considerable development effort, has never been made commercially viable.

TABLE 3

Routes to phosphorus oxychloride

(1) $2PCl_3 + O_2 \rightarrow 2POCl_3$

(2) $3PCl_3 + 3Cl_2 + P_2O_5 \rightarrow 5POCl_3$

(3) $3NaCl + P_2O_5 \rightarrow POCl_3 + Na_3PO_4$

The first of these reactions, which is highly exothermic, may be conducted continuously, preferably using neat oxygen which is bubbled through a sufficient depth of $PCl_3$, care being taken to avoid inhibiting trace impurities. The second reaction, also exothermic, is employed by big producers of $P_2O_5$ and the pentoxide is mixed with $PCl_3$ to which is added chlorine in a large nickel reactor operated batchwise and is followed by fractionation with return of the unreacted $PCl_3$ to the system. The purity of commercial $POCl_3$ is very high. Care needs to be taken to minimise the bromine content of the chlorine used, to avoid poor coloured products.

Phosphorus oxychloride, which boils at 107.4°C and melts at 1°C, is a dense liquid (s.g. 1.686 at 15.5°C) which fumes in contact with moist air. The liquid can react with water with explosive violence to form phosphoric acid and HCl; this sometimes happens after an "induction" period and great care should be taken to avoid water contamination of this material.

$POCl_3$ is more corrosive to metals in practice than $PCl_3$ and in the U.K. it is always handled in bulk in nickel, lead lined or glass vessels and transportation is by road nickel tankers only. Mild steel drums with thick plastic liners are suitable for small container requirements. The threshold limit value for this material is taken to be 0.5 ppm in air.

## 5.1 Applications of $POCl_3$

Historical growth of demand for $POCl_3$ in the U.S.A. in the period 1968-78 has been 2.1% per annum[12]. The principal industrial application of phosphoryl chloride is for the manufacture of trialkyl and more particularly triaryl phosphate esters used mainly to confer flame retardancy to plastics and fluids sometimes together with other important physical properties. Typical products used to achieve this kind of application are shown in Table 4.

Triaryl phosphates such as tricresyl, cresyl diphenyl, isopropyl phenyl phenyl and tertiary butyl phenyl phenyl phosphates are all used in large quantities as flame retardant plasticisers for polyvinyl chloride systems and some other polymeric materials also incorporate these compounds.

TABLE 4

Commercially important organic phosphates made from phosphorus oxychloride

$$POCl_3 + 3C_8H_{17}OH \rightarrow OP(OC_8H_{17})_3 + 3HCl$$

triocytl phosphate

$$POCl_3 + 3ClCH_2.\underset{\diagdown O \diagup}{CH.CH_2} \rightarrow OP(OCH(CH_2Cl)CH_2Cl)_3$$

tris(dichlorpropyl) phosphate

$$POCl_3 + 3CH_3.C_6H_4OH \rightarrow OP(OC_6H_4CH_3) + 3HCl$$

tricresyl phosphate

$$POCl_3 + 2Ph.OH + isoPr\text{-}C_6H_4OH \rightarrow OP(OPh)(OC_6H_4\text{-}isoPr)(OPh) + 3HCl$$

isopropylphenyl diphenyl phosphate

$$POCl_3 + C_8H_{17}OH \rightarrow OP(OC_8H_{17})(Cl)(Cl) + HCl$$

$$OP(OC_8H_{17})(Cl)(Cl) + 2NaOPh \rightarrow OP(OC_8H_{17})(OPh)(OPh) + 2NaCl$$

octyldiphenyl phosphate

The same products are widely used as flame retarding hydraulic fluids in high temperature seals, underground machinery, and turbines. Major producers in Europe are Ciba-Geigy and Albright and Wilson and in the U.S. FMC and Stauffer. A kindred family of materials are the alkyl aryl phosphates such as octyl diphenyl phosphate made by Monsanto. The production of both of these groups of products worldwide is probably around 100,000 tons/year. They are sold in Europe for prices around £1000/ton.

## 6. Thiophosphoryl halides

The only commercially significant member of this family is thiophosphoryl chloride (phosphorothionyl chloride) which can be manufactured by either of the two routes shown below or can arise as a by-product in other methods of manufacture of the important insecticide intermediates O,O-dialkyl phosphorochloridothioates. This kind of intermediate is used in the manufacture of organophosphorus insecticides including the well known methyl Parathion, Diazinon and Fenitrothion[10].

$PSCl_3$ is made on a scale of several thousands of tons per annum but while there is some merchant market for it the bulk of the material appears to be used captively in the U.S.

TABLE 5

Chemistry of thiophosphoryl chloride $PSCl_3$

$PCl_3$ + S + ($AlCl_3$ catalyst) → $PSCl_3$

$P_4S_{10}$ + $6PCl_3$ + $6Cl_2$ → $10PSCl_3$

$PSCl_3$ + RONa → $(RO)PSCl_2$ + NaCl
O-alkyl phosphorodichlorido-thioate

$PSCl_3$ + 2RONa → $(RO)_2PSCl$ + 2NaCl
O,O-dialkyl phosphorochlorido-thioate

The material is offered for sale by Rhone Poulenc and Ciba Geigy in Europe.

7. Phosphorus pentahalides

Organic compounds with five ligands round the central phosphorus atom are generally called phosphoranes and are desired as synthesis intermediates[13]. The phosphorus pentahalides are the theoretical precursors of these but are of comparatively small commercial significance in this way. Some of the chemistry of the pentahalides can be carried out by making them in situ, adding the trihalide and halogen to the desired reaction.

Although phosphorus pentafluoride $PF_5$ (b.p. - 84.8°C) has been on commercial offer for some time in the U.S.A. and finds some use as a catalyst in ionic polymerisations and in semi-conductor applications it is not readily available in Europe. Only $PCl_5$ is substantial commercially; made by chlorinating $PCl_3$ (Figure 2)[14] it is a fuming solid with a melting point of 160°C, subject to ready hydrolysis and reverting to the trichloride and chlorine at temperatures above 150°C.

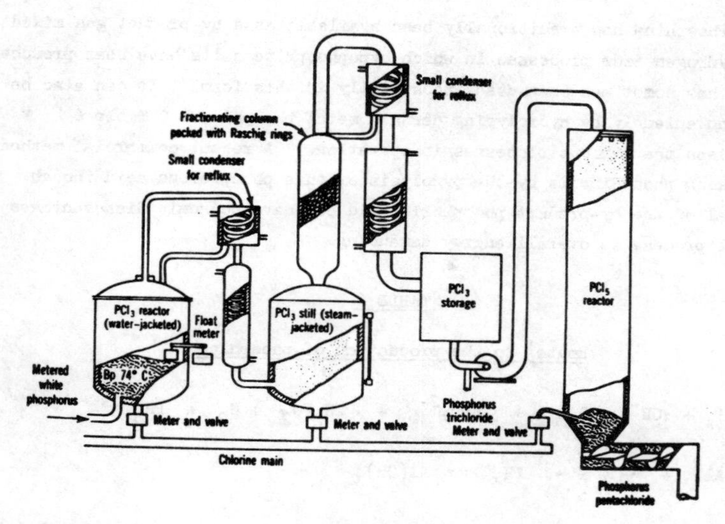


Figure 2 Plant for the continuous production of phosphorus trichloride and phosphorus pentachloride

Reproduced with permission from p.307 of ref. 14.

$PCl_5$ is used in industrial organic chemistry mainly to convert carboxylic acids to acid chlorides, sulphonic acids to sulphonyl chlorides and phosphonic acids to phosphonyl chlorides as well as undergoing addition reactions with olefins and aromatics. Its reactions with ammonia will be dealt with under phosphorus-nitrogen compounds.

$PCl_5$ reacts with sodium or potassium fluorides to give the corresponding hexafluorophosphate $XPF_6$, but these high fluorine content salts surprisingly do not appear to have found significant commercial application despite the commercial role achieved by their sister compound sodium monofluorophosphate. This latter compound is widely used as a toothpaste anti-decay additive and is made by the simpler process of reacting sodium fluoride with sodium metaphosphate in a melt:

$$NaF + NaPO_3 \rightarrow Na_2O_3PF$$

8. Phosphorus hydrides[15]

The simple hydride phosphine $PH_3$ is not readily made from its elements for both thermodynamic and kinetic reasons. The material is very easily oxidised and traces of phosphorus can render it spontaneously inflammable. It boils at $-89^{\circ}C$ and freezes at $-133.8^{\circ}C$ and it can be stored as liquid or as a compressed gas; its critical temperature is $51^{\circ}C$ and its critical pressure 64 bars.

Phosphine has traditionally been available as a by-product gas mixed with hydrogen from processes in which hypophosphite salts have been produced and it has sometimes been used industrially in this form. It can also be made conveniently by hydrolysing certain metal phosphides. Table 6 summarises the various processes in operation. A recent commercial method for making phosphine is by the pyrolysis of pure phosphorous acid though disposal of the by-product phosphoric acid can have economic disadvantages and the process is overall energy demanding.

TABLE 6

Routes to the production of phosphine

$$P_4 + 4OH^- + 3H_2O \rightarrow HPO_3^{2-} + 2H_2PO_2^- + H_2 + PH_3$$

$$AlP + 3H_2O \rightarrow PH_3 + Al(OH)_3$$

$$4H_3PO_3 \rightarrow PH_3 + 3H_3PO_4$$

$$2P_4 + 12H_2O \rightarrow 5PH_3 + 3H_3PO_4$$

The most soundly based process for large scale investment is that built on the hydrolysis of elemental phosphorus. A commercial plant of this kind is in operation in Canada run by Cyanamid and based on a process originally invented in the U.K.[16] Because of the corrosive conditions associated with this reaction, which are those of concentrated phosphoric acid containing phosphorus heated to 280°C, the only practical containing material is carbon. A fabricated carbon vessel is contained in metal and the energy input can be by electrode heating to optimise heat transfer.

The practical chemistry of phosphine is influenced by its ready oxidisability and its wide explosion limits with oxygen. While the gas is extremely poisonous it is not worse than some other industrial materials the continuous TLV being 0.3 ppm in air[8].

## 8.1 Applications of Phosphine

TABLE 7

Typical Organic reactions of phosphine of commercial significance

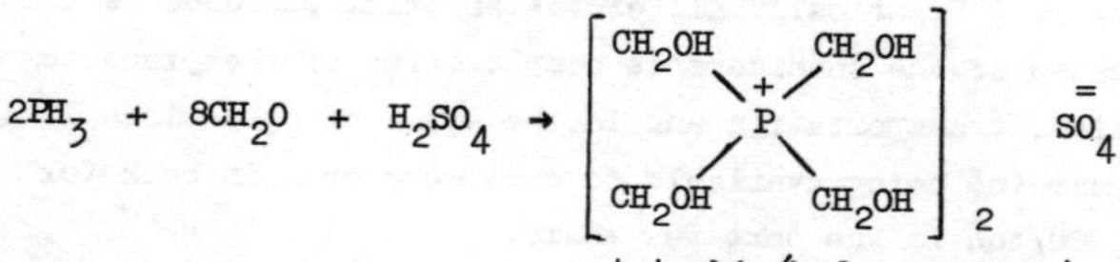

$$2PH_3 + 8CH_2O + H_2SO_4 \rightarrow \left[ (CH_2OH)_4P^+ \right]_2 SO_4^{=}$$

tetrakis (hydroxymethyl) phosphonium sulphate

$$PH_3 + 3C_4H_8 \rightarrow P(C_4H_9)_3$$

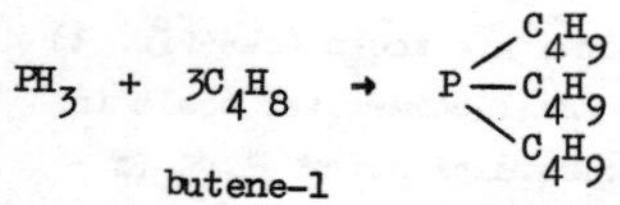

butene-1 tri-n-butylphosphine

$$PH_3 + 3C_8H_{16} \rightarrow P(C_8H_{17})_3$$

octene-1 trioctylphosphine

$$P(C_8H_{17})_3 + H_2O_2 \rightarrow OP(C_8H_{17})_3 + H_2O$$

trioctylphosphine oxide

A commercial organophosphine which cannot be produced from $PH_3$ is triphenylphosphine which is made :-

$$PCl_3 + 3PhCl + 3Mg \rightarrow P(Ph)_3 + 3MgCl_2$$

Phosphine is used mainly[15]:-

1. As a reactant to produce phosphonium salts, an example of which is tetrakis (hydroxy methyl) phosphonium sulphate, a major chemical for flame-proofing cellulosic fabrics. Phosphonium salts are more stable than ammonium salts in general and are effective as phase transfer catalysts.

2. To react with olefins to produce di- or tri-alkyl phosphines and their oxides or sulphides. These are used as metal extractants or as metal complexers for catalysts[17]. Tertiary phosphines form complexes with most transition metals like Pd, Pt, Rh, Co and Ni. They are able to form nonionic complexes soluble in organic solvents. They can also replace part of the Co in metal carbonyls. Tertiary phosphines are incorporated in a number of important hydroformylation and methanol homologation catalysts.

One popular tertiary phosphine used in such catalyst systems as $RhH(CO)(PPh)_3$ is triphenyl phosphine which cannot readily be made from phosphine and is made from $PCl_3$ by a Grignard route. It is not surprising that this material is on offer at prices around £6,000/ton.

The World production of pure $PH_3$ is probably little more than a thousand tons per year. Practically all of the material produced is for captive use and because of the considerable complexities of the processes involved in production, transportation and in the scale of operations it is difficult to see phosphine being available to customers even in bulk for less than about £5,000/ton in the next few years.

## 9. Phosphorus oxides

Although several lower oxides of phosphorus are known (see Fig. 3) and indeed $P_4O_6(P_2O_3)$ has been available on a small commercial scale in the U.K., only $P_4O_{10}(P_2O_5)$ is a substantial commercial material. In England Albright and Wilson is probably the biggest producer of separated dry $P_2O_5$ in the world since it operates one thermal phosphoric acid process by burning liquid $P_4$ at rates up to 600 kilos/hour with heat recovery of the very exothermic process and the operation contains a separate step in which giant condensers catch the $P_2O_5$ from where it is mechanically recovered and screwed either to the acid reactors or to a separation system.
Lower oxides can be rigorously excluded from the $P_2O_5$ though it retains trace quantities of arsenic oxide arising from the arsenic content of the phosphorus. Adequate control of the humidity of the burner air permits the commercial oxide to be made almost entirely in the form of fine crystals of the hexagonal (H) form though several other forms exist.

### 9.1 Uses of phosphorus pentoxide

Apart from its uses to make oxychloride and as a desiccant $P_2O_5$ is used to react with long chain alcohols to form a mixture of mono and diphosphates used as wetting agents. A special use, unique to the U.S.A. (Eastman-Kodak), is the reaction with by-product diethyl ether to give triethyl phosphate. $P_2O_5$ has been used in methods to manufacture the insecticide tetraethyl pyrophosphate. Equations summarising these are given.

TABLE 8

Typical organic reactions of $P_2O_5$

$$P_2O_5 + 3ROH \rightarrow (RO)P(=O)(OH)_2 + (RO)_2P(=O)OH$$

mixture of acid phosphates

where ROH may typically be an ethoxylated long chain alcohol.

$$P_2O_5 + 3(C_2H_5)_2O \rightarrow 2(C_2H_5O)_3P{=}O$$

triethyl phosphate

$$P_2O_5 + 4(C_2H_5O)_3P{=}O \rightarrow 3(C_2H_5O)_2P(=O){-}O{-}P(=O)(C_2H_5O)_2$$

tetraethyl pyrophosphate

Much work has been done over the years in attempts to make triaryl phosphates from phenols, $P_2O_5$ and phosphoric acid but realistic process conditions have been very difficult to achieve. It is known that attempts are currently being made to establish a commercial process based on this chemistry.

## 10. Phosphorus sulphides[18]

As in the case of the oxides the structure of the sulphides is based on the $P_4$ tetrahedral unit. However sulphides precisely analogous to the oxides do not always occur, the structures probably being conditioned by specific solid state effects in some cases. Figure 3 shows the structures properly characterised. Of the sulphides only two are commercially established though $P_4S_7$ has been attributed value in patents[19].

Phosphorus sesquisulphide is formed by reacting white phosphorus (liquified) with sulphur at around 350°C using a slight deficiency of phosphorus to prevent residual firing. The product is washed continuously to remove the more reactive higher sulphides leaving pure $P_4S_3$ a yellow solid which melts at 172°C and boils at 407°C. Although this compound is relatively stable at room temperature it burns at temperatures above 100°C. It is used industrially in a finely milled condition in bound mixtures with sodium chlorate to form the heads of strike anywhere matches.

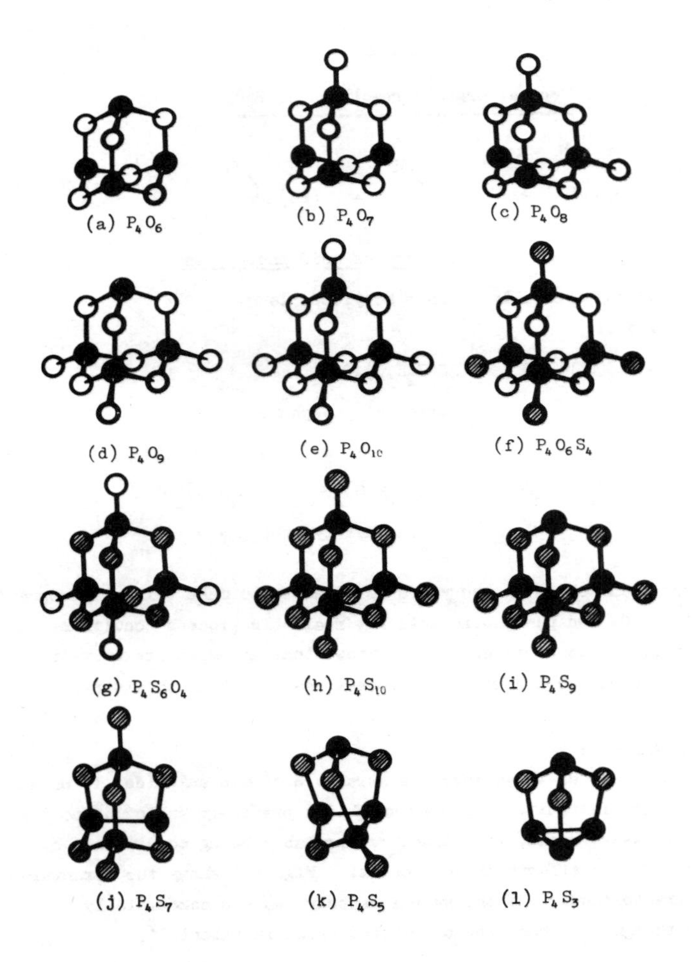


Figure 3 Molecular structures of oxides and sulphides of phosphorus

Reproduced with permission from p.57 of ref. 4.

Tetraphosphorus decasulphide $P_4S_{10}$ is normally designated phosphorus pentasulphide $P_2S_5$. It is a dense (s.g. 2.09) crystalline yellow triclinic solid having a melting point of 290°C and an atmospheric b.p. of 513°C.

The industrial manufacture of phosphorus pentasulphide involves the reaction above 250°C of molten phosphorus and molten sulphur which are pumped from storage tanks into a stirred reactor designed to permit substantial heat removal. There are a number of patents describing improvements[20, 21] to this basic process which may be followed by a controlled secondary reactor allowing products to complete reaction to ensure a uniform or particular chemical reactivity, the material being then cooled or flaked. Because of the potentially reactive nature of the material the solidifications and packing stages of production may be blanketed by carbon dioxide. For certain applications further stages of purification may be required.

Care is required to guard against firing, by traces of excess phosphorus, or the hydrolysis of the product by undue exposure to moist air. Commercial material is however conveniently handled in large aluminium tote bins and in polythene lined steel drums with good closures.

## 10.1 Applications of phosphorus pentasulphide

An extremely important reaction of $P_4S_{10}$ is that with aliphatic alcohols or substituted phenols to form the dialkyl or diaryl dithiophosphoric acids. The biggest single application of pentasulphide arises from this reaction followed by reaction of these acids with metal oxides to form the zinc or sometimes the barium salts of such acids. These salts are used to provide improved wear, detergency, anti-oxidancy and corrosion resistance under extreme pressure conditions in oils.

Sodium salts of the dialkyl dithiophosphoric acids are used as flotation agents in certain mineral recovery processes.

A very complex and large application of the dithio-phosphoric acids and their chlorinated derivatives is in insecticides.

TABLE 9

Typical reactions of phosphorus pentasulphide to produce insecticides[10]

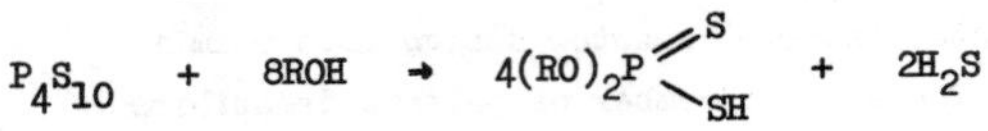

$$P_4S_{10} + 8ROH \rightarrow 4(RO)_2P(=S)SH + 2H_2S$$

O,O-dialkyl dithiophosphoric acid

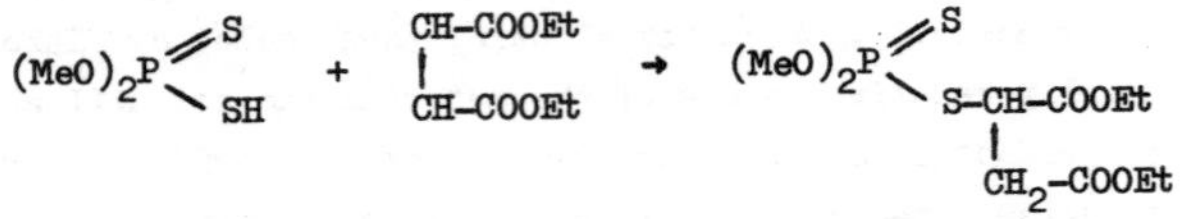

$$(MeO)_2P(=S)SH + CH(COOEt){=}CH(COOEt) \rightarrow (MeO)_2P(=S)S\text{-}CH(COOEt)\text{-}CH_2\text{-}COOEt$$

Malathion

$$2(RO)_2P(=S)SH + 3Cl_2 \rightarrow 2(RO)_2P(=S)Cl + 2HCl + S_2Cl_2$$

$$(MeO)_2P(=S)Cl + NaOC_6H_4{\cdot}NO_2 \rightarrow (MeO)_2P(=S)OC_6H_4{\cdot}NO_2 + NaCl$$

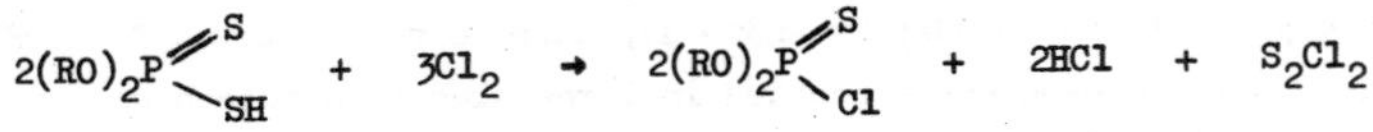

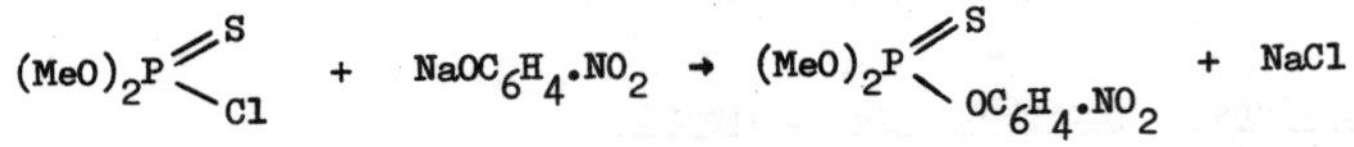

methyl Parathion

The growth in demand for phosphorus pentasulphide for all applications in the U.S.A. has been 5.5% in the period 1970–79.[22]

## 11. Some statistics

Useful parameters for the major industrial products are set out in the following two tables.

TABLE 10

Standard enthalpies of formation of major phosphorus intermediates[5]

| Compound | Formula | State | $\Delta H_f^o$ kJ mole$^{-1}$ |
|---|---|---|---|
| Phosphorus pentoxide | $P_4O_{10}$ | solid | - 3013 |
| Phosphorus trichloride | $PCl_3$ | liquid | - 306 |
| Phosphorus pentasulphide | $P_4S_{10}$ | solid | - 364 |
| Phosphorus oxychloride | $POCl_3$ | liquid | - 598 |
| Phosphine | $PH_3$ | gas | + 4 |

Synthesis of the first four compounds can be seen to be considerably exothermic.

TABLE 11

Four major inorganic phosphorus intermediates, capability and price in the U.S.A. in 1978 [11,12,22]

| | Plant capacity tons/yr | Price $/tonne |
|---|---|---|
| $P_4O_{10}$ | 7,000 | ~ 1,300 |
| $PCl_3$ | 130,000 | 880 |
| $POCl_3$ | 66,000 | 880 |
| $P_4S_{10}$ | 120,000 | 820 |

The major producers of this range of materials in the U.S.A. are FMC, Monsanto, Occidental and Stauffer. In Europe they are Albright and Wilson and Hoechst.

## 12. Phosphorus nitrogen compounds[23,24]

Amorphous materials arising from reactions between phosphorus vapour or phosphorus halides, oxy halides or thio phosphoryl halides and nitrogen or ammonia at elevated temperatures include PN, $P_3N_5$, $PN_2H$ (phospham), PNO and PNS. These products have been explored as slow release fertilisers but none has been found commercially practical.

Many reactions between phosphorus halides or oxyhalides and amines have been examined but few of these have produced commercial products though the reaction between oxychloride and dimethylamine to give hexamethyl phosphoramide produces a good complexing solvent with some role among industrial solvents.

A family of compounds arises from reaction between $PCl_5$ and $NH_4Cl$.

$$nPCl_5 + nNH_4Cl \rightarrow (PNCl_2)_n + 4nHCl$$

These materials $(PNCl_2)_n$ are known as chlorophosphazenes or phosphonitrilic chlorides. The main products of this reaction are the cyclic trimer and cyclic tetramer though rings up to the heptamer are formed. Corresponding fluoro compounds can be made from the chlorides by reaction with potassium fluoro-sulphite. The chloride may also be substituted by a number of groups such as amino, alkoxy, alkyl or aryl.

These materials have a fascinating chemistry but are of industrial interest almost entirely in respect of the polymers which they can form, illustrated in the figure 4.

There are hundreds of papers and patents describing phosphazene polymers for rubber, fluids or resins in many high temperature applications.

1. LINEAR-TYPE POLYMERS

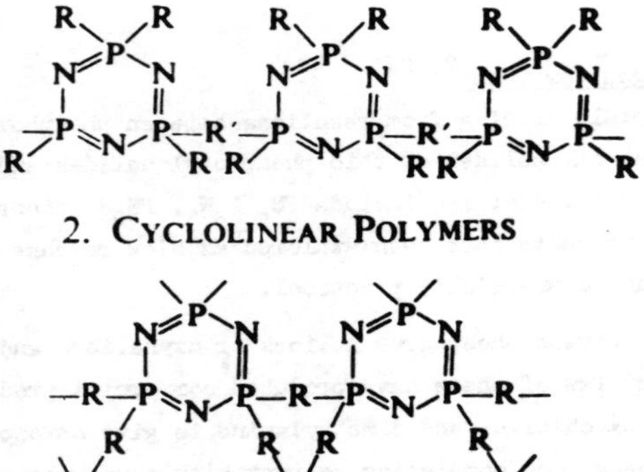

2. CYCLOLINEAR POLYMERS

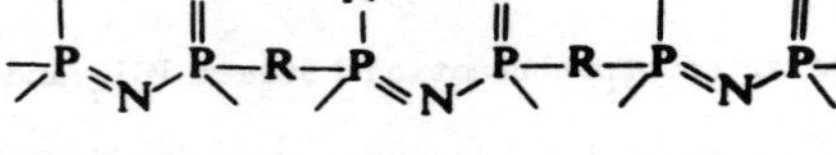

3. CROSS-LINKED CYCLOMATRIX POLYMERS

Figure 4
Phosphazene polymers.
Reproduced with permission from p.337-9 of ref. 23

Recently there has been a considerable amount of technical activity directed at the use of phosphazene polymers as flame retardants for wool, polyesters and closed cell foams. No commercial production of phosphazene intermediates is however current in the U.K. The extent to which any of the many ventures pursued in the U.S.A. have achieved what could be described as a commercial process is hard to assess. Several major companies such as Dupont or Firestone have developed polymers from fluorinated phosphazenes. Continued production of the materials is not demonstrated anywhere though both Ethyl Corpn. and Nippon Chemical offer the chemicals for sale.

## 13. Phosphorous Acid

This material is manufactured commercially both in the form of 70% solution and more particularly as pure anhydrous acid $H_3PO_3$. The operation is carried out in plant a large part of which is glass, hydrolysing phosphorus trichloride with the weak acid in a packed column with removal and recovery of hydrochloric acid. The acid is concentrated in both atmospheric and vacuum evaporators and the molten material is converted to product on a rotating drum flaker. World production appears to be about 6-8,000 tons annually as anhydrous acid equivalent.

Pure orthophosphorous acid is a colourless deliquescent solid melting at 70.1°C with a specific gravity of 1.65 at 21°C. The acid is highly soluble in water and behaves as a dibasic acid implying the predominant structural formula to be the phosphonic arrangement

HO     O
  \\  //
   P
  /  \\
HO     H

A small bulk user of phosphorous acid would expect to pay about £1,600/ton for product in kegs.

## 13.1 Applications of $H_3PO_3$

The longest established commercial application of $H_3PO_3$ is for the production of metal phosphites of which lead phosphite $PbHPO_3$ is the most versatile as a stabiliser for poly vinyl chloride and as a vulcaniser for rubber. Small quantities of phosphorous acid may be added to esterifications and condensation reactions to restrict colour formation; this use is well known in polyester manufacture.

Some modern applications are shown in Table 12.

TABLE 12

More recent commercial organic reactions of phosphorous acid

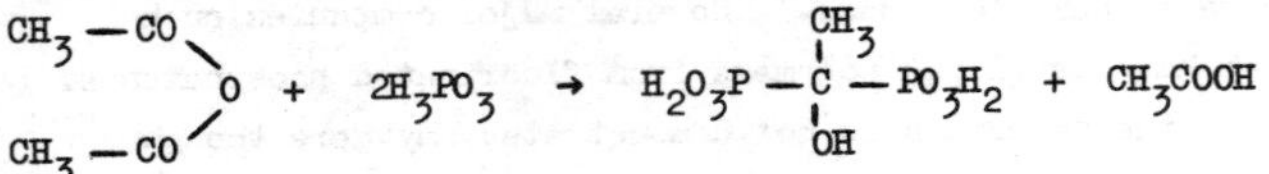

$$(CH_3CO)_2O + 2H_3PO_3 \rightarrow H_2O_3P-C(CH_3)(OH)-PO_3H_2 + CH_3COOH$$

acetodiphosphonic acid

ADPA is an important heavy metal sequestrant and scale and corrosion inhibitor made by Albright and Wilson and others.

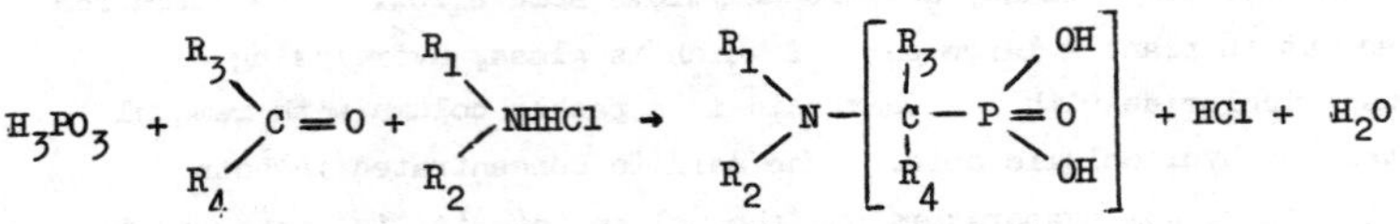

$$H_3PO_3 + R_3R_4C{=}O + R_1R_2NHHCl \rightarrow R_1R_2N-[C(R_3)(R_4)-P(=O)(OH)_2] + HCl + H_2O$$

amino alkylene phosphonic acids

A family of such products is made, primarily by Monsanto, to provide industrial sequestrants and inhibitors though some of these may be made from $PCl_3$ directly.

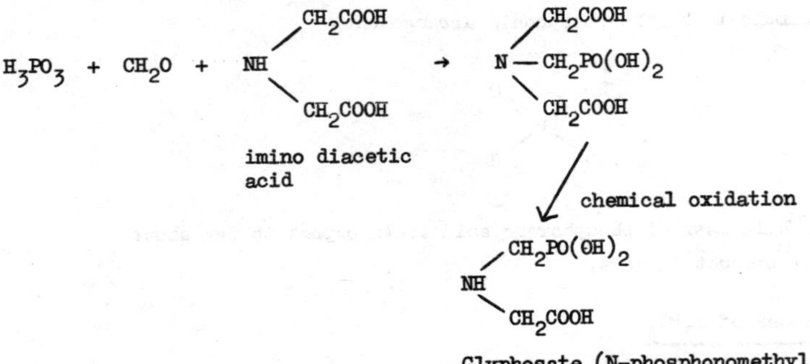

$$H_3PO_3 + CH_2O + NH(CH_2COOH)_2 \rightarrow N(CH_2COOH)_2-CH_2PO(OH)_2$$

imino diacetic acid

$$\xrightarrow{\text{chemical oxidation}} NH(CH_2PO(OH)_2)(CH_2COOH)$$

Glyphosate (N-phosphonomethyl glycine).

The isopropylamine salt of this material is the most important of this new family of broad spectrum herbicides manufactured by Monsanto[26].

## 14. Hypophosphorus Acid $H_3PO_2$ and its salts

Pure anhydrous hypophosphites of calcium and sodium have long been established as medicinal tonic materials. The initial process for these has traditionally been based on the alkaline hydrolysis of white phosphorus.

Today technical sodium hypophosphite monohydrate, which is increasingly used on a scale of some thousands of tons/year for the electroless plating of nickel on metals and plastics, is prepared by the reaction of white phosphorus with an aqueous solution of sodium hydroxide combined with the addition of sufficient calcium hydroxide to remove the by-product phosphite as calcium phosphite[27]. The co-product, impure phosphine, may be recovered for use. Equations outlining the reactions are set out below. Variants using non aqueous solvents have been described[28].

TABLE 13

Hypophosphites production

$$4P + 2Ca(OH)_2 + 3H_2O \rightarrow Ca(H_2PO_2)_2 + CaHPO_3 + PH_3 + H_2$$

Traditional route to medicinal hypophosphites.

$$PH_3 + 2NaOCl \rightarrow H_3PO_2 + 2NaCl$$

Clean up reaction.

$$4P + 3O\bar{H} + 3H_2O \rightarrow PH_3 + 3H_2PO_2^-$$

Controlled production of sodium hypophosphite using NaOH together with enough $Ca(OH)_2$ to precipitate the phosphite as calcium phosphite.

Pure hypophosphorus acid is prepared industrially from a technical grade of sodium or calcium hypophosphite by the use of an acid cation exchange column. The aqueous acid so produced is concentrated by evaporation and is normally sold as either the 30% or the 50% solution to British Pharmacopoeia standards. World production is probably little more than 1000 tons/year of 50% acid. Structure is $O{=}P\begin{smallmatrix}\diagup OH\\ -H\\ \diagdown H\end{smallmatrix}$ for the acid.

A number of highly purified salts have to be made from the pure acid, which is mono basic in its function; these include magnesium, calcium, manganese and iron salts which are sold for pharmaceutical applications and for use as antioxidants and catalysts in a variety of resin and synthetic fibre industries.

In 1979 Albright and Wilson commissioned what is probably the largest plant in the world for the manufacture of the pure hypophosphorus acid.

15. Conclusion

The transition between inorganic and organic phosphorus chemistry is difficult to define sharply. While the described chemistry of phosphorus has become almost overwhelming in the last twenty years yet the number of purely inorganic industrial phosphorus compounds has hardly changed at all. For example two far reaching industrial processes involving phosphorus and established in the last twenty years, the use of phosphorus compounds in homogeneous catalysis and the use of phosphorus ylids in vital product synthesis, have made only a little impact on the tonnage or type of inorganic intermediates used by industry.

Considering the vital and varied role played by phosphorus in the chemistry of life it is a little ironic that perhaps the most potentially subtle of all industrial applications, the processes in conjunction with enzymes or bacteria for the synthesis of proteins, calls for little more from the phosphorus producer than phosphoric acid. Nonetheless the inorganic phosphorus industry has experienced considerable impact from the politics of sodium triphosphate and has been pleased to feel a growing flush of enthusiasm in products such as phosphine, phosphorous acid, hypophosphites or monosodium fluorophosphate which were little more than fine chemicals or development products twenty years ago.


1. A.F.Childs, "Modern Inorganic Chemical Industry", Chemical Society Special Publication, London, 1977, p.375 et seq.
2. J.R.van Wazer, "Phosphorus and its Compounds", Interscience, New York, 1958, Vol. 1, p.96.
3. A.D.F.Toy, "The Chemistry of Phosphorus", Pergamon Texts in Inorganic Chemistry, Oxford, 1975, p.399, Vol.3.
4. D.E.C.Corbridge, "Phosphorus", Elsevier, Amsterdam, 1977.
5. S.B.Hartley et al, Chemical Society Quarterly Reviews, 1963, Vol. XVII, No.2.
6. A.Wilson, "The Metal Phosphides", in Mellor's Comprehensive Treatise on Inorganic and Theoretical Chemistry, Longmans, London, 1971 Vol. VIII, Supp. III.
7. R.H.Tomlinson, "Halides of Phosphorus" in reference 6 above.
8. Threshold Limit Values for 1978, Guidance Note EH 15/78, Health and Safety Executive, H.M.S.O.
9. G.M.Kosolapoff, "Organophosphorus Compounds", Wiley, New York, 1980.

10. C.Fest and K.J.Schmidt , "The Chemistry of Organophosphorus Pesticides", Springer-Verlag, Berlin, 1973.

11. Chemical Marketing Reporter, "Chemical Profile" Schnell Publishing Company, U.S.A., July 7, 1980, p.9.

12. Chemical Marketing Reporter, "Chemical Profile", Schnell Publishing Company, U.S.A., July 2, 1979, p.9.

13. S.Trippett, "Organophosphorus Chemistry", Specialist Reports, Chemical Society, London, 1970, 1970, Vol. 1 onwards.

14. Kirk-Othmer, "Encyclopedia of Chemical Technology", 2nd Edition, 1968, Vol. 15, p.307.

15. a) E.J.Lowe, "Phosphorus Hydrides" in reference 6 above.
b) American Cyanamid Company, Speciality Products publicity literature, 1980.

16. British Patent 990,918, (1965), to Albright and Wilson (Mfg.) and Hooker Chemical Corporation.

17. S.Vastog et al., J. Molecular Catalysis, 1979, 5, 189.
C.A.Tolman, Chemical Reviews, 1977, 77, 313.

18. H. Hoffman and M. Becke Goehring, "Phosphorus Sulphides", Topics in Phosphorus Chemistry, Ed. by Griffith and Grayson, Wiley, New York, 1974, Vol. 8, p.193.

19. U.S. Patent 3,560,597, (1971), to Stauffer Chemical Company.

20. British Patent 652,514, (1951), to Monsanto Chemical Company.

21. British Patent 1,050,137, (1966), to Knapsack Atiengesellschaft.

22. Chemical Marketing Reporter, "Chemical Profile", Schnell Publishing Company, U.S.A., January 2, 1978, p.9.

23. H.R.Allcock, "Phosphorus Nitrogen Compounds", Academic Press, New York, 1972, Chapter 16.

24. E. Fluck "Phosphorus Nitrogen Chemistry", Topics in Phosphorus Chemistry, Ed. by Griffith and Grayson, Wiley, New York, 1967, Vol. 4, p.291.

25. British Patent 1,023,785, (1966), to Monsanto Chemical Company.

26. "Pesticide Manual", British Crop Protection Council, 6th Edition, 1980.

27. Swiss Patent 322,980, (1957), to La Fonte Electrique.

28. U.S. Patent 2,977,192 (1961), to Food machinery and Chemical Corporation.

# Lithium Chemicals

By J. E. Lloyd
LITHIUM CORPORATION OF AMERICA INC., 26 ADELAIDE TERRACE, BLACKBURN BB2 6ET, U.K.

## 1. HISTORICAL BACKGROUND

The discovery of lithium is usually attributed to Arfvedson, a Swede working in the laboratory of Berzelius in 1817. Analysing rocks from an iron ore mine on the island of Utö, he isolated a sulphate which was not sodium, potassium or magnesium. The stone is thought to have contained the mineral petalite or spodumene. From the greek - 'lithos' - the 'new' element was first named Lithion. Sir Humphrey Davy electrolysed lithia to produce the first metal in trace amounts in 1818, and about the same time Gmelin observed the characteristic red flame colouration which is the present-day key to lithium analysis. Bunsen is reported as having isolated the metal in gram quantities in 1854 by electrolysis of the chloride. Lithium had been detected in natural mineral waters by Berzelius in 1825, and by the end of that century Lithia waters were being served in a venerable gentleman's club in Pall Mall as being beneficial to health - presumably as a cure for gout! In fact today a popular bottled water is sold in Zurich from a source named 'Lithenee'. Curative properties have been attributed to lithium and its compounds for many years, beginning around the 1840's. This may prove to be yet another example of human experience being ahead of understanding!

By 1923 there was an industrial use for the metal in a lead based alloy containing only 0.04% lithium. It was used as a bearing metal in German railway axles. This initiated the production of the carbonate, which was then being used by the glass industry, and of the metal at Langelsheim in the Harz. The benefits of lithium to the ceramic industry were the stimulus for tonnage production by the Maywood Chemical Company, New Jersey, U.S.A., in 1929. The element was no longer a curiosity, and now had an industrial future........

The industrial application of lithium has several notable 'milestones'. Two, however, may be considered to have special significance. The first was the demonstration by Cockcroft in 1932 of Einstein's theory of the equivalence of mass and energy by the reaction:-

$${}^{7}_{3}Li + {}^{1}_{1}H = 2\,{}^{4}_{2}He + 17.3 me.v.$$

which led to the understanding of the disintegration and fusion of atomic nuclei.

The second, the discovery by Dr. Cade, working in 1949 in a small township in Queensland, Australia, was that lithium really does have a beneficial effect upon the distressing human conditions of mania and depression.

## 2. SOURCES

The element does not of course occur in nature. Traces of lithium compounds are found in nearly all rocks, and in many brines and fresh water springs. Sea water contains about 0.17 ppm and the average in the earth's crust is estimated at about 60 ppm. The principal minerals of economic importance are the lithium aluminium silicates:-

| | | |
|---|---|---|
| Eucryptite | = | $Li_2O.Al_2O_3.2SiO_2$ |
| Spodumene | = | $Li_2O.Al_2O_3.4SiO_2$ |
| Petalite | = | $Li_2O.Al_2O_3.8SiO_2$ |

Although theoretically highest in $Li_2O$ content, eucryptite is rarely found in substantial quantities.

Other minerals of minor importance as sources of Li are:-

Lepidiolite - a complex lithium mica containing potassium and fluorine, and of variable Li content.

Amblygonite - a complex phosphate $(Li.Al)F.PO_4$, which has the highest $Li_2O$ of all ores, but only minor deposits are known.

These minerals are normally found within a matrix of other rocks, usually pegmatites, and a great deal of beneficiation is necessary to yield a concentrate which can be processed chemically. Because of this work, the recovery of lithium from brines is regarded as less costly than extraction from hard ores. This is, however, dependent upon the concentration of Li within the brines and the presence of other ions, in particular, magnesium.

More than 90% of the western world's lithium minerals are produced in the U.S.A., either as spodumene concentrate from North Carolina, or from the chloride brines of Nevada. The greater part of these ores are upgraded to chemicals or purified concentrates. The latter product is used in significant quantity.

This market was supplied from the Rhodesian petalite deposit until the imposition of the U.N. sanctions in 1972. A notable feature of this ore is its low iron content, which makes it a suitable material for direct use in the glass and ceramic industries. It is anticipated that the establishment of Zimbabwe will result in the return of substantial quantities of petalite to the market.

Brazil and Portugal also produce some Li minerals in small amounts. Lithium mineral concentrates are produced and converted to chemicals in both the U.S.S.R. and China, although very little is known about the nature or scale of these operations.

Table 1. MAJOR IDENTIFIED WORLD RESOURCES

| | Metric Tons $Li_2CO_3$ | Metric Tons Li |
|---|---|---|
| Tin-Spodumene belt, North Carolina, U.S.A. | 2,252,000 | 426,000 |
| Clayton Valley, Nevada, U.S.A. (Brine) | 3,690,000 | 698,000 |
| Great Salt Lake, Utah; U.S.A. (Brine) | 2,051,000 | 388,000 |
| Preissac-LaCorne, Quebec, Canada. (Spodumene) | 402,000 | 76,000 |
| Bernic Lake, Manitoba, Canada (Spodumene, petalite) | 328,000 | 62,000 |
| Bikita, Rhodesia (Spodumene, petalite, lepidolite) | 386,000 | 73,000 |
| Atacama, Chile (Brine) | 5,709,000 | 1,080,000 |
| Manono, Zaire (Spodumene) | 4,757,000 | 900,000 |
| | 19,575,000 | 3,702,000 |

## 3. STRUCTURE OF THE LITHIUM INDUSTRY

In the U.S.A. two producers supply approximately equal shares of the lithium market. Both mine and beneficiate spodumene ore from the pegmatites of North Carolina.

Foote Mineral Company, Exton, Pennsylvania
A part of Foote's concentrate is converted to carbonate at a new plant, near the King's Mountain, N.C., mine.
The rest of the concentrate is treated to reduce its iron content, and sold to the glass/ceramic industry.
Foote also produces carbonate from LiCl separated from the brines of Silver Peak, Clayton Valley, Nevada, which has a capacity of about 8,000 tes/year CE ('carbonate equivalents'). $Li_2CO_3$ is converted to other derivatives at facilities in Pennsylvania, Tennessee and Virginia.
Du Pont convert LiCl to lithium metal for sale by Foote.
A substantial part of Foote's $Li_sCO_3$ is supplied to West Germany.
This Company has recently announced that it will build a plant to process the Atacame brines, in co-operation with the Chilian authorities.
Foote Mineral Company, U.K. Branch, London is a sales office.

Metallgesellschaft A.G. Frankfurt, W. Germany
At Hans Heinrichhutte, Langelsheim (in the Harz Mountains), the first site of industrial production of lithium chemicals, M.G. now converts Foote carbonate to many lithium derivatives.
Degussa A.G. produce lithium metal for M.G.
Metallgesellschaft is the sole producer in Western Europe of the important product lithium hydroxide. The only significant competition in the European market has been from U.S. imports.
Recently, both the Russians and the Chinese have sold into the E.E.C.
M.G. also produces organolithium compounds, and has contributed significantly to their use by European industry.

Lithium Corporation of America, Gastonia, North Carolina
The ore from Lithco's Hallman-Beam mine is concentrated by froth flotation at the mine, and converted to chemicals and metal at the plant at nearby Bessemer City. This plant has a capacity of roughly 13/14,000 tes/year "C.E.'s". It produces over seventy

lithium-based products, many in very small quantities, as a service to the market.
In 1979, Lithco established:

Lithium Corporation of Europe Ltd.

'L.C.E.' is building a plant to produce speciality, lithium-based chemicals including butyl-lithium, at Bromborough, Wirral. This plant will be in operation in the latter part of 1980.
L.C.A. intend that Lithium Corporation of Europe will become the focus of its activities in Europe and related market areas.
These facilities will be complementary to its U.S. plant.

Asia Lithium Company, Japan

This is a joint venture company between Lithco and Honjo Chemical Company. The latter converts L.C.A. carbonate, and together the companies produce butyl-lithium.

There is a substantial network of agents of these companies throughout the world, and suppliers who convert and re-pack small quantities.
Also, there is some trade, including barter/exchange deals, in carbonate and hydroxide.

## 4. EXTRACTION OF LITHIUM

From spodumene, there are basically three process routes:

1. Acid Extraction

Spodumene concentrate is heated in a rotary kiln at 1030/1040°c.
The α-spodumene is transformed to the open β-phase.
β-spodumene is mixed with 93% $H_2SO_4$ and roasted at 200/250°c.
The hydrogen ion replaces the Li ion in the mineral, giving soluble lithium sulphate which is filtered from the ore residue. The impure $Li_2SO_4$ stream is purified by treatment with $Ca(OH)_2$ and $Na_2CO_3$, to remove calcium and magnesium, and concentrated by evaporation. Pure $Li_2CO_3$ is precipitated by soda ash. The sodium suphate is recovered by crystallisation of the decahydrate, which is dried to the anhydrous salt, and sold. Fig. 1.

FIGURE 1

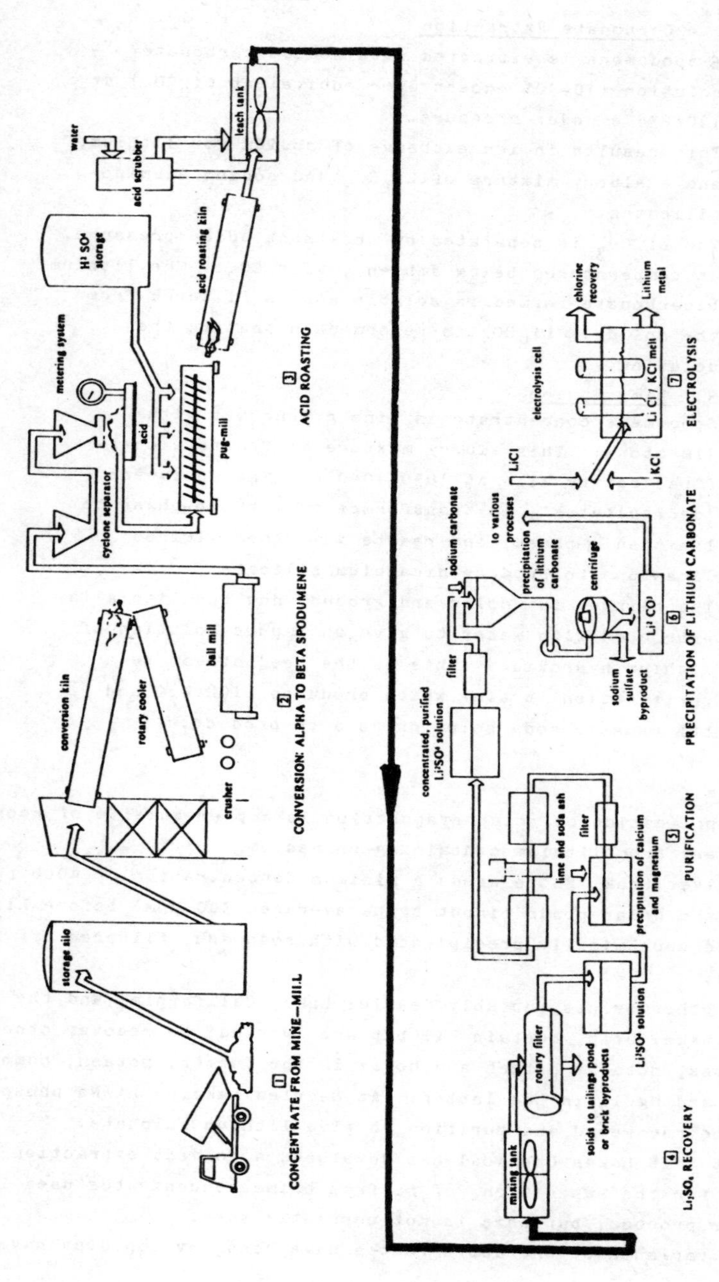
storage silo
conversion kiln
rotary cooler
crusher
ball mill
cyclone separator
metering system
acid
pug-mill
$H_2SO_4$ storage
acid roasting kiln
acid scrubber
water
leach tank
1 CONCENTRATE FROM MINE—MILL
2 CONVERSION: ALPHA TO BETA SPODUMENE
3 ACID ROASTING
mixing tank
rotary filter
solids to tailings pond or brick byproducts
$Li_2SO_4$ solution
lime and soda ash
precipitation of calcium and magnesium
filter
4 $Li_2SO_4$ RECOVERY
5 PURIFICATION
concentrated, purified $Li_2SO_4$ solution
filter
sodium carbonate
precipitation of lithium carbonate
centrifuge
sodium sulfate byproduct
$Li_2CO_3$
to various processes
6 PRECIPITATION OF LITHIUM CARBONATE
LiCl
KCl
electrolysis cell
LiCl / KCl melt
chlorine recovery
Lithium metal
7 ELECTROLYSIS

2. Carbonate Extraction

β-spodumene is extracted with sodium carbonate solution (10-30% excess over equivalent $Li_2CO_3$) at 190/235°c under pressure.

This results in ion exchange of sodium for lithium, and a slurry mixture of $Li_2CO_3$ and sodium alumino-silicates.

The $Li_2CO_3$ is separated by treatment under pressure, at temperatures below ambient, with $CO_2$. The lithium bicarbonate formed is soluble and is filtered from the solids. $Li_2CO_3$ is reformed on heating the solution.

3. Lime Fusion

Spodumene concentrate is fine ground wet with limestone. This slurry mixture is fed to a coal-fired rotary kiln at 1030/1040°c. The spodumene 'decrepitates', is transformed from the α-phase to the open β-phase, and reacts with the calcined limestone to produce dicalcium silicate.

The clinker is cooled and ground, and the lithia is extracted with water to give an impure solution of lithium hydroxide. This is the feed stream to a purification 'train' which produces $LiOH.H_2O$ and 50% caustic soda solution as a co-product.

Brines

The application of solar evaporation is a prerequisite of economic recovery from lithium-containing brines.

At Silver Peak, Foote needs a minimum concentration of 4000 ppm from the solar ponds (input brine averaged 300 ppm) before $Li_2CO_3$ can be economically precipitated with soda ash, filtered off and dried.

Some other brines, notably Searles Lake, California, and the Great Salt Lake, Utah, contain Li, but are 'worked' to recover other solutes, notably potash and borax in the former; potash, common salt and $MgCl_2$ in the latter. At Searles Lake, a Li/Na phosphate can be recovered and purified to give lithium sulphate.

Great Salt Lakes Chemical has developed a solvent extraction process for the separation of Li from brine concentrates used in their process, but this is not currently used.

Many variants of the above routes have been developed,as have alternatives to them, but these are not currently exploited by the industry.

## 5. COMPARATIVE CHEMISTRY OF LITHIUM

Lithium is the leading element of the Group 1a series, and, as such, exhibits in many of its properties the same characteristics as the common alkali metals: sodium and potassium. However, in some respects it shows similarities with the alkaline earth metals, in particular with magnesium. This is manifested by:

(a) the formation of a normal oxide, rather than peroxide, on reaction with oxygen;
(b) decomposition of the carbonates on heating;
(c) direct formation of nitrides and carbides from the elements;
(d) the very low solubility of the carbonates, fluorides and phosphates;
(e) the high degree of hydration of the ions;
(f) solubility of the salts in polar organic solvents, viz methanol and ethers;
(g) the solubility of the metal alkyls in non-polar organic solvents.

The low atomic weight of lithium results in its compounds bearing a higher percentage of the anion than other comparable cations. Thus:

- the perchlorate $LiClO_4$ and nitrate $LiNO_3$ generate a higher proportion of oxygen per unit weight of the compound;
- the peroxide $Li_2O_2$ and hydroxide LiOH will absorb more carbon dioxide;
- the hydride LiH yields more hydrogen per unit weight than any other;
- the hypochlorite LiOCl will generate more free chlorine per unit weight;
- Li on oxidation evolves more heat - 10.25 kcals/gm. cf.Na at 2.16 kcals/gm.
- the ionisation of Li gives the highest emf per unit weight of all metals.

Also, because of this effect, the lithium salts in either the fused state or the aqueous state deviate most from ideal behaviour. They depress the freezing points of fluid systems. In the fused state they are good fluxes. They reduce surface tension and

TABLE 2 - COMPARATIVE ATOMIC, IONIC AND MOLECULAR PROPERTIES OF THE ALKALI METALS

| | Li | Na | K | Rb | Cs |
|---|---|---|---|---|---|
| Atomic Number | 3 | 11 | 19 | 37 | 55 |
| Electronic configuration | 2,1 | 2,8,1 | 2,8,8,1 | 28,8,18, 8,1 | 2,8,18 8,1, |
| Atomic Weight | 6.94 | 22.99 | 39.10 | 85.47 | 132.91 |
| Heat of atomisation from standard kcal/mole | 39.0 | 25.9 | 19.8 | 18.9 | |
| Heat of formation of molecules from atoms, kcal/mole | -27.2 | -18.4 | -12.6 | -11.3 | -10.4 |
| Ionisation potential for gas, kcal | 123.8 | 117.9 | 99.7 | 95.9 | -89.4 |
| eV | 5.36 | 5.18 | 4.41 | 4.16 | 3.96 |
| Electroni affinity, eV | 0.54 | (3s)0.74 | (4s)0.7 | | |
| Normal electrode potential, V | 3.038 | 2.71 | 2.92 | 2.92 | 2.93 |
| Electronegativity | 1.0 | 0.9 | 0.8 | 0.8 | 0.7 |
| Ionic radius A | 0.68 | 0.97 | 1.33 | 1.47 | 1.67 |
| Covalent radius for C.N. 12,A | 1.58 | 1.92 | 2.38 | 2.53 | 2.72 |
| Internuclean distance in molecule A | 2.67 | 3.08 | 3.91 | | 4.55 |

PROPERTIES OF THE ALKALI METALS IN THE METALLIC STATE

| | Li | Na | K | Rb | Cs |
|---|---|---|---|---|---|
| Appearance | Silvery White solid | Silvery,white solid and liquid, purple vapour | Silvery white solid and liquid,green vapour | Silvery white solid | Silvery white solid |
| Lattice | Body centred cubic | | | | |
| Hardness (Mohs'scale) | 0.06 | 0.07 | 0.04 | | 0.02 |
| Specific Gravity @ 0° | 0.5 | 0.972 | 0.859 | 1.525 | 1.903 |
| m.p., °C | 179.5 | 97.8 | 63.5 | 38.7 | 29.8 |
| b.p., °C | 1336 | 883 | 762 | 700 | 670 |
| Heat of fusion kcal/gm atom | 0.69 | 0.63 | 0.57 | 0.53 | 0.50 |
| Effective number of free electrons per atom | 0.55 | 1.1 | 0.97 | 0.94 | 0.85 |

viscosity and because of this they bring reactants into contact, and enhance reaction rates.
The small ionic radius of the lithium atom means that its compounds with other small atoms and cations are strongly ionic in bond form. The high ionic potential results in a high energy of hydration - the ion is strongly solvated in aqueous solutions, and these show the widest deviations from the ideal. By reason of being the most electro-positive of elements, with its small size it exhibits strong covalency in many compounds. This confers a special place on lithium in the field of organic-metallic chemistry. It also leads to the solvation of the ionic compounds in organic solvents which exhibit high solubilities for the salts. Similarly, the organo-lithium compounds, for example BuLi, are readily soluble in non-polar solvents such as hexane and cyclohexane.
Even so, the solution has sufficient anionic strength to determine stereo-specific polymerisation of isoprene and butadiene.

## 6. TECHNOLOGY AND APPLICATION OF LITHIUM

As in the case of many specialised chemicals, the use of lithium by industry is as a means to achieving a desirable technological end.
The end user would, in many cases, never know that the product he buys contains lithium, or that the process by which it was made used lithium derivatives.
There are, of course, some exceptions: the man who buys lithium hypochlorite for his swimming pool is likely to be interested because he pays so much more for it than for other forms of available chlorine; the manic depressive will probably know that his pills of lithium carbonate (or citrate) are effective because they contain Li - his doctor will almost certainly have told him.

In order to develop this theme, there is some merit in attempting some division into categories. The divisions are not clear cut, and there could be many more categories; however, no apology for the imperfection is offered...

## 6.1. Market Products Containing Lithium

6.1(i) Glasses, glazes and enamels, glass-ceramics and ceramics. Glasses in many forms contain the element. This application is almost certainly the oldest industrial use. There are scientific papers by O. Schott detailing the benefits of lithium glasses, dated 1880.

Glass refers to the super-cooled liquid in the solid state. A glass used as a surface layer on a ceramic body is a glaze, and a glass used on a metal body is an enamel.

The majority of glasses are $SiO_2$ - $Na_2O$ based, and contain additional oxides which modify the base structure in manifold ways.

Substitution of the alkali $Na_2O$ by $Li_2O$ produces effects which are commonly termed 'contracting' the glass. The cause is a 'tightening' of the silica network commonly attributed to the high ionic field strength of $Li_2O$, and is manifest in a glass by:

- an increase in density
- an increase in surface hardness
- an increase in resistance to chemical attack
- an increase in resistance to thermal shock.

The lithium ion is said to have a powerful fluxing action. It is a good solvent in oxide, chloride and fluoride systems, which is a characteristic used in other applications.

The presence of lithia in the glass constituents fed to a furnace will lower the melting temperature, reduce the melting time, increase the throughput, and reduce the energy consumption. At the same time it will change the viscosity of the molten glass and its surface tension, and hence its 'working' properties.

Where electrical resistance heating of molten glass is used, Li will increase the conductivity of the melt.

$CaF_2$ is commonly used as a flux in melting glasses. Not all the fluorine appears in the glass. A great deal is discharged from the furnace, and may cause pollution.

Substitution in part or whole of $CaF_2$ by $Li_2O$ can reduce fluorine emission from glass tanks. This $Li_2O$ is commonly added as the beneficiated ore: petalite or iron-reduced spodumene.

This way the glass maker buys $SiO_2$ and $Al_2O_3$ in his mix with $Li_2O$.

Pure $Li_2CO_3$ is also used in this application - simplicity of melt control justifies the higher cost.

Oven-to-table cooking ware, developed from the famous 'Pyrex' borosilicate glasses, is, in fact, $Li_2O$ containing low-expansion glass. The lithia content permits the over-worked sales demonstration of plunging a red-hot frying pan into ice-cold water. Such ware may be additionally coated with a lithia-containing glaze, to give superior properties and improved appearance.

Photochromic glass, the glass which darkens in sunlight, is well-known in sun-glasses, and increasingly in spectacle lenses. The light-sensitive agent within the glass is a silver halide in an alumino-borosilicate structure. The more effective photochromic glass contains $Li_2O$ in order to control the solubility of the silver halide within the glass and the subsequent size of halide crystals on cooling and annealing.

U.V. photons dissociate the silver halide, and thus produce the silver ions in situ, which darken the glass. Removal of the stimulating light permits re-combination and return of the glass to transparency.

Very pure lithium carbonate is used for this application.

Many spectacle lenses are toughened by treatment in a bath of molten salts containing lithium, which replaces the sodium in the glass structure, thereby forming a compressive skin.

Sealed beam car headlight envelopes, using the same property, can suffer substantial thermal shock under some weather conditions.

Glazes on tableware and industrial ceramics, as well as vitreous enamels on steel and cast iron, are everyday applications of lithium-containing glasses. The lithium is used to impart attractive appearance, hardness, and resistence to machine washing to glazes on porcelain. By reducing a glaze melting point, output can be increased. The toxic lead content of old-fashioned glazes is removed, while retaining the brilliance of lead glazes.

Much domestic kitchenware of sheet steel is enamelled by spraying with a suspension of a glass-frit, which is then fired to melt and bond it to the surface. The presence of lithium contributes to matching the thermal expansion properties of the glass to the steel, thereby eliminating 'crazing'.

Again, melting-point control, surface tension and viscosity are important variables influenced by $Li_2O$, which is used to develop high-quality, durable finish to cookers, baths, fridges etc., at low cost.

Vitreous enamelled steel is widely used industrially. Recent developments include agricultural storage tanks; and in the future, corrosion-resistant car exhaust systems. In the States, domestic hot water systems are made in enamelled steel rather than copper as in the U.K. and Europe.

Regarding other surface coatings, it is possible to produce lithium poly-silicates solutions, with similar sol/gel properties to those of sodium and potassium. These have found use in two areas:

Firstly, they bond zinc-rich water-based suspensions of zinc powders, which are used to protect steel structures against corrosion. The polysilicate air dries to give an insoluble coating. A particular application is the painting of the oil-carrying tanks of supertankers. The use of solvent-based paints in such spaces is a potential hazard.

Secondly, such polysilicates are used to bond PTFE (Teflon) to aluminium kitchenware, to produce the well-known 'non stick pan'. The 'Lithsil', together with a fine PTFE suspension, is sprayed onto the sand-blasted surface and 'fired' at 400/450$^{\circ}$c to produce the coating.

The development of glass ceramics by Stookey of Corning was said to have begun 'the fourth golden age of glass'. Most of the glass systems which are commercially interesting contain lithia. Common glasses will re-crystallise ('devitrify') under heat treatment. Stookey found a way to produce a large number of small uniform interlocking crystals in the glass. This can be achieved by controlled catalyzed crystallisation of glasses containing nucleating agents - $TiO_2$ $ZrO_2$. These are added to the glass batch, melted, and a shape (casserole, frying pan or coffee pot) moulded. This when solid, is heat-treated, producing a crystalline ceramic in glass structure, which is very resistant to thermal shock and mechanically strong. It is also very transparent to

infra red.
The 'ceramic' cooker top is becoming increasingly common in Europe now.
There are defence and industrial applications for these materials.
Regarding ceramics, there is commercial interest in a range of $Li_2O$ based oxide ceramics, which incorporate $ZrO_2$, $Nb_2O_5$, $CoO_3$, $MoO_3$, $Al_2O_3$, $MnO_2$, $TiO_2$. Some of these are of interest to the glaze/enamel application. More complex systems are of value in speciality electronics applications, as ferrites, capacitances, etc.
Refractories are used when low thermal expansion is very desirable, for example in kiln furniture. E.g. supports used to hold metal items for furnace brazing $Li_2O$ find an application for refractories, but $Li_2O$ is not kind to conventional refractories because of its fluxing ability.
6.1(ii) Lubrication. Oil and grease are engineering imperatives for moving machinery. The property requirements of lubricating oils have to satisfy many conditions, and their formulations are frequently complex.
In a grease, a metal soap is used for form a matrix or 'sponge', which retains a lubricating oil in contact with a bearing surface. The oil is the lubricant and the soap keeps it in place under difficult operating situations. A wide range of metal soaps have been used for this purpose, notably Na, Ca, Mg, Al and Ba stearates, and, in particular, the 12-hydroxystearates derived from the saponification of hydrogenated castor oil. The properties required of such a soap are that it:

- .. has a high melting point (termed the 'drop' point)
- .. retains its thickening property under mechanical shear (measured by penetration of the grease by an object before and following mechanical agitation).
- .. holds a high volume of oil per unit of soap, and
- .. is resistant to water.

The sodium soaps are relatively cheap and have high drop points, 180/190°c, but are water soluble, whereas the calcium soaps are very insoluble but have a low melting point.
Lithium soaps were found to contain the best of both these: a high drop point, low solubility and, of course, will

contain a higher proportion of oil - more stearate chains per unit of soap. It has subsequently been found that the 12-hydroxystearate gives an even better performance than other stearate chains substituted in other sites, or substituted with different groups. This was explained by the presence in the grease of fibres of the 12-OH.St in the form of interlocking spirals. This was shown by electron micrographs.

Over 50% of all greases made are now based on Lithium 12-OH Stearate. The content of the soap within the grease is a function of the duty expected of it, and may be 5-10% by weight. A general purpose grease will contain about 8% of the soap, and it is one advantage of the Li grease that it is widely applicable for many conditions.

Grease manufacture is the largest single application of $LiOH_2H_2O$.

6.1(iii) Synthetic Rubbers. The discovery by F.W. Stavely that metallic lithium and many organo-lithium compounds initiate the polymerisation of isoprene to form an elastomer of nearly identical properties to those of natural rubber began a new era in rubber technology.

The production of such polyisoprenes and polybutadienes is the foundation of the modern synthetic rubber industry. Processes for anionic polymerisation have been developed by Phillips, Firestone and Shell, and are used to produce the greater part of the world's synthetic rubber polymers.

The styrene-butadiene co-olymers are the most common, and, used in tyre manufacture, account for the largest proportion. Other uses are growing rapidly as in the range of elastomers.

The 'castable' elastomers, such as the 'K' resins developed by Phillips, constitute a growth area for anionic catalysis using butyl-lithium.

In addition to producting solid elastomers by anionic polymerisation, it is possible to make a wide range of liquid polybutadienes of different molecular weights. These were developed by Lithium Corporation of America and given the name of Lithenes. These fluid polymers are solvent free, have outstanding resistance to water, acids and alkalis, with good adhesion to metals, high electrical resistance, and are air-drying. These properties make them eminently suitable as surface coatings.

They are made and marketed by Revertex Limited, who have an exclusive world licence from L.C.A.

There are a very broad range of polymerisations which can be initiated by lithium or its compounds. The most commonly used catalyst is butyl-lithium. The quantity of Li in the product is of course minute.

6.1(iv) <u>Films, Fibres and Paper.</u> Many uses of Li have been proposed in the patent literature relating to this area of technology. Relatively few have been commercialised - as is common when there is no sound scientific basis underlying the application.

Li catalyst are used in some processes for the manufacture of polyethylene terephthalate films and fibres, to which they bestow improved properties in terms of ease of extrusion, good moisture pickup, dyeability, and clarity of colour. Anti-static properties are also induced in textile fabrics by the incorporation of Li compounds.

A significant use of lithium hydroxide is in the production of acid dyes. The lithium salts of these dyestuffs are significantly more soluble in water than the sodium derivatives. They can be used in water solution rather than in suspension, and give a higher concentration in the dyebath. This can lead to improved performance and economics in such applications as through dye-printing of man-made fibre carpets, and in the dyeing of coloured tissue papers.

## 6.2. Uses of Lithium in Human and Animal Welfare

The part played by lithium and its derivatives in making products which are beneficial to man and farm animals is important in present day society.

The greater part of these applications lies in the synthesis of organic molecules which assist in the growth of humans and animals, cure or prevention of disease, and control of pests or fungi which restrict the growth of plant foods. Of these, the synthesis of the carotenes and of vitamin A, using Li metal as the transfer agent of the ethynyl group, is the most important single process.

A number of semi-synthetic penecillins use di-methoxybenzoic acid as an intermediate. This compound is produced by L.C.A. (amongst others) using butyl-lithium.

Several other commercial syntheses use this agent in the intermediate stages of the production of cardo-vascular drugs, contraceptives and herbicides. All these, of course, contain no lithium. There is a very large bibliography on the use of Li in organic syntheses.

More surprising, perhaps, is the rapid growth in the number of references in medical literature to the use of lithium in the treatment of manic depression in man. Here it is the lithium ion which is the active factor. As noted in the introduction, folk lore has indicated some medicinal benefit of Li, for many years. Manic depression is a common disease of modern man. It is a psychiatric illness whose symptoms range from suicidal despair to flamboyant elation, and violence. The treatment of identified sufferers from the disease may range from mild tranquilizers to hospitalisation and electric shock therapy. It seems obvious that, while extreme cases will be so treated, many of lower intensity will be given tranquilizers. However, large numbers of mild cases may never be diagnosed. The story is a fascinating one, which can only be told briefly here.

LiCl has been used in the U.S. as a NaCl replacement in cases requiring a salt-free diet. In 1949 there were a number of deaths amongst patients with certain other associated medical conditions. Death was attributed to gross over use of LiCl. About the same time, Dr. Cade published the results of his work in a distant country district of Queensland, Australia. He was able to demonstrate even with only a small number of patients that carefully controlled administration of Li was beneficial to cases of acute mania. This work stimulated a great deal of interest, and research on the topic grew dramatically. From 1840 to 1964, the total number of articles published in scientific journals was just over 500 - to date there are 6,000 references in reputable bibliographies. Lithium therapy is now recognised; it regained F.D.A. approval in 1969. By April 1980, a respected journal of psychiatry carried an article showing that the use of lithium in mental illness has saved the States $2.88 billion (1000 million) in the past ten years. An additional economic benefit is claimed for the increased production achieved by the nation as a result of better health.

There are some problems. Excess intake does give rise to side effects, and can be dangerous in some medicinal conditions. Correct therapy requires control of the Li content of the blood, and this means analysis by flame photometry during the initial period of treatment.

As $Li_2CO_3$ is not patentable and is readily available, it has not been widely promoted, and the amount of research into the biochemistry of its action is surprisingly small. In fact, even with the volume of literature published, the mechanism by which the ion influences the human body is not fully understood. Observed benefits in the treatment of alcoholism and drug addiction are still being studied.

## 6.3 Energy-related Application of Lithium

A number of diverse uses of Li and its compounds have emerged from the development stage in energy-consuming systems. There are substantial real prospects in energy storage, and a distant promise of energy generation in three or four decades.

6.3.(i) Li in Aluminium Production. In the classic Hall-Heroult (1894/1895) process for the production of aluminium, $Al_2O_3$ dissolved in cryolite ($3NaF.AlF_3$) is electrolysed, and the oxygen ion is discharged at the carbon anode with the production of $CO/CO_2$.

The overall reaction is:

$$2Al_2O_3 + 3C \quad = \quad 4Al + 3CO_2$$

Aluminium collects at the cathode which forms the base of the cell. From many years of operating experience have emerged practices which, whilst not fully understood scientifically, have permitted production and efficiency to grow steadily over 85 years.

The last 25/30 years have seen the emergence of scientific understanding of the process, resulting from study stimulated by economic and other pressures. In the last ten years, energy costs and capital for new plants have greatly increased the incentive to improve the traditional process. Li can contribute to such improvement just as Hall and Heroult (separately but at the same time) claimed in their patents that it should.

By reason of the interactions of the many variables involved in the smelting process, the influence of the chemistry of the bath of molten electrolyte is complex. It is not a simple solution of $Al_2O_3$ in $3NaF.AlF_3$. Experience has taught that certain other fluorides added to the bath can improve the operation. Some of these 'additives' are

contained in the $Al_2O_3$ feed, e.g. Ca Na. The primary objective of the additions is to permit a lower operating temperature for the electrolyte. The lower the temperature, the higher the electrical conductivity and hence the lower the power losses due to resistance heating. Common additives are $CaF_2$, $MgF_2$, $AlF_3$ (excess over cryolite stoichoimetric), and more recently LiF. The problem has been to locate optimum operating conditions. There is now strong evidence which shows that under some conditions this optimum includes LiF at 2.0/2.5%wt.LiF concentration in the bath. The Li is added to the cell as $Li_2CO_3$, and converted to LiF by reaction with $AlF_3$ in the electrolyte. Properly used, there is significant economic benefit, together with reduced power consumption per tonne Al. Benefits include:
..the output per cell which can be increased, giving better capital utilisation;
..the consumption of carbon and bath chemicals being reduced;
..the life of the carbon lining of the cathode being increased, meaning lower maintenance costs.
The are some disadvantages, being:
..that some Li is produced by the electrolysis and appears in the Al metal where it can cause some difficulties during subsequent processing ;
..the control of the smelting process is more difficult;
.. the cost.
However, the aluminium industry is the fastest growing consumer of $Li_2CO_3$, and this is potentially the largest single use. The incentive is the energy saving.
This is also the goal of the new process for aluminium production poineered by Alcoa at the 30,000 te.p.a. scale. ASP (Alcoa Smelting Process) electrolyses $AlCl_3$ dissolved in a bath of KCl:LiCl eutectic. The chlorine is recycled and an energy saving over the conventional process of 25/30% is claimed. There are some technological problems.

6.3.(ii) Other Electrolytic Processes Using Li. Li metal itself is of course produced by electrolysis of LiCl in KCl/LiCl melt. A process for the production of titanium by electolysis of $TiCl_4$ in solution in a KCl/LiCl melt is being developed in production scale cells by Howmet in the States. This gives a substantial saving of energy over the conventional sodium or magnesium processes. An improvement to the conventional magnesium electrolysis is possible by using LiCl in the electrolyte of the cell.

6.3.(iii) Lithium Batteries. Lithium is the most eletro-positive of the metals, with a Standard Electrode Potential of 3.045v compared with Na, 2.71v; Zn, o.76v; etc. Consequently, Li metal cells have the potential to store the largest amount of power/unit weight or volume. Concentrated power sources either as primary (disposable) or secondary (rechargeable) batteries are now commercially available and economic to use. They have significantly longer useful life in watches, heart pacemakers, flashlights, radios, tape recorders, shavers, calculators etc. There is a great deal of commercially directed activity in the field right now, as well as continuing effort in defence orientated areas. The greatest potential of course lies in the development of the electric vehicle, and off-peak power storage. In the U.S.A., a substantial proportion of the research in these fields is directed towards the use of lithium, whilst most of the work in Europe is concerned with sodium. Together with Li or Li alloy electrodes, the multitude of battery 'couples' proposed use electrolytes containing Li salts for their special qualities. The Li ion is highly mobile in the liquid and solid state system, which gives advantageous properties. The salts are soluble in both inorganic and organic solvents, which enables a wide range of electrolytes to be used. Ionic conductance in the solid state is of importance to a wider range of electronic applications than batteries, and here Li has a role to play. In the Na/S battery, the use of $Li_2O$ equivalent of β - alumina, in which the $Na_2O$ in the crystal structure is replaced by lithium oxide, leads to improvement in the performance of the cell.

6.3.(iv) Fusion Energy. Certainly a subject of future promises, but one consuming considerable current research and development effort.

The production of fusion energy will depend upon harnessing the reaction:

$$^{2}_{1}H + ^{3}_{1}H \longrightarrow ^{4}_{2}He\ (3.5MeV.) + n(14.1MeV.)$$

The deuterium would be obtained from heavy water separation, and tritium by:

$${}^{6}_{1}Li + n \longrightarrow {}^{3}_{1}H + {}^{4}_{4}He$$

and:

$${}^{7}_{1}Li + n \longrightarrow {}^{3}_{1}H + {}^{4}He + n$$

These reactions constitute the 'breeding' cycle of a fusion reactor. To work efficiently, the breeder blanker must be enriched in the ${}^{6}_{1}Li$ isotope. A practical design of reactor will almost certainly use Li metal as the coolant and heat transfer medium. A very large quantity of lithium metal will be required for a commercial ractor system. The are many problems to be solved, not the least of which is that of containment of the lithium metal in large fast-flowing streams at high temperature. Under the conditions which can be expected, the corrosive properties of Li metal are such that it will rapidly attack known materials of construction. Some of the chemistry of this attack is somewhat unusual. $Li_3N$, which is almost certain to be present in any commercial Li metal, readily attacks Ni, Cr, and Ti.

6.3.(v) Absorption refrigeration, heat pumps and dehumidification

These are a diverse collection of applications and concepts, which are dependent upon the absorption of heat (latent heat of evaporation) when a solution is concentrated by the removal of a solvent. A domestic refrigerator works by allowing a low boiling point liquid to evaporate in the cold compartment. The vapour is then compressed by a pump and condensed at a temperature above the surroundings, by cooling in the heat exchanger at the back of the unit.

There is another form of refrigerator, called an absorption machine, which used to be made in domestic sizes by Electrolux, but now is much less common. The absorption cycle uses two fluid streams in a totally enclosed system. One is the 'refrigerant', which provides the cooling effect; the other is the 'absorbent', which carries the refrigerant through part of the cycle. It is commonly a solution containing the refrigerant.

In operation, refrigerant vapour is generated by heating the solution, and is condensed by the outside air (or cooling water). The condenser and generator operate at the highest pressure in the system, and the condensed refrigerant is expanded into a lower pressure region, where it evaporates and absorbs heat from the surroundings, thereby cooling them. The vapour then passes back into contact with the absorbent, which was concentrated by the removal of the refrigerant vapour, before passing back to the boiler again.

Such an absorption refrigeration system is used for some large air conditioning units - the Houston Astrodome and J.F. Kennedy Airport are major examples. In these units, the working fluid is a solution of lithium bromide in water. The water is the 'refrigerant' and the solution the 'absorbent'. The operation of these machines is dependent upon the deviation from Raoults Law: LiBr exhibits one of the largest degrees of non-ideality. Such installations have proved to be most economic in large air-conditioning installations, but not successful in smaller units.

A heat pump is basically a refrigerator operated 'in reverse', pumping low grade heat to higher levels of temperature. An absorption system can be used as a heat pump, and there may prove to be a significant application of such units in connection with the use of solar energy. It is also possible to use the difference in vapour pressure of the solvent ($H_2O$) over solutions of different concentration of, say, LiBr, to produce a flow of vapour which in turn can be used to drive an engine of some form. In this way the system can be described as a 'thermal battery', akin to a steam accumulator.

Some of these aspects are being studied at Salford University in conjunction with solar heating systems.

Dehumidification is in some cases an adequate way of conditioning the surrounding air, both to improve comfort and, in some cases, for industrial purposes, for example in the production of chocolate confectionery, and pharmaceuticals in some climates.

Of growing importance is the handling of lithium foil for battery manufacture. This must be carried out at low moisture levels because of the reactivity of the metal.

Operating theatres and hospitals are a major user of dehumidification plants. These installations contact the air to be dried directly with a solution of lithium chloride. This has an extremely low vapour pressure, for example 3.5mmHg at 25$^{o}$c over 40% LiCl. In this way a reduction of water content of the air can be achieved at lower cost than if refrigeration were used. There is an additional benefit in that the LiCl solution destroys bacteria, thereby producing a sterilised environment from the outside air.

Such a unit can also be arranged to recover a high proportion of the heat in the air, and thereby contribute significantly to reducing the cost of heating buildings.
There is a Swedish development of this principle available, which makes use of the ability of LiCl to absorb moisture, and of the principle of the rotary disc heat exchanger. LiCl is carried on a porous surface formed into a honeycomb. A disc of such material rotates between the inlet duct of an air heating/circulating system carrying incoming cold/moist air, and the outlet duct carrying warm air leaving the building. The warm air evaporates water from the LiCl, removing latent heat, and also gives sensible heat to the material supporting the LiCl. As this part of the surface in turn faces the incoming moist air, it removes water from it, giving up heat of dilution and condensation, and supplying sensible heat from the heat transfer surface.

## 7. DEMAND AND SUPPLY, PRODUCTION CAPACITY, WORLD RESOURCES

The potential demand for lithium for use in secondary batteries for vehicles or power storage, and for fusion power generation has given rise, especially in the U.S.A., to concern in respect to the availability of sufficient reserves of ore and of the ability of the lithium industry to meet future requirements. To help understand this situation, production figures are given:

### 7.1. Lithium Chemicals: Demand and Supply

(1,000 metric tons as lithium carbonation equivalents)

| Year | Estimated Western World Demand | Basic U.S.A. Capacity | *Western World Supply |
|---|---|---|---|
| 1972 | 16 | 16 | 19 |
| 1974 | 22 | 19 | 21 |
| 1976 | 19 | 20 | 21 |
| 1978 | 22 | 24 | 26 |
| 1980 | 25 | 27 | 29 |
| 1982 | 27 | 29 | 34 |
| 1985 | 30 | 38 | 44 |

*Includes supply from U.S.A., U.S.S.R., and China.

Growth of the lithium market is projected to be between 4 and 5 per cent annually during the next five years. Lithium Corporation's expansion plan already in progress will add 8,000 metric tons of carbonate equivalents. In addition, Foote Mineral Company is considering a brine operation in Chile which could increase their capacity by another 6,000 metic tons. These increases are expected to provide adequate capacity for the next ten years. The lithium industry does not foresee being overtaken by a sudden burst in demand for lithium metal or compounds. It is a full-grown modern integrated mining and chemical industry. Its productive capacity has readily adjusted to demands by existing and new applications.

## 7.2. World Resources.

Increases in production capacity must be supported by adequate reserves and resources. Once thought to be limited, new resources for lithium continue to be uncovered by extensive worldwide exploration. A symposium on 'Lithium Resources and Requirements by the Year 2000' was held in Golden Colorado between January 22-24, 1976. The papers presented on all major aspects of these subjects are available as Geological Survey Professional Paper 1005 from the U.S. Government Printing Office.

Lithium sources may be categorised into three groups. Briefly, the following table summarises the currently available lithium resources.

Western World Reserves and Resources

| Type of Reserve | U.S.A | Other Western Sources |
|---|---|---|
| Class A Proven (Measured) | 375 | 1,656 |
| Class B Indicated (Not Measured) | 47 | 456 |
| Class C Inferred Geologically | 2,859 | 4,969 |
| Sub Total | 3,281 | 7,081 |
| TOTAL | 10,362 | |

* Lithium Sub-Panel Report to the National Research Council, August 10th, 1976.

It is assumed that Soviet Bloc and Eastern World countries are well developed in lithium resources, and are self-sufficient for current and future needs. A large deposit of lithium minerals has been reported to exist in the Kola Peninsula of northern european Russia. Other deposits are known in several areas of Siberia. The Chinese also produce lithium, and have unknown reserves of brine and lithium ores.

## 7.3 Major Commercial Consumers of Lithium

During the 1970's, a major growth in lithium consumption occurred in the aluminium industry. Within the next 2-4 years, the rapid increase in lithium demand in this field is expected to subside. The following table shows projected worldwide growth for lithium-based products, taken as a whole.

*Projected Western World Lithium Requirements (as lithium metal)

| | Metric tons/year for period shown | | | |
|---|---|---|---|---|
| | 1982/87 | 1987/91 | 1991/95 | 1995/2000 |
| Li in Batteries | 280 | 7,000 | 21,500 | 49,000 |
| Conventional Uses | 5,000 | 6,000 | 7,500 | 10,000 |
| Annual Li Production (required) | 5,300 | 13,000 | 29,000 | 59,000 |

* Argonne National Laboratory Report 3/1/79.

The projected growth for lithium in batteries made by A.N.L. is primarily for electric vehicle requirements. Successful development of lithium EV batteries would have a considerable impact on lithium demands.

Requirements for lithium in portable and peak load storage batteries is not expected to have a great effect on lithium production.

In fusion power reactors, lithium is used to breed tritium, which is recycled as a fuel source for the reactor. In addition, lithium serves as a coolant for the reactors. The development of fusion reactors is progressing rapidly; however significant demand for lithium in this energy field is not expected until after the year 2000.

## 7.4 Summary

Considering even the projected large requirement for the EV, the lithium industry could readily respond. Western world reserves and resources of all classes are estimated at 10.4 million metric tons of lithium, of which 3.3 million tons are located in the United States. The maximum demand for lithium metal through to the year 2000 would represent less than 5% of western world reserves and resources.

# Magnesium Compounds of Industrial Significance

By T. P. Whaley*
INSTITUTE OF GAS TECHNOLOGY, 3424 SOUTH STATE STREET, CHICAGO, ILLINOIS 60616, U.S.A.

## General History[1]

Magnesium minerals have been used by man for many centuries, largely because of their abundance in deposits of soapstone, serpentine, and asbestos that are readily accessible. Early man carved soapstone into bowls and other vessels, and asbestos has been used for lamp wicks for centuries, but the use of dolomite in early iron and steel production and of epsom salt as a laxative in the early days of medicine marked the beginnings of an industrial market for magnesium compounds. As construction of bessemer and open-hearth furnaces increased with the expanding steel industry during the later part of the nineteenth century, magnesite furnace linings also became an important business. In fact, magnesia or magnesite refractories have continued to be the major application of magnesium compounds to the present time.

## Magnesium Chemistry

The magnesium atom is one of the smallest atoms; with an atomic radius of 1.60 Å, it is only slightly larger than a lithium atom with an atomic radius of 1.52 Å. It has a nuclear charge of 12+ and an atomic weight of 24.312; its twelve electrons are distributed according to the sequence $1s^2\ 2s^2\ 2p^6\ 3s^2$, the $3s^2$ electrons being

*Formerly with International Minerals and Chemicals Corp.

removed quite readily to form a +2 oxidation state. No other oxidation states are formed, although the transitory existence of Mg+ ion has been suggested[2] in certain electrolysis reactions involving magnesium electrodes. With the loss of two valence electrons, magnesium forms a rather small ion compared with the isoelectronic sodium ion, but large compared with other divalent cations such as $Co^{++}$ or $Cd^{++}$. As is true with all of the alkali and alkaline earth metals, the $Mg^{+2}$ ion is colorless.

With an oxidation potential of 2.38 volts at 25°C, magnesium is one of the more electropositive elements. It reacts with water to form hydrogen, but the insoluble magnesium oxide film that forms protects the metal from further attack by water; this protective oxide film permits magnesium to be used as a structural metal. The insolubility of the oxide, or hydroxide, is also the reason why it can be extracted from seawater and why it is used as an antacid. Since solubility is determined by the Gibbs free energy required to break the crystal bonds and the degree of hydration of the resulting ion, the solubility of magnesium compounds compared with other alkaline earth metal compounds is often a function of its very strong hydration tendency in both crystals and solutions. The degree of hydration decreases with the size of the alkaline earth metal ion, i.e., from Mg to Ba. Magnesium ions, being small, hold on to water molecules very tightly and are difficult to remove; for example, magnesium perchlorate is a widely-used dessicant. Because of their strong electrostatic effect, magnesium ions also display a strong tendency to form complexes with oxygen-containing ligands, such as alcohols and sugars; this explains the use of magnesium compounds in sugar processing and the role of magnesium as the metal ion in chlorophyl

Solubility relationships of magnesium compounds, compared with those of other alkaline earth elements, are inconsistent but generally regular within a given group. For example, the solubility of alkaline earth metal sulphates, chlorides, iodates, nitrites, and chromates decreases from magnesium to barium, whereas the solubility of the hydroxides increases from magnesium to barium. The solubility values for the alkaline earth metal fluorides, oxalates, and carbonates reach a minimum at calcium, and are, therefore, irregular.

Occurrence[3]

The most widespread occurrence of magnesium is, of course, as the central metal atom in chlorophyll - the green coloring agent in plants. This accounts for the role that magnesium plays as one of the minor nutrients in fertilizers; after all, no magnesium - no chlorophyll, and no chlorophyll - no green plants. However, chlorophyll is scarcely a starting point for the production of magnesium compounds; this is to be left to the oceans, where magnesium is the third most abundant element, and the earth's crust, where it is the eighth most abundant element.

As the second most abundant metal in seawater (next to sodium), magnesium can be considered to be in virtually inexhaustible supply, since it averages 0.13%, by weight, of seawater. Within the earth's crust, a major source of magnesium compounds is represented by the oceanic deposits that were formed from ancient oceans and whose composition is reflected in the solubility relationships among the various sodium, potassium, magnesium, and calcium salts of halides and sulphates. As would be expected from such oceanic deposits, the magnesium compounds found there are water-soluble. The major magnesium minerals in the earth's crust, however, are water-insoluble compounds - generally, silicates, carbonates, oxides, and hydroxides Of these, the most important are the carbonates, magnesite ($MgCO_3$, 29% Mg) and dolomite ($CaCO_3 \cdot MgCO_3$, 13% Mg), followed by the silicate such as olivine ($Mg_2Fe_2\ SiO_4$, 19% Mg), talc or soapstone ($4\ Mg\ SiO_3 \cdot H_2SiO_3$), serpentine or asbestos ($H_4Mg_2Si_2O_9$), forsterite ($Mg_2SiO_4$), and sepiolite ($Mg_2Si_3O_8 \cdot 2H_2O$). Next, but of lesser importance, are the oxides and hydroxides such as spinel ($MgO \cdot Al_2O_3$, or other double oxides containing both $M^{+2}$ and $M^{+3}$ species) and brucite ($Mg[OH]_2$, 42% Mg).

Magnesium is also the second most abundant metal in meteorites thus indicating that it is an important element in other celestial bodies, as well as planet Earth. In the form of silicates, it is used to classify meteorites in terms of mineralogical and chemical content.

The most important single magnesium mineral is probably magnesite, $MgCO_3$, which is mined in many areas, but primarily in U.S.S.R (about 2 million tons per year), North Korea (about 1.65 million tons per year), China, Greece, and Austria (each about 1.1 million tons per year). Altogether, about 10 million tons per year of magnesite are mined through the world.

Next most important magnesium mineral is probably dolomite ($CaCO_3 \cdot MgCO_3$, 13% Mg), which is used for precipitating magnesium hydroxide from seawater, for production of dolomite refractories (about 2 million tons per year in the United States), and as the starting material for producing magnesium metal by the Pedgeon process of heating briquettes of ferrosilicon, dolomite, and feldspar to 2100°F.

Olivine (or chrysolite) is becoming increasingly more important in today's technology, not as a starting material for magnesium compounds but for its own special properties. Essentially a mixture of magnesium and iron silicates, olivine is generally represented by the formula $(Mg,Fe)_2 SiO_4$. A rather abundant material, it is important in foundry operations because of the superior finish on castings made with olivine molds and also because of the reduced silicosis hazards involved in using olivine, as compared with silica. It is also used as a flux in steelmaking operations.

Markets for Magnesium Compounds

Essentially all of the major markets for magnesium compounds can be accounted for by only five products, as shown in Table 1.

Table 1. MAJOR MARKETS FOR MAGNESIUM COMPOUNDS

| Magnesium Compound | Market, % of Total | Average Price (1980) |
|---|---|---|
| Refractory Magnesia | 70-75% | \$130/ton |
| Caustic-Calcined Magnesia | 13% | \$230/ton |
| $Mg(OH)_2$ | 9-6% | \$130/ton |
| $MgSO_4$ | 6-5% | \$210/ton |
| $MgCO_3$ | <1% | \$255/ton |

The remainder of the market for magnesium compounds is represented by 1) the less prominent compounds (acetate, formate, bromide, chloride, nitrate, nitrite, sulfide, etc.), and 2) the captive markets for magnesium chloride as the starting material for magnesium metal production.

In the United States - the largest single market area for magnesium compounds - the five major product categories listed in

Table 1 accounted for a total usage of 973,166 tons in 1977, representing a total value of $149,225,000. This volume does not include the import tonnage, which roughly equalled the export volume of 70-80 tons. Figure 1 shows a 25-year trend of the consumption and shipments of the two principal magnesia products in the United States, refractory magnesia and caustic-calcined magnesia.[4] As shown in Figure 1, refractory magnesia markets have been levelling off in the United States since about 1965, due primarily to the limited growth in U.S. steelmaking markets since that time and a levelling off in the construction of new BOF (basic oxygen furnace) units. The caustic-calcined magnesia markets have contined to grow at the modest rate shown in Figure 1 because it is tied to the chemical industry, pharmaceutical industry, pollution control, etc. where overall growth is still taking place. Students of America's stock market will note a rough correlation between the 1955-1977 consumption of refractory magnesia in the United States and the Dow-Jones stock market averages for that period. This is not surprising, because refractory magnesia consumption is tied to steel production, as is the entire economy of today's industrial countries.

World-wide, magnesia consumption has continued to grow, particularly in countries such as Japan where steel production is still growing, from 3000 tons in 1955 to an estimated 6500 tons in 1980.[5] In 1955, one-sixth of the world's total magnesia production came from seawater magnesia plants, while in 1980, one-third of the total world production of magnesia is expected to come from seawater extraction plants. Of the seawater-derived magnesia, the percentage going into different markets has remained approximately constant during the 25-year period, with 77 to 81% going into refractories, 13 to 16% going into magnesium metal, and 5 to 7% going into all th

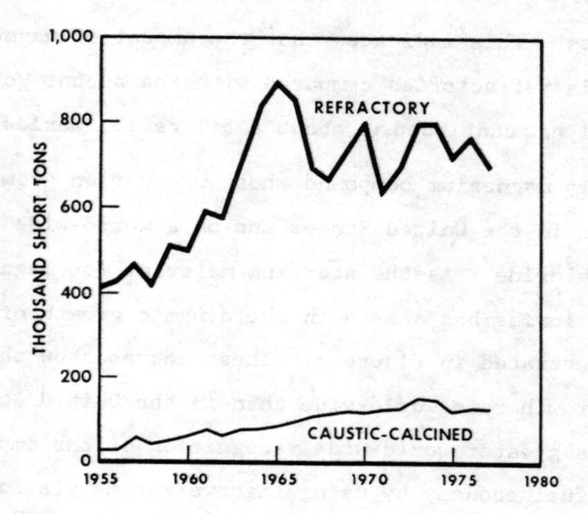


Figure 1. Consumption and shipments of magnesia in the United States.[4]

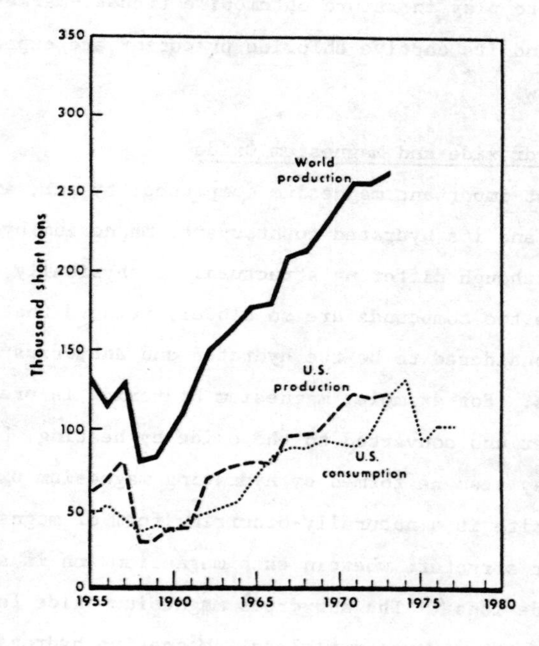


Figure 2. Production and Consumption of Primary Magnesium.[4]

other markets. This indicates that the amount of magnesia going into magnesia refractories compared with the amount going into magnesium metal has continued at about a 5:1 ratio, worldwide.

The only magnesium compound whose production growth has continued, both in the United States and on a world-wide basis, is magnesium chloride. As the starting material for magnesium metal, magnesium chloride has shared in the dynamic growth of the metal itself, illustrated in Figure 2. These curves show the much greater magnesium growth rate world-wide than in the United States,[6] probably because of a greater world-wide recognition of the importance of increasing fuel economy by using lightweight metals in automobiles. For this reason, and the increasingly greater role that fuel economy is expected to play in future automobile trends, markets for magnesium metal and its captive chloride precursor are expected to continue to grow.

Magnesium Hydroxide and Magnesium Oxide

The most important magnesium compounds, by far, are magnesium oxide, MgO, and its hydrated counterpart, magnesium hydroxide, $Mg(OH)_2$. Although differing structurally, physically, and chemically, these two compounds are so closely related that they should really be considered to be the hydrated and anhydrous forms of the same species. For example, magnesium hydroxide is precipitated from seawater and converted to the oxide by heating, but magnesium hydroxide may also be formed by hydrating magnesium oxide. The mineral brucite is a naturally-occurring form of magnesium hydroxide with a layer structure wherein each magnesium ion is surrounded by six hydroxide ions.[7] The anhydrous magnesium oxide form is found in nature as the mineral periclase. Magnesium hydroxide, when heated strongly, starts to lose water at about 350°C and is virtually completely changed to the anhydrous MgO at 600°C. However, it remains chemically reactive, dissolving readily in dilute acids and hydrated readily, even in cold water, if the calcination temperature remains below 900°C. This reactive form of magnesium oxide is called "caustic-burned" or "caustic-calcined" magnesia. If the calcination takes place above 900°C, the magnesium oxide is converted to a very unreactive form, called "dead-burned" or "sinter" magnesia, which is virtually insoluble in water and does not hydrate in contact with water.

The specific gravity or density of magnesium oxide prepared from $Mg(OH)_2$ depends on the temperature at which it was calcined, the density-temperature curve being essentially linear from 600°C (sp.gr. 2.94) to 1000°C (sp.gr. 3.39) but showing a decided break at 1200°C (sp.gr. 3.48) and gradually approaching the x-ray value of 3.58 as the fusion temperature is approached. This is particularly important in the production of periclase for refractories, because the need for high-density, low-porosity material requires high calcining temperatures to achieve the maximum density values. From 1400°C to 1800°C, the specific gravity values of 3.52 to 3.57 are only slightly below the 3.58 value for fused MgO.[1] Table 2 gives properties of MgO and $Mg(OH)_2$.

Production of Magnesium Hydroxide, $Mg(OH)_2$

Although a limited amount of magnesium hydroxide is obtained from the naturally-occurring mineral brucite and by slaking a reactive MgO, most of the $Mg(OH)_2$ is obtained from seawater or magnesium brines. Seawater composition varies somewhat around the world, but on the average is as follows in Table 3.

Table 2. PROPERTIES OF $Mg(OH)_2$ AND MgO

| | $Mg(OH)_2$ | MgO |
|---|---|---|
| Melting point | d. 350°C | 2800°C |
| hardness, Mohs | 2.5 | 5.5-6.0 |
| solubility, g/100 ml $H_2O$ | 0.0009 (18°C) | 0.00062 (20°C) |
| | 0.0004 (100°C) | 0.0086 (30°C) |
| density, $g/cm^3$ | 2.38 | 3.58 (x-ray) |
| index of refraction | 1.559, 1.580 | 1.736 |
| crystal form | trigonal | fcc |
| lattice parameter, Å | a = 3.11 | a = 4.20 |
| heat of fusion, cal/mole | | 18,500 (at 2642°C) |
| heat capacity (T=°K), cal/deg. mole (273-2073K) | | $10.86 + 0.001197T - \frac{208700}{T^2}$ |
| thermal conductivity at 100°C, cal/sec·$cm^2$ (°C/cm) | | 0.086 |
| electrical resistivity, μΩ-cm | | $2 \times 10^{14}$ (850°C) |
| | | $3 \times 10^{13}$ (980°C) |
| | | $4.5 \times 10^{8}$ (2100°C) |
| heat of formation, kcal/mole | | -143.84 (25°C) |
| color* | white | white |

* Both magnesium hydroxide and magnesium oxide are white solids, the reflective power of MgO in the visible spectrum being so great that it is used as a white color standard.

Table 3. SEAWATER COMPOSITION, g/kg

| | | | |
|---|---|---|---|
| Na | 10.77 | Cl | 19.354 |
| Mg | 1.290 | $SO_4$ | 2.712 |
| Ca | 0.412 | Br | 0.067 |
| K | 0.399 | B | 0.0045 |

At the average Mg concentration of 1.29 g /kg , about 3 g magnesium hydroxide can be produced from each liter of seawater; this means that 300 tons of seawater must be processed for every ton of MgO separated from it.

The underlying principle in precipitating magnesium from seawater involves the addition of hydroxyl ion in the form of calcium hydroxide so that insoluble $Mg(OH)_2$ will precipitate according to the equation:

$$Mg^{++} \text{ (in seawater)} + Ca(OH)_2 \rightarrow Mg(OH)_2\downarrow + Ca^{++}$$

Seawater impurities must be removed by pre-treatment. This was formerly accomplished by a pre-treatment of the raw seawater with lime to precipitate a portion of the magnesium as $Mg(OH)_2$, which occluded silica, ferric oxide, and $CaCO_3$ and was removed as a sludge. Unfortunately, the $CaCO_3$ that resulted from the reaction of the added lime with dissolved $CO_2$ formed stable supersaturated solutions which subsequently resulted in a $CaCO_3$ contamination of the $Mg(OH)_2$ product. One scheme for eliminating this problem involves the acidification of the seawater to pH 4 with $H_2SO_4$ to decompose any carbonate or bicarbonate ions, and degassing it to remove any carbon dioxide that might have remained in the water.[5]

After the decarboxylation step, a measured amount of lime (or calcined dolomite) is added to the treated seawater as a slurry of 10 to 15 lbs of CaO per $ft^3$ of water. The choice of adding either lime or calcined dolomite is determined by which of the two is most readily available. In the Gulf Coast area of the United States, for example, oyster shells are readily available and, since they are primarily $CaCO_3$, they can be converted to lime by simple calcining. Where dolomite is more accessible, as it is near the UK site at Hartlepool, the calcining and precipitation reactions are as follows:

$$CaCO_3 \cdot MgCO_3 \text{ (dolomite)} \xrightarrow{\text{calcine}} CaO \cdot MgO + 2CO_2\uparrow$$

$$Mg^{++}\text{(seawater)} + CaO \cdot MgO + 2H_2O \rightarrow 2Mg(OH)_2\downarrow + Ca^{++}$$

When roughly 70% of the magnesium values has been precipitated with lime, the seawater becomes saturated with respect to $CaSO_4$. If all the magnesium were precipitated, the $Mg(OH)_2$ product would be contaminated with $CaSO_4$. Fortunately, calcium sulphate forms stable supersaturated solutions and careful processing can avoid all product contamination except minor components that can be removed with good water washing.

Boron is another component of seawater that represents a potential problem in magnesium processing because 1) boron impurities in $MgCl_2$ feed to electrolytic magnesium reduction plants can cause a serious loss of efficiency, and 2) boron impurities in magnesia refractories can cause a serious loss of hot strength. However, it is possible to eliminate the boron contamination by special control of reaction conditions.

The large volume of seawater that must be processed in order to produce significant quantities of magnesium requires very large thickening and processing equipment. As shown in the first part of the flow diagram of Figure 3, the screened seawater and lime slurry are mixed in large flocculators, transferred to large thickeners or settling tanks with diameters on the order of 250 to 300 ft, and, after washing with fresh water to remove impurities, transferred as a 10 to 15% $Mg(OH)_2$ slurry underflow to large rotary vacuum filters of about 14 ft diameter and 18 ft length. The filters concentrate the slurry from 45 to 55% $Mg(OH)_2$.

A major problem in processing $Mg(OH)_2$ slurries is the small particle sizes and slow rates of settling, which increases both capital costs and operating costs. Consequently, a great deal of development work has been done to improve settling rates, increase particle size, etc., although this often comes at the expense of product purity if additives must be used to improve settling rates.

The filter cake from the vacuum filters can be sold, as such, or dried to a $Mg(OH)_2$ product, or further calcined to form MgO, eith caustic-calcined or dead-burned.

Production of Magnesium Oxide, MgO[1]

Magnesium oxide, MgO, is usually produced by calcining the naturally-occurring magnesite (primarily $MgCO_3$) or magnesium hydroxide, according to the equations:

$$MgCO_3 \xrightarrow{>600^\circ C} MgO + CO_2$$

$$Mg(OH)_2 \xrightarrow{>600^\circ C} MgO + H_2O\uparrow$$

If the calcination takes place below 900°C, the product is the reactive form of MgO, caustic-calcined magnesia, which can hydrate to $Mg(OH)_2$ in cold water and is soluble in acids. A particularly

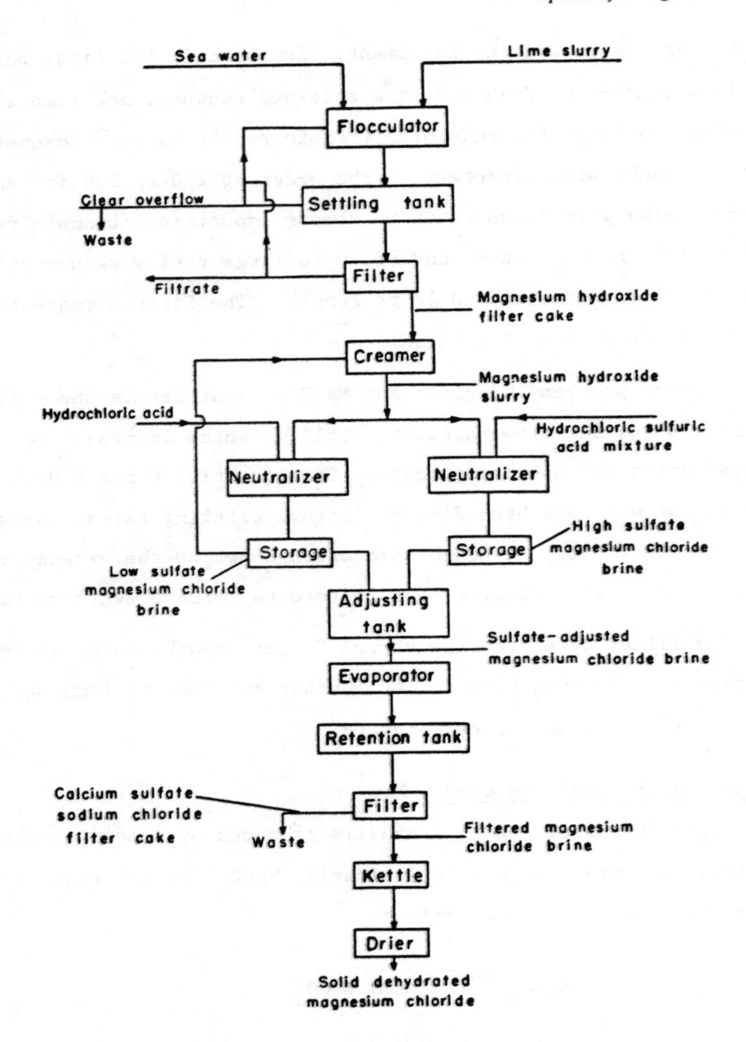


Figure 3. Process For Recovering Magnesium Compounds[3]

reactive form of MgO for use in the rubber and pharmaceutical industries is produced by calcining the basic carbonate, $5MgO \cdot 4CO_2 \cdot 6H_2O$, or $Mg(OH)_2$ below 900°C in small, externally-heated rotary kilns. Closer control of calcining temperatures is possible with externally-heated rotary kilns than with internally-fired rotary kilns, where temperatures in the firing zone may exceed 900°C and reduce the overall reactivity of the caustic-calcined magnesia. Generally, the most reactive MgO products are produced from magnesium carbonates and basic magnesium carbonate, with purities in the range of 99%. Caustic-calcined magnesias produced from $Mg(OH)_2$ are generally less reactive.

Externally-fired rotary kilns are also used for close control of temperatures in the selective calcining of dolomite, $MgCO_3 \cdot CaCO_3$. The thermal decomposition of pure calcium carbonate starts at about 825°C, whereas decomposition of $MgCO_3$ begins below 400°C, so carefully controlled heating of some dolomite samples between these two temperatures permits a selective decarboxylation and production of a mixture of $CaCO_3$ and caustic-calcined MgO, which can sometimes be separated by mechanical methods. Some dolomite samples behave as a true double compound and do not exhibit the same sharp break in the decomposition curve that is shown by other dolomite samples that behave more like a mechanical mixture of magnesium and calcium carbonates. Only the latter types of dolomite permit selective calcination into MgO and $CaCO_3$.

Dead-Burned Magnesia

When calcination temperatures exceed 900°C, magnesium oxide is converted to an unreactive, or dead-burned, form that does not hydrate in cold water (unless very finely divided) and is not soluble in acids. This is the type of MgO destined for basic refractory bricks. The dead-burned magnesias for refractory use fall into two categories, 1) brick grade, or periclase, with a 90 to 98% MgO content, and 2) maintenance grade, grain magnesite, with a 65 to 90% MgO content.

Two types of kilns are used to produce dead-burned magnesia, 1) large gas, oil, or coal-fired rotary kilns ranging in size from 6 to 12 ft in diameter and in lengths up to 400 ft, and 2) specially designed vertical gas-fired shaft kilns. Both types of kilns are lined with magnesia refractory bricks to withstand the operating temperatures of 1400 to 2000°C. Careful control of the kiln operating conditions must be maintained, which is somewhat easier to do with rotary kilns than with shaft kilns. Shaft kilns were more popu lar in earlier days of refractory-grade MgO processing than they are today, and large rotary kilns currently predominate, even though rotary kilns have a higher fuel consumption than shaft kilns.[5] Rotary kiln calcination is controlled by adjusting the angle of inclination of the kiln, feed rate, and speed of rotation. Flow is countercurrent, and the maximum calcining temperature is reached jus before discharge. In recent years, finely divided MgO and other particulate matter are removed from the effluent gases to avoid atmospheric contamination, usually by electrostatic precipitators.

The sintering temperature of pure MgO is above 2000°C, so impur ities such as iron oxides, alumina, silica, etc. must be present to achieve sintering at lower temperatures. Natural magnesites contain enough of these impurities to permit good quality, dead-burned periclase to be produced at calcining temperatures as low as 1400°C. The higher-purity, seawater-derived magnesium hydroxide, however, is too pure to sinter at these lower temperatures, and small amounts of ferric oxide were formerly added in order to achieve sintering at temperatures comparable to those used to calcine natural magnesite. This practice permitted seawater magnesia producers to more closely approximate the lump form, low-porosity, higher apparent density (up to 3.2 g/ml) material produced by dead-burning natural magnesite, but the marketplace demanded higher purities as well as high densities and a new technique was developed. The customary direct dead-burning of the seawater $Mg(OH)_2$ in rotary kilns simply could not produce a grain size greater than about 60 mm in diameter.[5] However, by first lightly calcining the $Mg(OH)_2$ to a reactive MgO and then pelletizing into almond-shaped briquettes under pressure, before dead-burning at temperatures of 1600 to 2000°C, permitted seawater-derived MgO with a purity greater than 95% to be dead-burned to a dense, lumpy product comparable to that produced from natural magnesite. This densification technique permitted acceptable periclase to be made in either rotary or shaft kilns, and higher grades of seawater periclase are now made by this technique.

A new seawater periclase process has been developed in recent years by the Steetley Company which avoids the light-burn and pelletizing steps, and thereby is a more energy-conserving process. This process uses a high-pressure tube press, developed by English China Clays and built and marketed under licence by Alfa Laval, to produce a $Mg(OH)_2$ filter cake with over 70% solids that can be dead-burned directly to a lumpy high-density refractory magnesia with the required physical characteristics.[5]

Magnesium oxide can also be produced from magnesium chloride and magnesium sulfate, although these processes entail considerably more difficulties than the $Mg(OH)_2$ or magnesite-based processes. Generally speaking, these raw materials would not be used for high temperature calcination routes to MgO unless warranted by a rather special set of circumstances. Both magnesium chloride and magnesium sulfate are found as major components in oceanic deposits, together with potassium chloride and potassium sulfate, as well as other sodium and calcium salts. Most of these oceanic deposits, however, are worked primarily for their potash values, and magnesium chloride and magnesium sulfate become by-products of the potash mining. It was under such a set of circumstances that International Minerals and Chemical Corp. developed processes for producing MgO from $MgCl_2$ and $MgSO_4$ contained in the potash waste brines at their Carlsbad, New Mexico (U.S.A.) mining operation, and used them to produce MgO during the period from 1953 to 1961.[1]

The magnesium chloride process, illustrated by the flow diagram in Figure 4, involved an initial precipitation of KCl, NaCl, and $KCl \cdot MgSO_4 \cdot 3H_2O$ (kainite) from the potash waste brines by a complex evaporation and recycling process, leaving behind a relatively pure solution of $MgCl_2$ which was concentrated by boiling at 182°C to 49.5% $MgCl_2$. After cooling and flaking, the hydrated product (a mixture of $MgCl_2 \cdot 6H_2O$ and $MgCl_2 \cdot 4H_2O$) was partially dehydrated from 62 to 64% $MgCl_2$ and then decomposed at 510 to 540°C to MgO and HCl in a 7 x 70 ft rotary decomposition kiln. The crude MgO product was abou 80% pure and contained $MgSO_4$, NaCl, KCl, and $MgCl_2$ as impurities; despite the low temperature of decomposition, the MgO was quite unre active compared with conventional caustic-calcined magnesia. The crude product was then mixed with either carbon black or fuel oil, pelletized, and calcined in a calcining kiln at about 1620°C, at

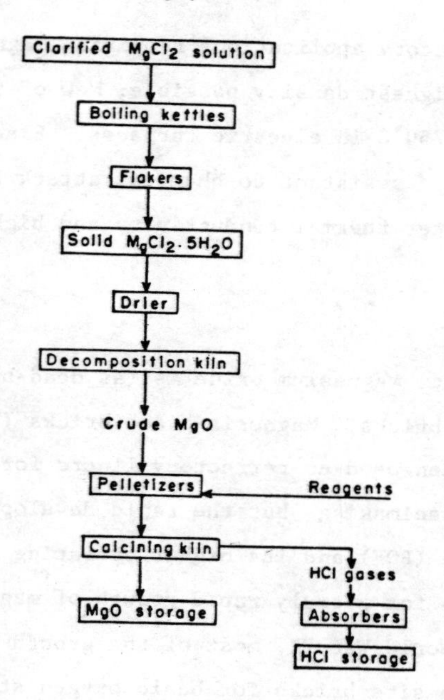


Figure 4. Preparation of MgO From $MgCl_2$ Liquor[3]

which temperature NaCl and KCl impurities vaporized and both $MgCl_2$ and $MgSO_4$ decomposed.

Magnesium oxide can be prepared from magnesium sulfate, the dissociation to MgO, $SO_3$, $SO_2$, and $O_2$ starting at temperatures just below 900°C and being complete at 1100° to 1200°C, but at much lower temperatures in the presence of a reducing agent ($H_2$, C, or CO). Mixed with carbon, $MgSO_4$ can be decomposed at 750°C according to the equation

$$MgSO_4 + C \rightarrow MgO + SO_2 + CO$$

Heated to 850°C in a stream of natural gas, $MgSO_4$ is 95 to 99% converted to MgO in about 1 hour.

For those refractory applications requiring magnesia of the highest purity and highest density possible, MgO of 96 to >99% purity is fused at 2750°C in electric furnaces. Electrically fused MgO is harder and more resistant to chemical attack than regular periclase, with greater thermal conductivity and higher electrical resistance.

Uses

The major use for magnesium oxide is (as dead-burned magnesia) in basic refractory bricks. Magnesia-based bricks (known as magnesite bricks), had been used as refractory liners for furnaces since the early days of steelmaking, but the rapid development of the basic oxygen furnace (BOF) and its higher operating temperatures provided the impetus for a truly rapid growth of magnesia-based refractories since World War II, most of the growth occurring betwee 1955 and 1965. Magnesite bricks for basic oxygen steelmaking furnaces are either tar (coal tar pitch) impregnated or tar-bonded, th pitch residue in the bricks improving their slag resistance.[8] Dead burned magnesite is also mixed with chrome ore to produce magnesite chrome and chrome-magnesite bricks that are either 1) unfired, chemically-bonded, or 2) fired at temperatures above 1650°C. The fired bricks are called direct-bonded bricks, with bonds formed at high temperature between components by solid state diffusion;[9] dire bonded bricks are used to line electric furnaces and open-hearth fu naces. Bricks with a preponderance of magnesite are known as magnesite-chrome bricks and are used largely in the glass and steel industries; chrome-magnesite bricks have a greater proportion of chrome ore and are used in copper refining.

Electrically fused magnesia is used extensively as a refractor for high temperature crucibles and, in pulverized form, as insulati for heating elements in electric furnaces and electric ranges (Calrods).

Caustic-calcined magnesia has a much greater variety of uses than the dead-burned product, but total tonnage is only about 13% of the total tonnage of magnesium compounds. Since the average cost of caustic-calcined material is nearly double the average cost of dead-burned magnesia, the total annual value of the latter is only about three times greater than the value of the material calcined at lower temperatures. The reactive MgO is used in the chemical industry for the production of various magnesium compounds and in the preparation of catalysts used in the production of organic chemicals. It is used in the rubber industry as a vulcanization accelerator and neutralizing agent, particularly in neoprene compounding. It finds use in uranium ore processing, where it functions as an absorbent and catalyst in the carbonate leach circuit, and in the dry-cleaning industry for decolorizing solvents. It is an ingredient in the formulation of dentifrices and in cosmetic powders. It is used to provide the essential minor element Mg to fertilizers. The approximate breakdown of caustic-calcined magnesia uses is shown in Table 4.

Magnesium hydroxide has its own unique uses, also, although caustic-calcined magnesia can often be used for the same applications as $Mg(OH)_2$, simply by hydrating the MgO. The principal use for $Mg(OH)_2$ is in medicine, where it has served as an antacid and laxative for many years. The well-known milk of magnesia is prepared by diluting the 30% $Mg(OH)_2$ filter cake that is obtained in the vacuum filtration step of the seawater magnesia process. It may also be produced by precipitating $Mg(OH)_2$ from $MgSO_4$ solution with caustic soda.

Table 4. USES OF CAUSTIC-CALCINED MAGNESIA[3]

| | | |
|---|---|---|
| 1. Chemical, Metallurgical, and Related Industries | | 59% |
| o Electric heating rods | 8% | |
| o Pulp and paper - for $SO_2$ recovery | 12% | |
| o Rubber compounding - vulcanizing accelerator | 9% | |
| o Petroleum additives | 9% | |
| o Other: Uranium process<br>Other magnesium compounds<br>Drycleaning solvent decolorizer<br>Catalysts for organic chemicals<br>Reagent in inorganic chemicals | | |
| 2. Agricultural | | 32% |
| o Animal feed | 21% | |
| o Fertilizer | 8% | |
| 3. Medicinals | | 0.5% |
| 4. Cements | | 8.5% |

Magnesium Carbonate

Magnesium carbonate forms three hydrates, $MgCO_3 \cdot 5H_2O$, which is stable below 13.5°C, $MgCO_3 \cdot 3H_2O$, stable between 13.5°C and 50°C, an $MgCO_3 \cdot H_2O$. The pentahydrate and trihydrate have both been identified in naturally-occurring deposits, although such deposits are rare; however, hydrated magnesium carbonate is truly stable only in an atmosphere of carbon dioxide, because of the tendency to form th more stable basic carbonate ($5MgO \cdot 4CO_2 \cdot xH_2O$) on exposure to air. Literature references on the composition of different magnesium car bonate and basic magnesium carbonate hydrates are often somewhat contradictory, particularly in the older technical literature. However, many of these compositions can probably be explained as being mixtures of the different magnesium carbonate hydrates, basic magnesium carbonate, and $Mg(OH)_2$.

Basic magnesium carbonate, as prepared by heating a solution or suspension of hydrated magnesium carbonate, may be either a pentahydrate or hexahydrate, depending on the temperatures of the precipitation and subsequent drying. The low (5 to 8 $lb/ft^3$) bulk densities of commercial $5MgO \cdot 4CO_2 \cdot xH_2O$ are due to trapped air bubbles in the solid flakes.

In addition to the various hydrates, magnesium carbonate forms many double salts with other alkaline earth metal and alkali metal halogenides and carbonates, the most common being dolomite, $MgCO_3 \cdot CaCO_3$. Some of the other double salts include $MgCO_3 \cdot Na_2CO_3$, $MgCO_3 \cdot Rb_2CO_3 \cdot 4H_2O$, $MgCO_3 \cdot (NH_4)CO_3 \cdot 4H_2O$, $MgCO_3 \cdot RbHCO_3 \cdot 4H_2O$, $MgCO_3 \cdot K_2CO_3 \cdot 8H_2O$, $MgCO_3 \cdot KHCO_3 \cdot 4H_2O$, $MgCO_3 \cdot MgCl_2 \cdot 7H_2O$, $MgCO_3 \cdot MgBr_2 \cdot 7H_2O$, and $2MgCO_3 \cdot MgBr_2 \cdot 8H_2O$. At a $CO_2$ pressure of 18 atmospheres, an unstable crystallized $Mg(HCO_3)$ is formed.

Properties

The most important properties of magnesium carbonate are probably the aqueous solubility and thermal stability, as discussed earlier. The temperature-solubility relationship of $MgCO_3$ is generally given only under a specified pressure of $CO_2$ because of the degree of hydration involved. Under a $CO_2$ atmosphere, magnesium carbonate displays retrograde solubility, i.e., solubility decreases with increasing temperatures, as shown in the following solubility data for $MgCO_3$ at 1 atmosphere pressure of $CO_2$,[1]

| Temperature, °C | 3.5 | 18 | 25 | 30 | 40 | 50 |
|---|---|---|---|---|---|---|
| Solubility, g $MgCO_3$/ 100 g solution | 3.56 | 2.21 | 1.87 | 1.58 | 1.18 | 0.95 |

The dissociation of $MgCO_3$ to MgO and $CO_2$ starts at about 350°C, but takes place rapidly above 500°C, as shown by the dissociation pressures,[1] which increase from 0.1 mm at 400°C to 100 mm at 500°C and 747 mm at 540°C. Other physical properties are given in Table 5.

Table 5. PHYSICAL PROPERTIES, $MgCO_3$ AND $MgCO_3 \cdot 3H_2O$

| | $MgCO_3$ | $MgCO_3 \cdot 3H_2O$ |
|---|---|---|
| Solubility, 25°C, g/100 g $H_2O$ | 0.0034 | 0.129 |
| Density, $g/cm^3$ | 3.037 | 1.850 |
| Crystal form | trigonal | rhombic |
| Index of refraction | 1.717, 1.515 | 1.495, 1.501, 1.526 |
| Thermal conductivity, $cal/sec \cdot cm^2(°C/cm)$ | 0.00023 (100°C) | |
| | 0.00025 (300°C) | |
| Lattice parameter, Å | a=5.61 | |

Production of Magnesium Carbonate

The most common of naturally-occurring magnesium compounds, magnesite, is primarily an impure form of magnesium carbonate that can be beneficiated by various minerals separation schemes to yield a sufficiently pure form of magnesium carbonate for refractories use However, most of the minerals beneficiation methods are generally not adequate for products intended for chemical application; these involve dissolution and precipitation steps.

The most common mineral impurities associated with magnesite are talc and silica, much of which can be removed by mechanical beneficiation steps such as screening and hydraulic separation; som deposits, such as those on the Greek Islands, have been beneficiate primarily by hand sorting - a process generally called "hand-cobbin Froth flotation involving the use of impurity depressors, such as sodium silicate, and magnesite collectors, such as oleic acid or oleates is used, as is the floating of impurities (with creosote oi by the process of "reverse floatation." Heavy media separation involving liquids with a density intermediate between that of magnesite or dolomite and its principal impurities has been used to beneficiate both magnesite and dolomite ores; this technique works because the product and impurities have different densities, and one of them floats while the other one sinks to the bottom of the heavy medium liquid.

One of the older, but still effective, methods involves the processing of dolomite ores to produce magnesium carbonate by first calcining the dolomite and then carbonating the calcine to produce magnesium bicarbonate according to one of the following two schemes (See Figure 5):

1. $$CaCO_3 \cdot MgCO_3 \xrightarrow{\text{calcine}} CaO \cdot MgO \xrightarrow{CO_2} Mg(HCO_3)_2 + \downarrow CaCO_3$$

$$Mg(HCO_3)_2 \xrightarrow{\text{heat}} 5MgO \cdot 4CO_2 \cdot XH_2O$$

2. $$CaCO_3 \cdot MgCO_3 \xrightarrow{\text{calcine}} CaCO_3 \cdot MgO \xrightarrow{CO_2} Mg(HCO_3)_2 + \downarrow CaCO_3$$

$$Mg(HCO_3)_2 \xrightarrow{\text{aerate}} 5MgO \cdot 4CO_2 \cdot XH_2O$$

This carbonation to the bicarbonate was first developed in England by H. F. Patterson in 1841, and Equation 1 above is still practised in modified form as the Patterson process. In both processes, the dolomite is calcined in rotary kilns or stack kilns (if mixed with coke). The calcine is slurried with water and carbonated with stack gases. Carbonation is carried out at 75 psi in the Patterson process, but carbonating at atmospheric pressure is used in the other. The Patterson pressure carbonation process involves the heat-treating of the bicarbonate solution with live steam to precipitate the basic carbonate. The precipitate forms

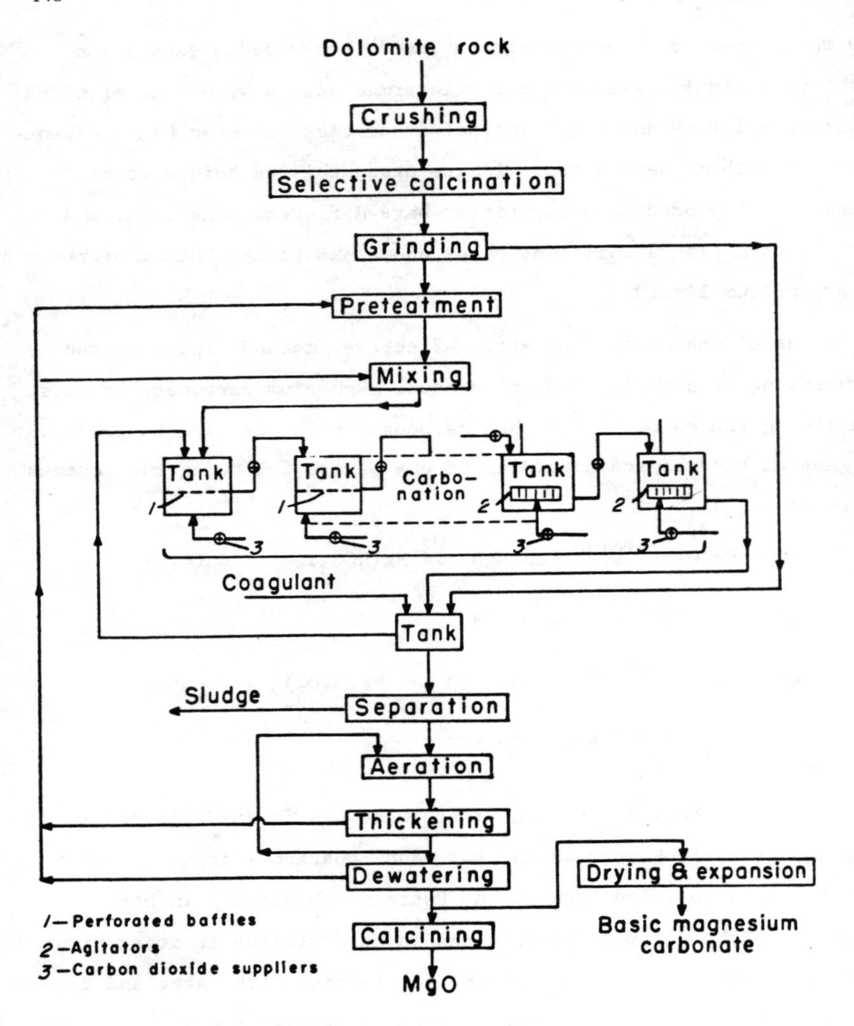


Figure 5. Manufacture of Magnesium Carbonate.[3]

heavy scale on the heated surfaces, which makes the maintenance of constant temperature in continuous units difficult; consequently, the Patterson process is often carried out on a batch basis. The other process involves atmospheric pressure carbonation and an initial precipitation of $MgCO_3 \cdot 3H_2O$ by aeration of the $Mg(HCO_3)_2$ solution at ambient temperatures; this is then converted to the basic carbonate or MgO, and the mother liquor recycled. This variation involves circulating large volumes of liquid; however, power requirements are reportedly lower than in the Patterson pressure carbonation process.

Seawater magnesia is converted by direct carbonation of the MgO slurry to $MgCO_3 \cdot 3H_2O$, which can be converted by boiling to the basic magnesium carbonate. The temperature of the carbonation determines whether the final product is to be the $MgCO_3 \cdot 3H_2O$ or the basic carbonate. If the initial precipitation of $Mg(OH)_2$ from seawater has been carried out with impure precipitating agent, the impure magnesia is converted to $Mg(HCO_3)_2$ and the solution treated as in the Patterson process.

Magnesia insulation, or other applications not requiring high purity, is produced by carbonating $Mg(OH)_2$ suspensions directly under careful control of concentrations and temperatures. This produces a precipitate of $MgCO_3 \cdot 3H_2O$, which is transferred to molds and permitted to set by forming the basic carbonate. Fillers such as asbestos fibers and calcium carbonate are added to the magnesium hydroxide suspension prior to carbonation to provide subsequent structural strength to the insulation. When the $MgCO_3 \cdot 3H_2O$ is converted to $5MgO \cdot 4CO_2 \cdot XH_2O$ in the molds, the fillers produce a lightweight but strong insulation for steam pipes that has an apparent density of 10 to 15 $lbs/ft^3$.

Magnesium carbonate is used as a filler in paints, paper, plastics, rubber, etc., the application dictating the required purity. Lower purity technical grades are generally adequate for use in many of the applications, particularly in fireproofing, fire-extinguishing and polishing compositions. Greater purity is required for $MgCO_3$ used in chemical and pharmaceutical applications such as antacids, various powder formulations, and as a starting material for other magnesium compounds. Small percentages of $MgCO_3$ are added to table salt to reduce caking.

## Magnesium Chloride

Magnesium chloride, $MgCl_2$, is an industrially significant magnesium compound primarily for its major role as the starting materia for magnesium metal production by electrolysis. However, the magnesium chloride market, as such, for this use is entirely captive, its production being integrated into the electrolytic metal production process as indicated in Figure 6. Other $MgCl_2$ markets, as indicated later in this section, are relatively modest.

Magnesium chloride is found in oceanic deposits as a component of double salts such as carnallite, $KCl \cdot MgCl_2 \cdot 6H_2O$. Such deposits from ancient seas led to statements in the earlier literature that magnesium is present in seawater as magnesium chloride. Modern chemistry, however, recognizes that seawater contains only the ioni species $Na^+$, $K^+$, $Mg^{++}$, $SO_4^{=}$, $Cl^-$, $Br^-$, etc., and the nature of the deposits from evaporated ancient seas is dictated by the relative solubilities of the salts and their hydrates. These solubilities and their phase relationships were investigated extensively in the

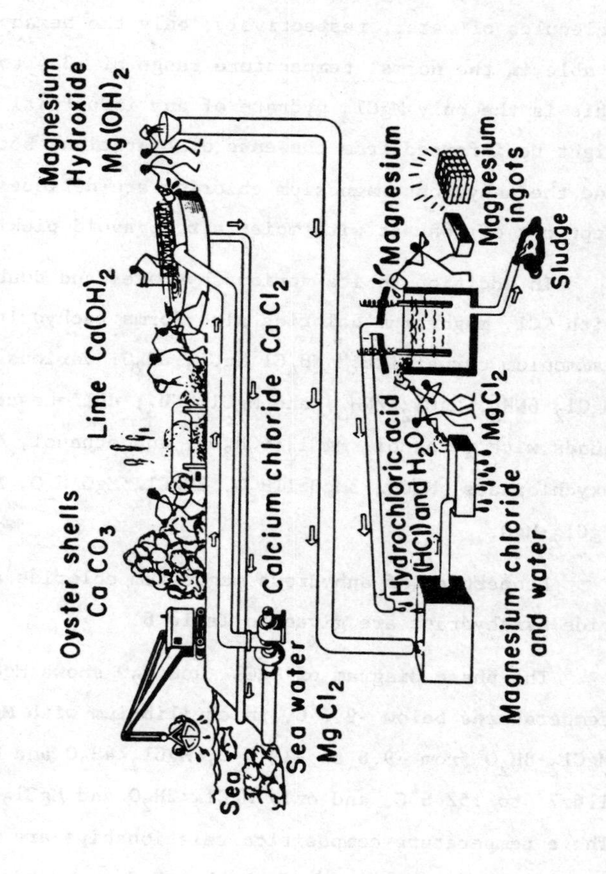


Figure 6. Dow Sea Water Process of Magnesium Production.[3]

classic studies of Van't Hoff and Meyerhoffer[1] and D'Ans,[1] undoubte prompted by the commerical development of the Stassfurt deposits i Germany.

Although $MgCl_2$ forms several hydrates, with 2, 4, 6, 8, and 1 molecules of water, respectively, only the hexahydrate $MgCl_2 \cdot 6H_2O$ stable in the normal temperature range of -3.4 to 116.7°C. Thus, this is the only $MgCl_2$ hydrate of any industrial significance. As might be inferred from the ease of hydration, both the hexahydrate and the anhydrous magnesium chloride are deliquescent and must be kept out of contact with moist air to avoid picking up moisture.

In addition to its various hydrates and double salt formation with KCl, magnesium chloride also forms tachydrite, $2MgCl_2 \cdot CaCl_2 \cdot 2$ "ammonium carnallite," $NH_4Cl \cdot MgCl_2 \cdot 6H_2O$; various ammoniates, $MgCl_2 \cdot 6NH_3$, $MgCl_2 \cdot 2NH_3$, and $MgCl_2 \cdot NH_3$; deliquescent addition compounds with methanol, $MgCl_2 \cdot 6CH_3OH$, and ethanol, $MgCl_2 \cdot 6C_2H_5OH$; oxychlorides, $MgCl_2 \cdot 3MgO \cdot 11H_2O$, $2MgCl_2 \cdot MgO \cdot H_2O$, $MgCl_2 \cdot MgO \cdot H_2O$, and $MgCl_2 \cdot MgO$.

Properties of anhydrous magnesium chloride and magnesium chlo ride hexahydrate are given in Table 6.

The phase diagram of $MgCl_2$ and $H_2O$ shows $MgCl_2 \cdot 12H_2O$ existing temperatures below -9.6°C, in equilibrium with $MgCl_2 \cdot 6H_2O$ or $MgCl_2 \cdot 8H_2O$ from -9.6 to -19.4°C, $MgCl_2 \cdot 4H_2O$ and $MgCl_2 \cdot 6H_2O$ from 116.7° to 152.6°C, and only $MgCl_2 \cdot 2H_2O$ and $MgCl_2 \cdot H_2O$ above 186°C. These temperature-composition relationships are utilized in the commercial dehydration[10] of the hexahydrates which finds both the hexahydrate and tetrahydrate losing two $H_2O$ molecules at 95° to 115°C, the dihydrate losing one $H_2O$ molecule at 135° to 180°C, an the monohydrate decomposing above 230°C.

Table 6. PROPERTIES OF $MgCl_2$ AND $MgCl_2 \cdot 6H_2O$

| | $MgCl_2$ | $MgCl_2 \cdot 6H_2O$ |
|---|---|---|
| Density, $g/ml^3$ | 2.316 | 1.56 |
| m.p., °C | 708 | d. 116-118 |
| b.p., °C | 1442 | d. |
| Lattice parameter, Å | 6.22 | 9.90, 7.15, 6.10 |
| Crystal form | hexagonal | monoclinic |
| Index of refraction | 1.675, 1.59 | 1.495, 1.507, 1.528 |
| Solubility, g/100 g. aq. soln. | 34.5 (0°C) | |
| | 37.9 (60°C) | |
| g/100 g. solvent | | |
| $CH_3OH$ | 15.5 (0°C) | |
| | 20.4 (60°C) | |
| $C_2H_5OH$ | 3.61 (0°C) | |
| | 15.89 (60°C) | |

## Production

Magnesium chloride has been produced commercially from three principal sources: 1) carnallite deposits, 2) underground brines, and 3) seawater magnesia, although the latter is the predominant source today.

In producing $MgCl_2$ from carnallite, the primary emphasis was placed on the production of KCl, with the by-product mother liquors concentrated until the solution contained up to 28% $MgCl_2$; precipitating crystalline NaCl, KCl, and $MgSO_4$ in the process. The remaining brine was then further evaporated until the density of the hot brine reached 1.435 g/ml, and permitted to cool, the cooled, solidified product being a glassy hydrate that was called "fused magnesium chloride." This by-product of KCl production was either sold as such or further processed.

Underground brines and seawater are both solutions containing magnesium and chloride ions, in addition to other cations and anions However, the process for producing $MgCl_2$ varies in accordance with the composition and the primary emphasis of the process. For example, Michigan brines contain little potassium and essentially no sulfate.[10] If bromine is to be extracted, the bromine extraction step is a chlorination step with $Cl_2$ to oxidize bromide ion to $Br_2$. Magnesium may be recovered by crystallizing out tachdyrite, $2MgCl_2 \cdot CaCl_2 \cdot 2H_2O$, as was formerly carried out by Dow Chemical Co. in processing Michigan underground brines, or it may be recovered as MgO by adding lime or dolomite to precipitate $Mg(OH)_2$, such as is used in precipitating magnesium from seawater. The precipitated magnesium hydroxide is calcined lightly to a chemically reactive MgO, which is converted to anhydrous $MgCl_2$ by chlorination in short shaft furnaces in the presence of carbon (Magnesium Elektron process or carbonaceous material; the $Mg(OH)_2$ may also be dissolved in hydrochloric acid and $MgCl_2 \cdot 6H_2O$ crystallized out by evaporation of the solution followed by careful dehydration of the hexahydrate (Dow Process).

Conversion of hydrated magnesium chloride to the anhydrous salt is usually carried out on a large scale by a two-stage process wherein $MgCl_2 \cdot 6H_2O$ is converted successively to $MgCl_2 \cdot 4H_2O$ and $MgCl_2 \cdot 2H_2O$ by heating in air at intermediate temperatures, followed by heating the dihydrate in an atmosphere of HCl or $Cl_2$ to avoid the formation of $MgO \cdot MgCl_2$ or MgOHCl and convert any oxychloride that formed during the first stage. As practised commercially, the first stage of the process involves either drum flaking or spray drying of the concentrated $MgCl_2$ solution to produce a mixture of the lower hydrates,

$MgCl_2 \cdot 2H_2O$ and $MgCl_2 \cdot H_2O$, which is compacted under pressure and then crushed for introduction into the second stage furnaces. The second stage involves heating the partially dehydrated material in rotary or shaft kilns in an atmosphere of HCl at temperatures in excess of 230°C; it can also be mixed with carbonaceous material and heated in chlorine in short shaft kilns, as in the Magnesium Electron process.

Wet hydrogen chloride is a very corrosive material, particularly at elevated temperatures, so both temperature and oxygen are controlled in the final dehydration step involving the HCl atmosphere. Temperatures are generally maintained below 325°C to avoid excessive corrosion, unless oxygen is removed from the crude HCl gas by passing it over red hot coke or charcoal, which permits operating temperatures approaching 450°C without excessive corrosion and a corresponding increase in dehydration rate.

Other dehydration steps include the reaction of ammonia with an aqueous solution of $MgCl_2$ and $NH_4Cl$ to form $MgCl_2 \cdot 6NH_3$, which may be decomposed at about 430°C, or the formation and subsequent thermal decomposition of ammonium carnallite ($NH_4Cl \cdot MgCl_2 \cdot 6H_2O$). However, neither of these methods is used commercially.[1]

Uses

The major use for magnesium chloride, except for its most important role as starting material for magnesium metal production, is in the preparation of oxychloride (sorel) cement for use in flooring cements and wall plaster. The cement is based on the equation –

$$3MgO + MgCl_2 + 11H_2O \rightarrow 3MgO \cdot MgCl_2 \cdot 11H_2O \text{ (sorel)}$$

and is made up on the job by carefully measuring and mixing the proper proportions of $MgCl_2$, $H_2O$, and a dry mix containing caustic-calcined MgO, fibrous fillers, and fine aggregates. Within a few hours, the cement forms a hard, dense, smooth-textured, sparkproof surface for floors or walls; the cement also serves as a base for tile or terrazzo floors.

Magnesium chloride also finds minor applications as a dust binder (similar to calcium chloride) in roads and mines, a fireproofing agent for wood, refrigerant brines, fire-extinguishing agent, an as a starting material for other magnesium compounds. It is the starting material for Grignard reagents and has been used for producing the gasoline anti-knock compound tetramethyl-lead, $(CH_3)_4Pb$.

## Magnesium Sulphate

Historically, magnesium sulphate is one of the most significant magnesium compounds because of its use as a laxative and cathartic in the early days of medicine. However, its place as a significant magnesium compound in today's modern commerce is relatively minor compared with the oxide, hydroxide, and carbonate.

Magnesium sulphate is an important member of the salts that constitute the various oceanic salt deposits, such as those at Stassfurt (Germany) and Carlsbad, New Mexico (U.S.A.). The principal simple salts found in nature are the mineral kieserite, $MgSO_4 \cdot H_2$ and epsomite, $MgSO_4 \cdot 7H_2O$. Double salts include: langbeinite ($K_2SO_4 \cdot 2MgSO_4$), schoenite ($K_2SO_4 \cdot MgSO_4 \cdot 6H_2O$), kainite ($2KCl \cdot MgSO_4 \cdot 6H_2$ leonite ($Na_2SO_4 \cdot MgSO_4 \cdot 4H_2O$), vanthoffite ($3Na_2SO_4 \cdot MgSO_4$), loewite ($Na_2SO_4 \cdot MgSO_4 \cdot 2.5H_2O$), bloedite ($Na_2SO_4 \cdot MgSO_4 \cdot 4H_2O$), and polyhalite ($2CaSO_4 \cdot K_2SO_4 \cdot MgSO_4 \cdot 2H_2O$).

Although magnesium sulphate forms many hydrates, the most important ones are the monohydrate and the heptahydrate (epsom salt

The pentahydrate and a monoclinic form of the heptahydrate are not stable and the hexahydrate is stable only in the temperature range of 48.2° to 67.5°C. The monohydrate is stable above 67.5°C, at which point it exhibits a solubility of 56.6 g $MgSO_4$ per 100 g $H_2O$; its solubility above this point decreases with increasing temperature to 50.4 g at 100°C, 23.9 g at 150°C, and 1.9 g at 195°C. Interestingly enough, the solubility of other magnesium sulphate hydrates increases normally from -3.8°C (27.95 g $MgSO_4$/100 g $H_2O$) to 67.5°C. Most of the $MgSO_4$ solutions show marked supercooling tendencies.

Magnesium sulphate forms other solvate systems comparable to the hydrate systems, such as the ammines ($MgSO_4 \cdot NH_3 \cdot 3H_2O$, $MgSO_4 \cdot 2NH_3 \cdot 4H_2O$, and $MgSO_4 \cdot 2NH_3 \cdot 2H_2O$) that result from the action of $NH_3$ on $MgSO_4 \cdot 7H_2O$ and the acid sulphates ($MgSO_4 \cdot H_2SO_4$, $MgSO_4 \cdot H_2SO_4 \cdot 3H_2O$, and $MgSO_4 \cdot 3H_2SO_4$) that crystallize out of solutions of $MgSO_4$ in $H_2SO_4$. As can be readily seen, the $MgSO_4 \cdot H_2SO_4$ is stoichiometrically equivalent to the bisulphate, $Mg(HSO_4)_2$. Oxysulfate salts, comparable to the oxychloride salts in the $MgCl_2$ system, are formed by the reaction of caustic-calcined MgO and $MgSO_4$ or by the thermal decomposition of magnesium sulphate hydrates; these include $3MgO \cdot MgSO_4 \cdot 11H_2O$, $5MgO \cdot MgSO_4 \cdot 8H_2O$, and $5MgO \cdot MgSO_4 \cdot 13H_2O$.

Comparable to the corresponding chloride salt, anhydrous $MgSO_4$ is quite hygroscopic and continues to pick up moisture from the air until the heptahydrate is formed.

Physical properties of magnesium sulphate and its two major hydrates are shown in Table 7.

Table 7. PHYSICAL PROPERTIES OF $MgSO_4$, $MgSO_4 \cdot H_2O$, and $MgSO_4 \cdot 7H_2O$

| | $MgSO_4$ | $MgSO_4 \cdot H_2O$ | $MgSO_4 \cdot 7H_2O$ |
|---|---|---|---|
| Density, g/ml | 2.66 | 2.517 | 1.636 |
| m.p., °C | d. 1124 | | $-6H_2O$ @ 150°C |
| Heat capacity, cal/deg./mole | 26.7 (296-372 K) | | 89 (291-319 K) |
| Solubility, g/100 ml $H_2O$ | 26 (0°C) 73.8 (100°C) | 68.4 (100°C) | 71 (20°C) |
| Crystal form | | monoclinic | rhombic |

Production

Magnesium sulfate is produced either from naturally-occurring oceanic salt deposits or by the neutralization of MgO, $Mg(OH)_2$, or magnesium basic carbonate with $H_2SO_4$. Technical grades are produced from kieserite, langbeinite, and various waste streams in potash processing. Higher purity material, such as epsom salts used in chemical or pharmaceutical applications, are produced by neutralizing MgO or $5MgO \cdot 4CO_2 \cdot xH_2O$ with sulfuric acid after precipitating iron and other impurities with an excess of MgO. After filtration, the solution is evaporated to a specific gravity of 1.35 and the epsom salt crystallizes out as the solution cools. Epsom salt may be dehydrated as shown in Figure 7.

Uses

A significant quantity of $MgSO_4$ is sold in the fertilizer market as the double salt, langbeinite, $K_2SO_4 \cdot 2MgSO_4$, which is isolated from the various oceanic salt deposits. This provides both magnesium and sulfur to the soil - both of them being necessary for plant growth - as well as a non-chloride form of potash. However, its use as a

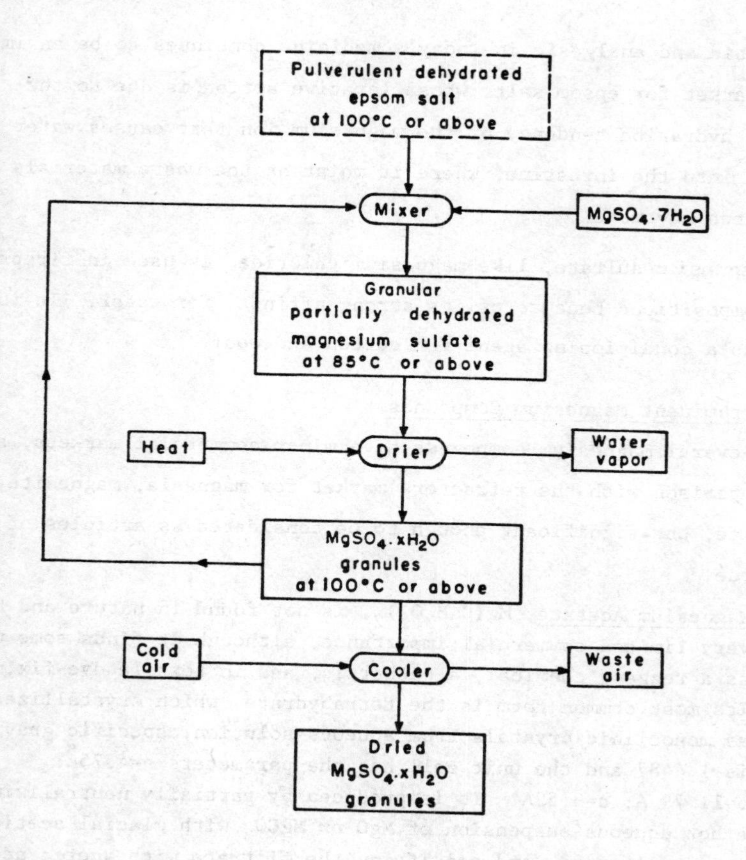


Figure 7. Process For Drying Magnesium Sulfate.[3]

cathartic and analgesic in today's medicine continues to be an impor tant market for epsom salt, whose laxative action is due to the strong hydration tendency of the magnesium ion that causes water to return into the intestine, where it moistens the waste materials and initiates movement.

Magnesium sulfate, like magnesium chloride, is used in fireproof ing compositions because of its strong affinity for water. It is used as a conditioning agent for cotton and wool.

Less Prominent Magnesium Compounds

Several magnesium compounds have minor commercial markets, smal by comparison with the refractory market for magnesia, magnesite, an dolomite, but significant enough to be considered as articles of commerce.

- Magnesium Acetate, $Mg(C_2H_3O_2)_2$, is not found in nature and has very limited commercial importance, although it finds some use as a reagent chemical, a cathartic, and in textile dye-fixing. Its most common form is the tetrahydrate, which crystallizes as monoclinic crystals from aqueous solution; specific gravity is 1.4487 and the unit cell has the parameters a=4.75Å; b=11.79 Å; c=8.52Å. It is produced by partially neutralizing a hot aqueous suspension of MgO or $MgCO_3$ with glacial acetic acid, filtering, and acidifying the filtrate with acetic acid (MgO dissolves in magnesium acetate solutions). The solution is evaporated, cooled, and the tetrahydrate crystals are separated from solution.
- Magnesium Alkyl Chloride, $C_nH_{2n+1}MgCl$, solution in ether is the well known Grignard reagent. It is prepared by the reaction of $C_nH_{2n+1}Cl$ with magnesium metal in ether, a preparation well known to all organic chemists since it is invaluable in organic syntheses. For this reason, its only use is in specialty organics and at least one firm produces the Grignard reagent commercially. It has been produced and utilized

<u>in situ</u> in an electrolytic process for producing tetramethyl-lead.

- <u>Magnesium Bromide</u>, $MgBr_2$, is found in oceanic salt deposits and may be prepared by methods analogous to those used for $MgCl_2$. Since seawater is the major bromine source in the world, its production being carried out by the oxidation of bromide ion with $Cl_2$, magnesium bromide can be produced from seawater or from natural brines such as those in Michigan and Arkansas (U.S.A.). The chemistry of $MgBr_2$ is similar to that of $MgCl_2$; it is soluble in alcohols and forms solvates or addition compounds with many organic compounds. It is used to prepare the electrolyte paste used in magnesium dry cell batteries, and formerly found a substantial use in medicine in treating nervous disorders; however, the latter use has been largely taken over by other tranquilizers.

- <u>Magnesium nitrate</u>, $Mg(NO_3)_2$, is an industrially significant compound almost exclusively for its captive use in concentrating nitric acid, although it finds minor industrial uses as a reagent chemical and in the preparation of catalysts. Its use in concentrating nitric acid is based on the strong affinity for water of $Mg(NO_3)_2$ solutions and the ternary $Mg(NO_3)_2$-$HNO_3$-$H_2O$ system that provides for a nitric acid composition of 90 to 95% $HNO_3$, thus circumventing the concentration limitation posed by the 66% $HNO_3$ azeotrope with water. Magnesium carbonate is neutralized with 60% nitric acid and the solution evaporated to a 72% $Mg(NO_3)_2$ concentration. This strong magnesium nitrate solution has an affinity for water comparable to sulfuric acid and much greater than that of 60% nitric acid; consequently, it is able to remove water from the 60% nitric acid when they are brought together in stripping columns and thus permit 90 to 95% nitric acid to distill over. The 170° to 180°C magnesium nitrate solution at the bottom of the column has a concentration of 60% and is then sent to a vacuum evaporator for removal of water and concentration to the original 72% $Mg(NO_3)_2$ solution. The 90 to 95% $HNO_3$ can be further concentrated in a rectifying column to 99.5%, with 80 to 90% $HNO_3$ exiting at the bottom. Magnesium nitrate is not sold commercially for this use, but is produced and used as needed.

o <u>Magnesium Sulfite</u>, $Mg(SO_3)_2$, is industrially significant only for its role in the "magnefite" process that is used in the manufacture of paper pulp. Magnesium bisulfite solution is produced by absorbing $SO_2$ in a 30% slurry of MgO and used to cook the wood chips for producing the pulp. After the pulp is washed, the combined liquors that include lignins, carbohydrates, and magnesium sulfite are mixed with recycled MgO, concentrated by evaporation and then sprayed into a furnace at 235°F. At this temperature, magnesium sulfite decomposes to MgO and $SO_2$, which are both recycled; the lignins and other organics in the liquors are consumed in providing fuel for the steam. It is claimed that this process revolves the troublesome problem of disposing of the black liquor from the paper-pulp operation, at the same time utilizing or recycling the components of the liquor. The "magnefite" process is possible because of the low decomposition temperature of magnesium sulfite. Magnesium sulfite is not produced and sold for this application, but is a captive material that is produced and used as needed in the process. Recycle efficiency for both MgO and $SO_2$ is rather good, with 75 to 88% of the MgO and 66 to 70% of the $SO_2$ being recovered. Makeup is 20 to 50 lb. of MgO and 70 to 100 lb. of sulfur (for producing $SO_2$) per ton of pulp.

## REFERENCES

1. A.F. Boeglin and T.P. Whaley, "Magnesium Compounds," in Kirk-Othmer Encyclopedia of Chemical Technology, 2nd ed., (Interscience, New York, 1966), Vol. 12, pp. 708-736.

2. M.A. Paul, E.J. King, and L.H. Farinholt, "General Chemistry," (Harcourt, Brace and World, Inc., New York, 1967).

3. H.B. Comstock, "Magnesium and Magnesium Compounds," U.S. Bur. Mines Inform. Circ. 8201, U.S. Government Printing Office: 1963.

4. B. Petkof, "Magnesium Compounds," in Minerals Yearbook, 1977, Vol. I, Metals and Minerals, pp. 609-616, U.S. Dept. of Interior, U.S. Government Printing Office, Washington: 1980.

5. W.C. Gilpin and N. Heasman, "Recovery of Magnesium Compounds from Sea Water," Chem. and Ind., No. 14, 16 July 1977, pp. 567-572.

6. B. Petkof, "Magnesium," in Minerals Yearbook, 1977, Vol. I, Metals and Minerals, pp. 601-608, U.S. Dept. of Interior, U.S. Government Printing Office, Washington: 1980.

7. R.D. Goodenough and V.A. Stenger, "Magnesium, Calcium, Strontium, Barium and Radium," in Comprehensive Inorganic Chemistry, (Pergamon Press, Oxford, 1973), Vol. 1, p. 641.

8. W.T. Bakker, G.D. MacKenzie, G.A. Russel, Jr., and W.S. Treffner, "Refractories," in Kirk-Othmer Encyclopedia of Chemical Technology, 2nd ed. (Interscience, New York, 1968), Vol. 17, pp. 227-267.

9. K.A. Baab and C.R. Beechan, "Refractories," The Encyclopedia of Chemistry, Third Edition, C.A. Hampel and G.G. Hawley, Eds., 960-963, Van Nostrand Reinhold Co., New York.

10. C.L. Mantell, "Magnesium Chloride," in Chlorine, ACS Monograph No. 154 (J.S. Sconce, ed.) American Chemical Society, Washington: 1962, pp. 574-583.

# Speciality Inorganic Aluminium Compounds

By K. A. Evans and N. Brown
THE BRITISH ALUMINIUM COMPANY LTD., CHALFONT PARK,
GERRARDS CROSS, BUCKINGHAMSHIRE SL9 0QB, U.K.

## Introduction

Aluminium compounds have been of service to man since at least Greek and Roman times, when they were used as mordants for dyeing, as fire retardants, for grinding, in ceramics and in medicines. Today aluminium compounds find their way into almost every sphere of human activity and several are manufactured on a vast industrial scale. This article will deal with the production and use of a number of the more important commercially available compounds, namely the hydroxide, oxide, sulphate, chloride, nitrate, phosphate, carboxylates and sodium aluminate.

## Aluminium Hydroxide

Aluminium hydroxide is the precursor of many of the inorganic aluminium compounds produced. People frequently refer to aluminium hydroxide as if it were one single compound; it does, however, exist in a number of different forms and to add to the confusion different names are used for the same mineral in different countries. Table 1 shows some of the names used.

Production of aluminium hydroxides: With the exception of small tonnages of boehmite and bayerite, the aluminium hydroxide normally produced is gibbsite, the $\alpha$-aluminium trihydroxide. The most important commercial route for the production of gibbsite is the Bayer process starting from bauxite[1].

TABLE 1

Aluminium Hydroxide: synonyms and forms

| | |
|---|---|
| $Al(OH)_3$ or $Al_2O_3.3H_2O$ | alumina hydrate |
| aluminium trihydroxide | hydrate |
| aluminium alpha trihydroxide | ATH |
| aluminium beta trihydroxide | nordstrandite |
| aluminium gamma trihydroxide | orthoaluminic acid |
| trihydrate of alumina | hydrate of alumina |
| hydrated alumina | alumina trihydrate |
| randomite | |
| gibbsite (aluminium alpha trihydroxide) | |
| hydrargillite (aluminium alpha trihydroxide) | |
| bayerite (aluminium beta trihydroxide) | |

Deposits are termed bauxites when they have at least 40% alumina present. They are formed as the result of weathering of aluminium silicates. The major impurities present in bauxite are usually iron oxide (2-20%), titania (2-8%) and silica (1-10%). Other impurities present in minor quantities are the oxides of calcium, magnesium, gallium, chromium, manganese and phosphorous[1].

Most of the bauxites which are currently being processed in the western hemisphere contain alumina primarily in the form of gibbsite whereas bauxites in the eastern hemisphere - Europe, Asia and Africa - generally contain appreciable quantities of boehmite which is more difficult to process.

In the Bayer process, crushed bauxite is treated with caustic aluminate solution containing 100-300 g $dm^{-3}$ of soda. The dissolution reaction is generally carried out under pressure at elevated temperatures ranging from 140°C to 280°C. The caustic solution reacts only with the alumina so that the impurities can be separated by settling and filtration leaving a clear solution.

To recover the dissolved aluminium hydroxide, the solution is diluted, if necessary, to 100-150 g $dm^{-3}$ soda and cooled to 50° to 70°C. The solution is then mixed with large quantities of recycled seed particles of gibbsite (up to four times by weight of the amount dissolved) and agitated in large crystallisers for periods of up to three days. The product slurry is then passed through a classification system in which the coarse fraction is separated and passed forward to the next step in the process while the finer fraction is recycled as seed for subsequent crystallisation. Filtered solution, depleted of about half of its alumina goes back to the extraction system via the evaporators. Figure 1 shows the broad principle of the Bayer process.

It is essential to control the process so that the product obtained is of the chemical and physical character required and the seed surface area is maintained. The addition of previously crystallised gibbsite as seed provides a measure of control over the particle size distribution of the product and is in fact the basis of the original Bayer patent[2]. The temperature at which crystallisation takes place, the degree of agitation and the quantity of seed added also influence the particle size distribution and purity of the gibbsite obtained.

During crystallisation, the particle size of the original seed undergoes changes due to crystal growth, production of new crystals by secondary nucleation, and cementing of particles into consolidated agglomerates/aggregates[3,4]. The extent of particle size enlargement that takes place depends upon the starting alumina supersaturation of the solution, the temperature and the quantity of seed added. Figure 2 shows a typical particle of gibbsite produced in the Bayer process.

Under the conditions of temperature and soda concentration employed in the Bayer process, the product from the crystallisers is always gibbsite irrespective of the nature of the alumina present in the starting bauxite. At lower temperatures, however, bayerite starts to be formed. Autoprecipitation, in the absence of seed crystals at $20^{o}C$, or neutralisation of sodium aluminate solution with carbon dioxide gas, leads preferentially to the formation of bayerite. However, because of the relatively rapid rate of crystallisation under these conditions, the product particles are usually of colloidal size and low crystalline order[5].

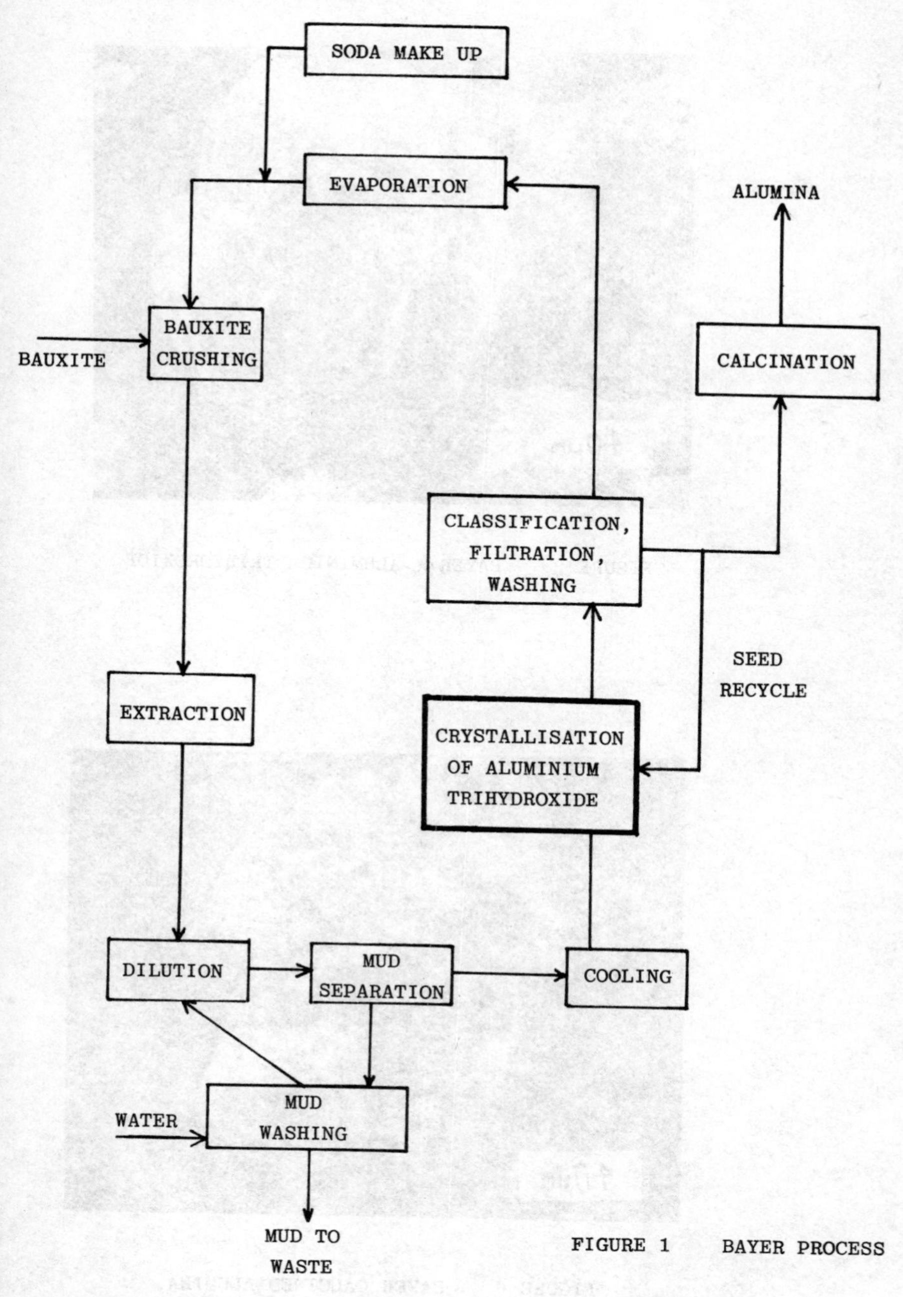


FIGURE 1 BAYER PROCESS

FIGURE 2 BAYER α-ALUMINIUM TRIHYDROXIDE

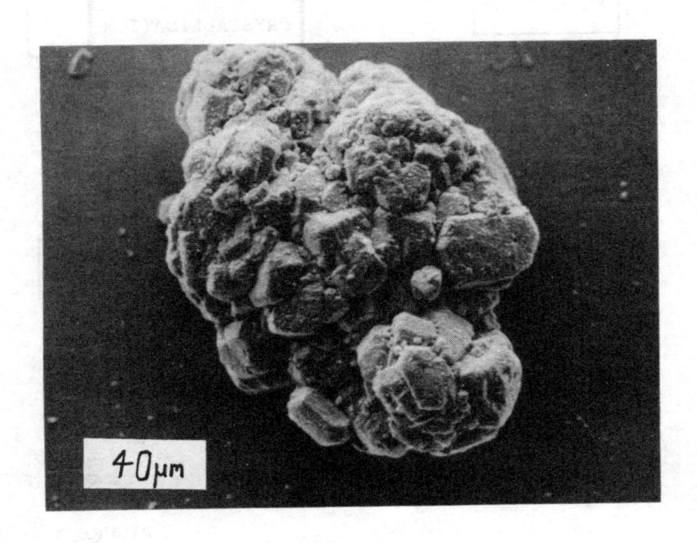


FIGURE 6 BAYER CALCINED ALUMINA,
SMELTER GRADE

When producing aluminium hydroxide from solution of pH < 7, the initial product is usually a gel which however will transform to crystalline trihydroxide on aging if the hydroxyl ion concentration is subsequently kept sufficiently high. In the absence of alkali ions, the product is bayerite. The first X-ray crystalline form to occur in the aging sequence is usually boehmite. Gelatinous boehmite itself is an important technical product[5].

Properties of Aluminium Hydroxide: Aluminium hydroxide is amphoteric and dissolves readily in solutions with a pH < 4 and > 9. Between these values its solubility is extremely low (solubility product at 20°C $(A\ell^{3+}) \times (OH^-)^3$ $1.25 \times 10^{-33}$)[6]. Mohs' hardness 2½-3½, refractive index 1.57.

Applications of Aluminium Hydroxide: Some 0.5-0.7 million tonnes of aluminium hydroxide are sold annually. The major proportion of this is used for the manufacture of other aluminium containing compounds, eg. aluminium sulphate, aluminium fluoride, aluminium nitrate, aluminium chloride, sodium aluminate and catalysts. The other major markets are toothpaste and fire retardants with much smaller tonnages going into glass[7], ceramic glazes, titanium pigment coating, paint[8,9] and pharmaceuticals.

When used in glasses or ceramic glazes, aluminium hydroxide adds sparkle and improves the resistance to chemical attack and thermal conductivity.

Fine grades of aluminium hydroxide (< 1 μm) have been found to be highly cost effective in increasing the brightness of paper[10,11,12].

Toothpaste: Many thousands of tonnes of aluminium hydroxide are now used annually for this application and a number of toothpastes on the market contain more than 50% by weight of aluminium hydroxide. The filler in a toothpaste should be sufficiently abrasive so as to remove particles adhering to teeth without causing damage to the tooth enamel or any exposed dentine. A particular advantage of aluminium hydroxide is that dentifrices containing it have very good cleansing properties but cause relatively less damage to tooth enamel than other commonly used fillers such as chalk and dicalcium phosphate[13]. Aluminium hydroxide has a Mohs' hardness of 2½-3½ and its abrasivity falls between that of chalk and dicalcium phosphate. To some extent the degree of abrasivity and cleaning properties can be controlled by judicious choice of the particle size of aluminium hydroxide used. Aluminium hydroxide's cleaning properties together with its non-toxicity, insolubility and compatibility with fluorides make it a very effective filler for toothpaste.

It is important that the active ingredients added to reduce dental caries (sodium fluoride, stannous fluoride and sodium monofluorophosphate) are not rendered ineffective by the filler added. The use of chalk in toothpaste has declined in recent years because of the increasing trend to adopt fluoride compositions; chalk can react with the fluorides added and so obviate their effectiveness. With other fillers the soluble fluoride level declines to some extent on aging but tests have shown that aluminium hydroxide and sodium monofluorophosphate have good compatibility and toothpastes containing them have been found by clinical trials to remain effective in reducing dental caries[14,15].

Co-milling the aluminium hydroxide with long chain fatty carboxylic acids can improve the retention of the fluoride[16] and with acid salts and organic acids can render the toothpaste non-corrosive[16]; this can be particularly important if the toothpaste is packed in unlined aluminium tubes.

Boehmite[18] and alumina[19,20] can also be used in toothpastes or polishes where a mere abrasive filler is required. The pastes and polishes incorporating alumina are very effective in cleaning tobacco stained teeth.

Fire Retardancy: Aluminium compounds have long been known for their ability to act as fire retardants. The annals of Claudius recorded that the storming towers used in the attack on Piraeus in 83 BC were protected against fire by treatment with alum. One of the earliest records[21] of aluminium hydroxide as a fire retardant is to Sir William Henry Perkin who late in the nineteenth century was attempting to reduce the flammability of cotton flannelette; he screened a large number of inorganic compounds and found aluminium hydroxide capable of rendering cellulose "non-flammable". A further investigation of 31 potential fire proofing agents for textiles some years later also identified the usefulness of aluminium hydroxide as a fire retardant[22].

Aluminium hydroxide's ability to improve the fire resistance of rubbers was patented in 1921[23] and in the past 20 years there have been innumerable papers and patents highlighting its fire retardant and smoke suppressant properties in different polymeric systems[24-30].

The current usage of aluminium hydroxide as a flame retardant and smoke suppressant is some 20,000 tonnes per annum in Europe and approximately 200,000 tonnes per annum in the USA. In the USA some 50% of this is in plastics, the remainder being divided between cellulose fibre loft insulation and in fire retardant latex carpet backings or underlay.

Aluminium hydroxide interferes with the primary stage of the combustion process. On heating to above 200°C it decomposes to give alumina and water vapour.

$$2A\ell\ (OH)_3 \rightarrow A\ell_2O_3 + 3H_2O$$

This reaction is strongly endothermic absorbing 1.97 kJ/g (470 cal/g) of hydrate and heat is thereby removed from the reaction zone of the polymer, thus inhibiting the production of gaseous products. On decomposition aluminium hydroxide releases 34.6% of its weight as water; this evolved water also serves to suppress burning as it dilutes the combustible gases. Its efficacy as a flame retardant for any particular polymeric system depends upon its endothermic decomposition process occurring over the correct temperature range for the polymer involved. Table 2 shows decomposition temperatures for a number of polymers, and Figure 3 shows the differential thermal analyses (DTA) of two samples of aluminium hydroxide differing in particle size distributions.

TABLE 2

Decomposition temperatures for various Polymers

| Polymer | Decomposition Temperature Range ($^{o}$C) |
|---|---|
| Polyethylene | 335 - 340 |
| Polypropylene | 328 - 410 |
| Polyvinyl chloride | 200 - 300 |
| Polystyrene | 300 - 400 |
| Polymethyl methacrylate | 170 - 300 |
| Nylon 6 and 6/6 | 310 - 380 |
| Polyvinyl acetate | 213 - 325 |
| Polyethylene terephthalate | 280 - 320 |

Most of the water vapour is evolved over the temperature range 200-400$^{o}$C which coincides with the decomposition temperature of many polymers. For these samples a heating rate of 10$^{o}$C/min was used. For the coarser material, BACO FRF 5 (median particle size by Coulter Counter 65 μm), three peaks were observed, whereas for the finer material, BACO FRF 80 (median particle size 6.5 μm), the first and third peaks are considerably diminished in size.

Weight for weight aluminium hydroxide is a less effective flame retardant than a number of other flame retardants on the market. It is, however, relatively cheap and has the nearly unique property of behaving as a smoke suppressant in addition to acting as a fire retardant[30]. Some of the other fire retardants commonly used actually increase the amount of smoke evolved when ignition does occur and also give off toxic products when they decompose. The importance of reducing the smoke hazard can be appreciated when it is realised that of the deaths that occur in domestic fires, some 75% are a consequence of the smoke and toxic fumes. Figure 4 shows the effect of progressively increasing the aluminium hydroxide content on the smoke generation characteristics of a burning polyester (tested to ASTM D 2863).

Major markets for aluminium hydroxide as a flame retardant are polyesters, particularly Sheet Moulding and Dough Moulding Compounds, rubber latex for carpet backing, PVC cables and conveyor belting, epoxies and cellulosic fibre loft insulation.

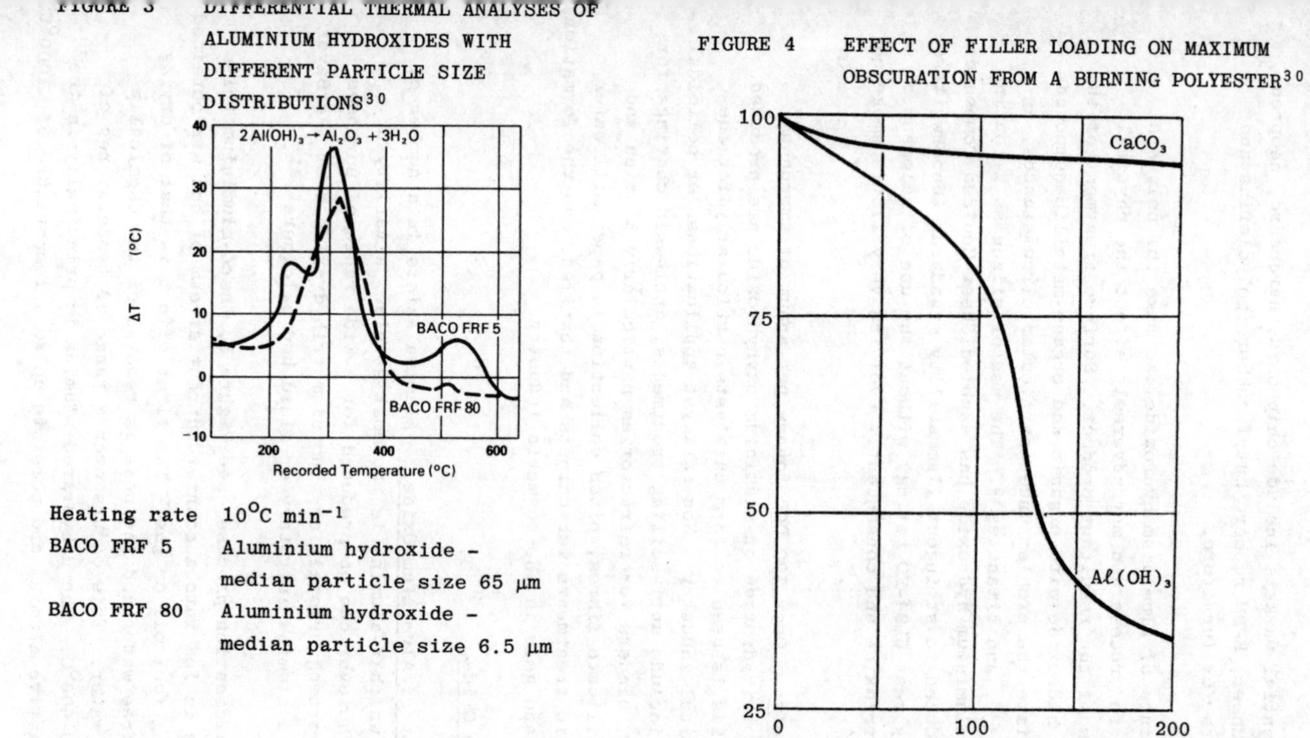


FIGURE 3 DIFFERENTIAL THERMAL ANALYSES OF ALUMINIUM HYDROXIDES WITH DIFFERENT PARTICLE SIZE DISTRIBUTIONS[30]

Heating rate $10^{\circ}C\ min^{-1}$
BACO FRF 5 Aluminium hydroxide - median particle size 65 μm
BACO FRF 80 Aluminium hydroxide - median particle size 6.5 μm

FIGURE 4 EFFECT OF FILLER LOADING ON MAXIMUM OBSCURATION FROM A BURNING POLYESTER[30]

Other potential markets include chipboard, hardboard, neoprene, nitrile rubber, EPDM rubber, butyl rubber, polyolefins and interliners for furniture.

High loadings of aluminium hydroxide can make the polymers difficult to process and may adversely effect the physical properties of the resultant product. Surface treatment of the aluminium hydroxide with organic and organo-metallic compounds can alleviate the problem under particular circumstances, eg. silanes[31,32], and titanates[33]. The use of silane as a coating agent on aluminium hydroxide has enabled Sheet Moulded Compounds to be produced to stringent flammability standards (Underwriters Laboratory test UL94-VO rating) without the use of halogens or antimony trioxide and consequently evolving very little smoke on burning[34].

Alumina gels: Over 100,000 tonnes per annum of amorphous aluminium trihydroxide or aluminium oxyhydroxide are produced. Most of this is used to form catalysts particularly for the petrochemical industry. Non-catalyst applications for colloidal aluminas include anti-soiling treatments, anti-skid coatings for packaging, binders for refractories particularly alumina and alumino silicate fibres, pitch collection in paper mill water, anti-static treatments for carpets and textiles, and the formation of foams and gels in the cosmetic industry.

## Aluminium Oxide

Production of Aluminium Oxide: Alumina exists in a number of forms of which $\alpha$-alumina is the most stable. A variety of aluminas however can be produced for a wide range of purposes when converting crystalline aluminium trihydroxide from the Bayer process to alumina in rotary or fluid bed calciners[1,35].

In the calcination process, see Figure 5, the $\alpha$-aluminium trihydroxide is fed into a counter-current stream of hot air obtained by burning fuel oil or gas. The first effect is that of drying off the free water and the next is removal of the chemically combined water. This occurs over a range of temperatures of about 180-600°C. The dehydrated alumina is principally in the form of active alumina and persists up to a temperature of ~ 1000°C.

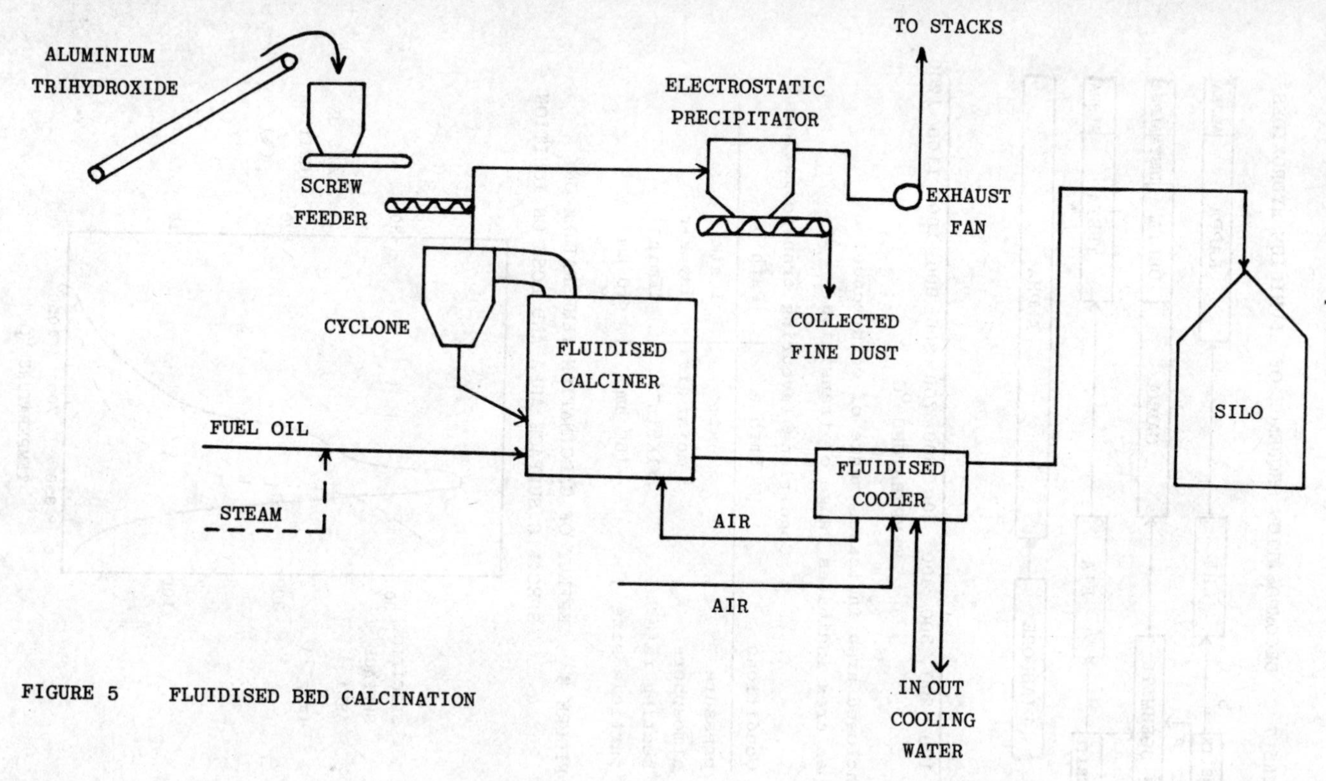


FIGURE 5 FLUIDISED BED CALCINATION

FIGURE 7 DECOMPOSITION SEQUENCE OF ALUMINIUM HYDROXIDES[5]

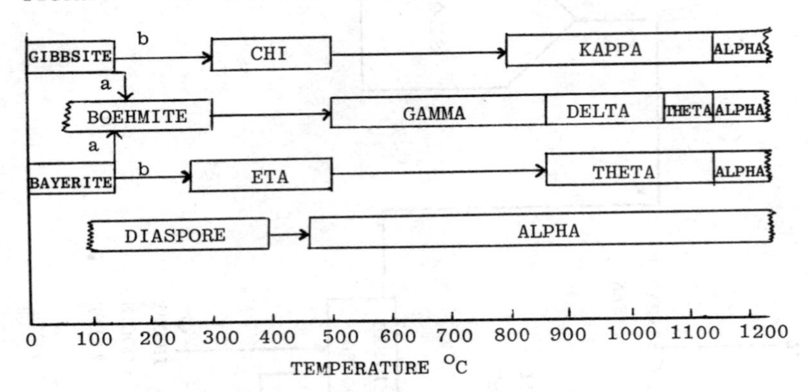


Enclosed area indicates range of occurrence;
open area indicates range of transition.

| Conditions | Conditions favouring transformations | |
|---|---|---|
| | Path a | Path b |
| pressure | >1 atm | 1 atm |
| atmosphere | moist air | dry air |
| heating rate | $>1 Kmin^{-1}$ | $<1 Kmin^{-1}$ |
| particle size | >100 μm | <10 μm |

FIGURE 8 EFFECT OF CALCINATION TEMPERATURE ON SPECIFIC SURFACE AREA AND LOSS ON IGNITION

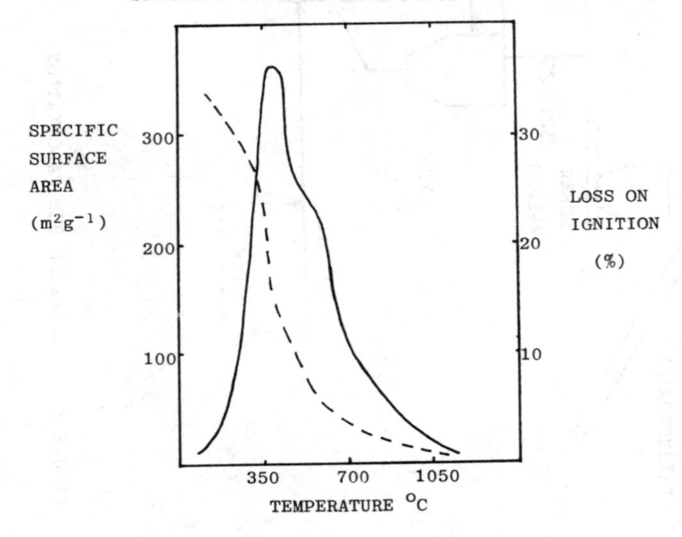

Further calcination at temperatures > 1000°C converts this to the non-absorbent α-form[1,35,36].

Smelter grade aluminas generally retain the physical nature of the starting aluminium trihydroxide; see Figure 6.* The conversion to the α-form however is usually no more than 25% and the specific surface area is relatively high at > 50 $m^2g^{-1}$ due to the presence of transition aluminas. The number and variety of transition aluminas depend on the atmosphere within the calciner, the amount of moisture present and the heating rate, see Figure 7. The relationships between calcination temperature; specific surface area and loss of ignition are illustrated in Figure 8[5].

In the production of speciality aluminas, physical changes involving recrystallisation of the alumina occur; see Figures 9 and 10. Higher temperatures are used and additives are often added to accelerate the conversion to α-alumina and its subsequent recrystallisation. Boron and fluorine containing compounds are the most frequently used additives. The specific surface area of the products lies within the range 0.5-20 $m^2$ $g^{-1}$ while the α-content is greater than 75%.

Because of the very small physical size of alumina, the exit gases from the calciner have a high dust content. Elaborate dust treatment is therefore essential to prevent loss of alumina. The gases are first passed through cyclones and then electrostatic precipitators. The dust is usually recycled back through the calciner so that the recirculating load can be as much as 2 to 10 times the output from the calciner[37].

Soda is the main impurity at a level commonly of between 0.05-0.6% and is normally present in the form of β-alumina ($Na_2O.11A\ell_2O_3$)

The annual production of aluminium oxide at over 30 million tonnes per annum[38] puts it amongst the major commodity chemicals; Table 3 shows the breakdown by country. Nearly 90% of this tonnage is, however, smelter grade alumina and is subsequently used for the production of aluminium metal.

*On p.168.

FIGURE 9 BAYER ALUMINA CALCINED AT 1250°C WITHOUT MINERALISER

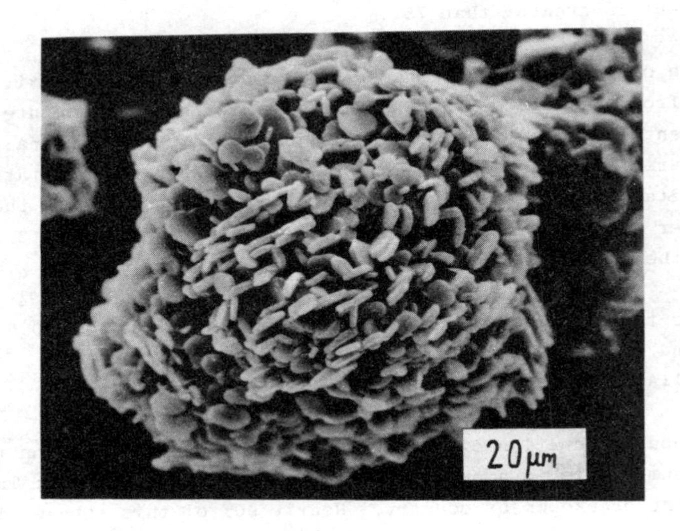


FIGURE 10 BAYER ALUMINA CALCINED AT 1250°C WITH FLUORINE ADDED

TABLE 3

Production of aluminium oxide[38]

1978

| | Thousand tonnes |
|---|---|
| Europe (ex USSR) | 4,656 |
| Asia | 2,372 |
| Africa | 622 |
| America | 11,224 |
| Australia | 6,776 |
| Eastern Countries | 5,152 |
| Total World | 30,802 |

TABLE 4

World alumina markets (excluding smelter grade), %[39]

| Market | 1975 |
|---|---|
| refractories (includes fused-cast and refractories made with sintered tabular grains) | 62 |
| abrasives (fused and polishing aluminas) | 16 |
| ceramics, electronic and technical (80+% $Al_2O_3$, includes spark plug insulators) | 11 |
| whitewares (includes electrical insulators below 80% $Al_2O_3$) | 9 |
| glass and enamels | 1 |
| miscellaneous | 1 |

Applications for aluminium oxide: The major markets for non-smelter or speciality calcined aluminas are refractories, abrasives and polishes, glass and enamels and ceramics (often divided between those with an alumina content greater than 80% - technical ceramics - and those with lower alumina contents - whitewares) are shown in Table 4. In addition to having lower surface areas and higher $\alpha$-alumina contents, many of these speciality aluminium oxide grades have carefully controlled particle size distributions and particle morphologies. The aluminas produced after calcining the aluminium hydroxide are sometimes used directly but more frequently they are ground by impact-milling, dry or wet ball-milling.

The properties of alumina which account for its widespread use are its high heat stability (m.pt. 2053°C), hardness (9 on Mohs' scale), inertness to chemical attack, abrasion resistance, mechanical strength and ability to withstand thermal and mechanical shock[5].

<u>Refractories</u>: Alumina use in refractories is widespread where its ability to withstand chemical attack and excellent thermal stability are utilised[40]. It is used in the manufacture of both high alumina bricks and castables and in order to reduce cost it is normally used in compositions with silica. Refractoriness decreases in almost direct proportion to the amount of alumina replaced by silica; the melting point of the refractory falls gradually from 2050°C for 100% alumina to 1545°C at a eutectic containing 5.5% alumina:94.5% silica. The refractories used to line tank and furnaces for glass production can be made from bricks cast from molten alumina. As the capital cost of manufacturing plants soar, the importance of maximising plant utilisation has accelerated the trend towards higher alumina refractories which, whilst more expensive, have a longer service life. The important refractory mullite, $3Al_2O_3.2SiO_2$, can be made synthetically from calcined alumina and silica rather than the traditional route from Kyanite.

High-alumina refractory bricks and castables are inert to most corrosive fumes and slags, and will resist both reducing and oxidizing atmospheres. In addition, they exhibit increased load bearing ability, spalling resistance, resistance to thermal shock and flame impingement.

<u>Polishing and grinding</u>: Naturally occurring alumina in the form of corundum and also emery (when mixed with magnetite) have been used as abrasives since at least 500 BC. Only a few materials are harder than corundum - diamond, boron carbide, silicon carbide - and alumina is widely used for polishing materials as diverse as rocks, ferrous and non-ferrous metals, car paint finishes and plastic spectacle lenses. The alumina can either be mixed with wax, grease, oil or water to form a block, paste or liquid suspension. A multitude of factors influence the usefulness of an alumina as a polishing material including the particle shape, particle size distribution, $\alpha$-alumina content, and internal porosity.

Several hundreds of thousands of tonnes of alumina are used each year for the production of fused alumina for manufacturing grinding wheels. In this process a fusion grade of alumina is melted in an electric-arc furnace using graphite electrodes. After solidifying, the fused alumina block is broken up into a fine grain and then formed into grinding wheels using either resin or glass as a bonding agent. Small additions of the oxides of titanium, chromium and zirconium are occasionally added to give improved performance. Even larger tonnages of fused grain are made directly from bauxite; the bauxites used are normally the high quality ($Al_2O_3$ > 85% on a calcined basis) South American ones.

Technical Ceramics: Alumina ceramics find application in almost every industry because of their outstanding resistance to heat, wear and chemical attack[41]. These properties are summarised in Table 5.

TABLE 5

Properties of high alumina bodies

High tensile strength
High compression strength
Hard and abrasive
Resistant to abrasion

Resistant to chemical attack
by a wide range of chemicals
even at elevated temperatures

High thermal conductivity
Resistant to thermal shock
High degree of refractoriness

High dielectric strength
High electrical resistivity
even at elevated temperatures
Transparent to radio frequencies
Low neutron capture area

Raw material readily available
and price not subject to
violent fluctuation

One of the most familiar examples of a high-alumina ceramic is the sparking plug insulator which typically contains 88-95% alumina. High-alumina content insulators were developed during the Second World War to meet the needs of high compression aero-engines for which the existing mullite plugs (~ 55% $A\ell_2O_3$) were inadequate[42]. The high silica content of these mullite insulators made them susceptible to attack from the lead in the fuel and other additives in the lubricants. Alumina was able to meet all the necessary properties: ability to withstand attack by lead tetraethyl and the other compounds in the fuel or lubricating oils; ability to withstand temperatures from sub-zero to over 1000$^{o}$C; reasonably high thermal conductivity in order to conduct heat from the nose of the plug to the engine block and thereby avoid pre-ignition; ability to withstand voltages of the order of 20,000 volts at high temperature; and high strength. Generally low soda aluminas ($Na_2O$ < 0.05%) are used for the manufacture of sparking plug insulators but there is now a tendency to move to a medium soda alumina ($Na_2O$ level < 0.25%) in a slightly lower alumina content plug.

Other applications where high alumina ceramics are widely utilised include: thread guides in nylon spinning where good abrasion resistance is necessary without causing damage to the nylon fibre; dies for wire drawing; rotors and stator blades in jet engines; laboratory ware such as crucibles; kiln furniture; thermocouple sleeves; as components for pumps used for corrosive substances; transparent tubes in high pressure sodium lamps; nosecones for jet aircraft and rockets; grinding media; substrates for electronic components; and insulators for high voltage power transmission.

For particularly critical applications very high alumina bodies may be utilised; alumina contents of 99.8% are readily achievable. These very high alumina ceramics are most readily manufactured using thermally reactive aluminas, as they can be sintered without fluxes at temperatures between 1450-1750$^{o}$C instead of the 1800-1850$^{o}$C for non-reactive calcined aluminas. These thermally reactive aluminas are produced by dry ball-milling in the presence of milling aids down to sub-micron particles. These milling aids are adsorbed onto the surface of the crystals during milling thereby overcoming the problem of mill packing and reducing the energy required for milling.

Hotelware: Alumina is also added to chinaware as a replacement for flint or quartz and gives the body greater mechanical strength and improves thermal shock resistance[43,44]. This is widely used for "hotelware" intended for canteens, restaurants and hospitals where the crockery is subjected to particularly hard use.

Glass: Calcined alumina is used in optical and other special glasses requiring a high degree of purity in raw materials, especially as regards impurities that can affect the colour of the glass such as iron and chromium. Up to 7% alumina is used in the production of low expansion borosilicate glasses for scientific equipment and ovenware. Alumina helps to stabilise the glass against devitrification and imparts to the glass improved toughness and chemical resistance.

Tabular alumina: Tabular or sintered alumina is the name given to large (50-300 µm) tablet shaped crystals of $\alpha$-$Al_2O_3$. It is produced by heating the alumina to a temperature slightly below its fusion point. Tabular alumina is used in particularly high grade refractory brick or monolithic liners for critical applications in the metal and glass industries. Other applications include linings for petroleum refining equipment, ceramics, catalyst supports, kiln furniture, and the filtering of molten aluminium prior to casting[45,46].

Activated alumina: The transition aluminas referred to earlier are widely used as adsorbents, catalysts and catalyst supports. Activated aluminas are very effective as desiccants for a very wide range of gases or liquids including air, nitrogen, hydrogen, oxygen, carbon dioxide, chlorine, natural gas, freon, ethylene, ammonia, alcohol, petroleum products, vegetable and animal oils, etc.[45] They are much more effective in removing moisture than commonly used desiccants such as anhydrous calcium chloride and silica gel[47] and under suitable conditions they can reduce the moisture content in a gas to less than 1 part per million by volume. They can be reactivated repeatedly with a comparatively small loss in adsorptive efficiency by heating to 175-200°C for several hours. Activated aluminas have a high affinity for fluorine and have been used to defluoridate potable waters[48].

Activated aluminas are produced by careful choice of calcination conditions and may have surface areas up to 500 $m^2g^{-1}$ with carefully controlled pore size distributions. They are frequently sold in a granulated form to facilitate handling.

Large tonnages of activated aluminas are used for supporting catalysts, especially for petrochemical use. The active catalyst such as molybdenum, chromium, nickel, or cobalt is impregnated onto the alumina and even though the alumina may have little or no effect as a catalyst on its own, it can significantly improve the efficiency of the catalyst system. Processes where activated alumina catalyst supports are used include reforming and cyclization, dehydrogenation, hydrogenation, isomerization, desulphurization and cracking of petroleum chemicals[45].

Activated aluminas behave as catalysts in their own right in the dehydration of alcohols, the isomerization of olefins and the recovery of sulphur from hydrogen sulphide in natural gas (the Claus process).

Ceramic fibres: The widespread concern about the use of asbestos has catalysed the development of synthetic ceramic fibres. They are acceptable substitutes for asbestos in many applications and are available as bulk fibres, rope, strip, spray, pre-formed shapes, mats, felts and blankets. Their thermal stability is in general superior to asbestos but depending upon their alumina content they may be more prone to chemical attack. Alumino-silicate fibres (Kaowool, Fiberfrax, Kerlene, Mackechnie) can be utilised at temperatures up to 1400°C and are attacked by hydrogen fluoride, phosphoric acid, concentrated alkalis and hot reducing atmospheres. Alumina fibres (Saffil, Fibral), however, can be used under hot reducing conditions and are resistant to virtually all chemicals; they can be employed at temperatures up to 1600°C.

Aluminium Sulphate

Some 50 acidic, basic and neutral hydrates of aluminium sulphate have been reported[49]; only a few of these, however, have been well characterised. Traditionally the formula $A\ell_2(SO_4)_3.18H_2O$ has been written for commercial dry aluminium sulphate but the material is generally amorphous and probably does not represent a distinct crystalline species. There is considerable confusion concerning the name and this material is referred to as "paper maker's alum", "coke alum", "patent alum" or simply "alum". The double salts potassium aluminium sulphate and ammonium aluminium sulphate are frequently called "common alum" or "ordinary alum".

Aluminium sulphate is sold as either "commercial" purity (with an iron level of up to 0.5%) or "iron free" aluminium sulphate. It is marketed in both solid ($A\ell_2O_3$ content 17/17.5%, 16% & 14%) and as a liquid (typically 6-8% $A\ell_2O_3$ content).

The worldwide market of "commercial" grade aluminium sulphate is over 2 million tonnes per annum whilst the iron free grade is about a quarter of this total, Table 6.

TABLE 6

Aluminium sulphate production[50]

1976

| | |
|---|---|
| USA | 989,000 |
| Europe (ex USSR) | 334,000 |
| South America | 82,000 |
| Asia | 783,000 |
| World (ex USSR) | 2,188,000 tonnes |

Commercial grade 17-17.5% $A\ell_2O_3$

Production of aluminium sulphate: Commercial grade aluminium is usually produced by the reaction of sulphuric acid (30-60%) with bauxite or clay[51]. Before use, the clays are normally calcined to destroy the organics and to increase the availability of the alumina for extraction. The optimum conditions for roasting the clay and the optimum strength of the sulphuric acid depend upon the particular source of the clay, its mineralogy and impurity content. The solid residue from the dissolution reaction is removed by sedimentation followed by a final polishing filtration although the neutral aluminium sulphate solution may have to be diluted first. The digestion takes place at a temperature of 100-120$^{\circ}$C.

Iron removal is difficult and "iron free" grade ($Fe_2O_3$ < 0.005%) is normally produced by the reaction of sulphuric acid with aluminium hydroxide; no heat input is necessary to sustain the reaction.

Applications of aluminium sulphate: Fifty percent of the aluminium sulphate produced is used by the paper making industry. Here the principal use of the aluminium sulphate is to promote the combination of resin and cellulose fibres to improve both the strength of the paper and its resistance to water and ink[52]. In addition the aluminium sulphate is used to neutralise the acidity of the wood pulp, soften hard water and to treat the resulting effluent.

Effluents from a wide variety of other industries can also be successfully treated with aluminium sulphate. Aluminium sulphate can be utilised to remove both coarse suspended particles and colloidal particles. It functions by reacting with the carbonates present in the water to produce a voluminous gelatinous precipitate of aluminium hydroxide which coagulates and carries down the suspended matter present. Additionally, the positive aluminium hydroxide species are attracted to the colloidal impurities which are normally negatively charged thereby forming larger and heavier flocs. Combinations of aluminium sulphate and polyelectrolytes are particularly effective in treating oil contaminated effluents; in these cases aeration is used to form flocs on the surface of the effluent which can be removed by skimming.

Potable waters can also be treated with aluminium sulphate to remove suspended impurities. Other uses of aluminium sulphate include: waterproofing textiles, dyeing, tanning, photography, for the production of colloidal aluminium hydroxide for gastric disorders, as a fire retardant in cellulosic fibre loft insulation and in fire extinguishers[51,52].

## Anhydrous Aluminium Chloride

Production of anhydrous aluminium chloride: Anhydrous aluminium chloride can either be prepared from aluminium metal or from aluminous materials such as clay or bauxite[53,54]. Its production from aluminium metal involves bubbling gaseous chlorine through molten aluminium and condensing the aluminium chloride vapour.

The reaction is exothermic and therefore requires careful control to maintain the reactor temperature of 600-750°C. To lower the melting temperature, small amounts of copper of magnesium are frequently added. The aluminium chloride is then condensed in a condensing chamber before being removed, crushed, sized and packaged under an atmosphere of dry air or nitrogen. The overall purity of aluminium chloride produced is > 98% with the main impurities being Cℓ, Aℓ, Fe, Si, Ti and Ca.

Chlorination of aluminous ores has been extensively investigated principally as a route to produce aluminium by subsequent electrolysis of aluminium chloride. This route to aluminium has a lower energy requirement than the Hall-Héroult process (by approximately one third), it should be environmentally cleaner but there are severe corrosion difficulties to overcome. The early patents involve feeding bauxite calcined at ~ 800°C into a reactor and introducing, at the required rate, chlorine and oxygen. The aluminium chloride vapour was condensed as described above (product purity ~ 95%).

Recently, the production of aluminium from non-bauxite aluminous ores has received considerable attention and one of the processes proposed consists of treating calcined clay or bauxite with chlorine at 350-800°C in the presence of a reducing gas, eg. carbon monoxide.

Applications of aluminium chloride: The catalytic properties of anhydrous aluminium chloride were discovered by Friedel and Crafts in 1877 and are now employed in a wide variety of reactions: dehydrogenation, decarboxylation, oxidation, alkylation, polymerization, acylation, desulphurisation and amination.

Other uses include the removal of dissolved gases and magnesium from aluminium alloys during secondary smelting and as a component of the flux used in arc welding titanium alloys[54]. The production of anhydrous aluminium chloride in the USA in 1978 was 75,000 and the amount imported into the UK was ~ 10,000 tonnes.

## Hydrated Aluminium Chloride

Aluminium chloride hexahydrate, $AlCl_3.6H_2O$, is a deliquescent powder which is soluble in both alcohol and water. It is commercially available as a 28% by weight solution and as a crystalline solid.

Aluminium chloride hexahydrate, $AlCl_3.6H_2O$, is usually made by dissolving aluminium trihydroxide in concentrated hydrochloric acid at atmospheric boiling. When the acid is used up, aluminium chloride hexahydrate is crystallised from the solution by first of all cooling to $0^{o}C$ and then sparging with $HCl$ gas. Any impurities from the starting aluminium trihydroxide remain in solution so that a purified $AlCl_3.6H_2O$ crystallises out. The crystals are readily filtered from the liquor and can be washed with ethyl ether before drying.

Aluminium chloride hexahydrate has a large number of applications in industry being used in textile finishing to impart crease resistance and non-yellowing properties to cotton fabrics, anti-static properties to polyester, polyamide and acrylic fibres and improved flammability resistance of nylon.

Basic aluminium chlorides are used in deodorants, anti-perspirants, fungicidal preparations, as water treatment agents, thickeners in paint, earth stabilisation to prevent water permeation during oil drilling and as the starting material for the production of alumina fibres. They have the general formula $Al_2(OH)_{6-n}Cl_n$ and are prepared by reacting an excess of aluminium with 5-15% hydrochloric acid at 65-100$^{o}C$.

## Sodium Aluminate

Production of sodium aluminate: The use of sodium aluminate, $NaAlO_2$, as a commercial product began in the nineteen twenties. It is now being used increasingly in the water treatment and paper industries amongst others. The commercial product is available as either a liquid containing about 40% by weight of sodium aluminate or as a solid product which is essentially anhydrous.

Sodium aluminate is of course the circulating liquid medium of the Bayer process for the production of alumina from bauxite. The process liquor, however, tends to be too contaminated with impurities so that commercial quantities of sodium aluminate are normally made by dissolving aluminium trihydroxide from the Bayer process in sodium hydroxide solution at atmospheric boiling. To obtain a colourless solution or to obtain a white product completely soluble in water and free from colour, it is important to start with a low-iron aluminium trihydroxide (< 0.010% $Fe_2O_3$). Even so, colloidal iron associated with certain bauxites, eg. Guyana, can, via the trihydroxide, impart a brown colouration to the sodium aluminate solution. This, however, can be removed with a small amount of active carbon.

Commercial grades of sodium aluminate generally contain some water of hydration and an excess of soda. In solutions, the high pH retards the reversion of sodium aluminate to the insoluble aluminium hydroxide, liquid sodium aluminate generally requires the higher $Na_2O$:$A\ell_2O_3$ ratios (up to 1.5:1) for good stability and storage properties. Commercial sodium aluminates are therefore not accurately represented by any of the commonly used formulae $NaA\ell O_2$, $Na_2O.A\ell_2O_3$ or $Na_2A\ell_2O_4$.

There are many other processes known for preparing sodium aluminate that do not require the addition of water[55]. For example, sodium aluminate can be prepared as a hygroscopic white crystalline solid by fusing equimolar amounts of sodium carbonate and aluminium acetate at a temperature of 800°C[56].

Applications of sodium aluminate: Sodium aluminate has a large number of uses as a coagulating agent. It is also used in conjunction with other coagulating agents such as aluminium sulphate, ferric salts, clays and polyelectrolytes[57]. In the paper industry, sodium aluminate improves sizing and filler retention. When added to titania paint pigment, the non-chalking behaviour of outdoor paints is improved. In the processing of acrylic and polyester synthetic fibres, the use of sodium aluminate enhances dyeing, antipiling and antistatic properties of the fibres. Sodium aluminate is often used in the preparation of alumina based catalysts. The reaction of sodium aluminate with silica or silicates can produce porous crystalline alumino-silicates which are used as absorbents and catalyst supports.

## Aluminium Nitrate

Aluminium nitrate nonahydrate, $Al(NO_3)_3.9H_2O$, is a white crystalline material, m.p. 73.5°C, soluble in cold water, alcohols and acetone. It is used primarily as a salting-out agent in the extraction of actinides. It is also a source of alumina used in the preparation of insulating papers, in transformer core laminates and in cathode ray tube heating elements. It is a stable compound but of minor commercial significance.

In general, aluminium nitrate nonahydrate is prepared by dissolving aluminium trihydroxide in dilute nitric acid and crystallising from the resulting aqueous solution. Other methods which have been patented involve treating bauxite or calcined clay with nitric acid. Liquor purification generally requires the removal of dissolved iron from the leach liquor before crystallisation takes place. This is generally achieved by solvent extraction using, for example, di-(2 ethyl hexyl) phosphoric acid[58]. There is negligible co-extraction of aluminium.

## Aluminium Phosphates

Aluminium forms a number of phosphates: the orthophosphate, $AlPO_4$; the primary and secondary orthophosphates, $Al(H_2PO_4)_3$ and $Al_2(HPO_4)_3$; the pyrophosphate, $Al_4(P_2O_7)_3$; the acid pyrophosphate, $Al_2H_6(P_2O_7)_3$; and the metaphosphate, $Al(PO_3)_3$[59].

The orthophosphate can be prepared by dissolving aluminium hydroxide in a stoichiometric amount of phosphoric acid. If an excess of phosphoric acid is used then the acid orthophosphates will be formed. An alternative route to produce the orthophosphate is to mix a salt of lower-boiling acid, such as sulphate, chloride or nitrate, with the appropriate proportion of phosphoric acid[60].

The principal use of aluminium phosphates is as a high temperature binding agent. The aluminium polyorthophosphates form viscous, adhesive solutions in water at room temperature which are highly effective in permanently bonding refractories even at high temperatures[61]. The aluminium polyortho- and metaphosphates are employed in glasses, ceramics and in catalysts[60]. They are incorporated into special glasses when particularly high chemical resistance is required.

## Aluminium Carboxylates

The aluminium salts of carboxylic acids are known as aluminium carboxylates. They are derived from aluminium trihydroxide by successive replacement of hydroxyl by carboxylate anions.

The aluminium formate series would for example comprise of the following:

| | |
|---|---|
| Normal aluminium formate | $A\ell(OOCH)_3$ |
| Monobasic aluminium formate | $(HO)A\ell(OOCH)_2$ |
| Dibasic aluminium formate | $(HO)_2A\ell(OOCH)$ |

The general methods for preparing mono- and dibasic aluminium carboxylates include:

(a) direct reaction of aluminium as the catalytic electrode with an aqueous solution of the acid[62,63];

(b) reaction of the metal with the acid in the presence of mercury or mercuric chloride in catalytic amounts[64];

(c) reaction by double replacement between aluminium alkoxide and the sodium salt of the organic acid[65].

Normal aluminium carboxylates are prepared by the direct reaction of the acid with aluminium chloride in an organic solvent or by a double displacement reaction using a soluble aluminium salt of the organic acid[66,67].

About twenty of the aluminium carboxylates are of industrial importance and have commercial applications. The applications fall into three general areas:

(a) finishing agents for the waterproofing of cloth and as mordants in the dyeing of textiles,

(b) pharmaceutical preparations because of their antiseptic, astringent and basic properties,

(c) the manufacture of cosmetics due to the gelling properties of some of the higher molecular weight aluminium carboxylates.

References

[1]T. G. Pearson, "Chemical Background of the Aluminium Industry", Royal Institute of Chemistry, London, 1955.

[2]U.K. Patent No. 10,093, 1887.

[3]N. Brown, J.Crystal Growth, 1972, 12, 39.

[4]N. Brown, J.Crystal Growth, 1972, 16, 163.

[5]K. Wefers and G. M. Bell, Technical Paper No. 19, Aluminium Company of America, 1972.

[6]Z. G. Szabo et al, Z. Anal. Chem., 1955, 146, 401.

[7]B. Schweig, Glass, 1945, 22, 263.

[8]J. J. Koenig, Tappi, 1965, 48, 123.

[9]J. E. Williams, Paint Varn. Prod., 1967, 57, 54.

[10]J. J. Koenig, "Paper Coating Pigments", 4th ed., Tappi Monograph No. 38, edited by R. W. Hagemeyer, Tech. Assoc. of the Pulp and Paper Industry, Atlanta, 1976, p.6.

[11]"Hydral 705 Paper Filler Pigments", brochure, Aluminium Company of America, Pittsburgh, 1967.

[12]"Blue-white Brightness at Lower Cost with Paperad", brochure, Reynolds Metal Company, Richmond, 1971.

[13]U.K. Patent No. 1,188,353 (Unilever) 1968.

[14]R. J. Andlaw and G. J. Tucker, Brit. Dent. J., 1975, 138, 462.

[15]P. M. C. James et al, Com. Dent. Ord. E pid., 1977, 5, 67.

[16]United States Patent 4,098,878 (Colgate Palmolive) 1978.

[17]U.K. Patent No. 1,537,823 (British Aluminium) 1979.

[18]U.K. Patent No. 2,037, 162 (Unilever) 1980.

[19]U.K. Patent No. 840,161.

[20]U.K. Patent No. 1,442,396 (Unilever) 1976.

[21]J. W. Lyons, "The Chemistry and Uses of Fire Retardants", Wiley, New York, 1970.

[22]M. N. Conklin, Color Trade J., 1922, 11, 171.

[23]U.K. Patent No. 183,922 (Sovereign Mills), 1922.

[24]W. J. Connolly and A. M. Thornton, Mod. Plastics, 1965, 43, 154.

[25]P. V. Bonsignore and J. M. Manhart, 29th Annual Conference Reinforced Plastics/Composites Institute, The Society of Plastics Industry, 1974.

[26]R. C. Hopkins, Polymer Age, 1975, 130.

[27]E. A. Woycheshin and I. Sobolev, 30th Annual Conference Reinforced Plastics/Composites Institute, The Society of Plastics Industry, 1975.

[28]R. C. Nametz, Ind. & Eng. Chem., 1967, 59, 99.

[29]A. Agnew, Proceedings of 2nd European Conference on Flammability and Fire Retardants, Copenhagen, Alena Enterprises, Ontario, 1978.

[30]"BACO FRF", brochure, B. A. Chemicals Ltd., Gerrards Cross, 1977.

[31]E. P. Plueddemann and G. L. Stark, Mod. Plastics, 1977, 54, 76.

[32]K. E. Atkins et al, 32nd Annual Conference Reinforced Plastics/ Composites Institute, The Society of Plastics Industry, Washington, 1977.

[33]S. J. Monte and G. Sugerman, 33rd Annual Conference Reinforced Plastics/Composites Institute, The Society of Plastics Industry, Washington, 1978.

[34]K. A. Evans and R. J. Shaw, Reinforced Plastics, 1980, 24, 272.

[35]W. M. Fish, paper presented at the AIME annual meeting, Dallas, 1975.

[36]B. A. Scott and W. H. Horsman, Trans. Br. Ceram. Soc., 1970, 69, 37.

[37]A. N. Adamson, The Chemical Engineer, June 1970.

[38]Metal Statistics, 1978, Metallgesellschaft AG, Frankfurt.

[39]G. MacZura, K. P. Goodboy and J. J. Koenig, "Aluminium Oxide" in Kirk-Othmer Encyclopaedia of Chemical Technology, 3rd edition, Vol. 2, Wiley, New York, 1978.

[40]B. Jackson, Euroclay, 1974, 2, 27.

[41]G. Richards, Design Engineering, 1973, 54.

[42]J. S. Owens et al, Ceramic Bulletin, 1977, 56, 437.

[43]C. R. Austin, H. Z. Schofield and N. L. Holdy, J.Amer. Ceram. Soc., 1946, 29, 341.

[44]W. E. Blodgett, Bull. Amer. Ceram. Soc., 1961, 40, 74.

[45]Calcined, Reactive, Tabular Aluminas and Calcium Aluminium Cement, brochure, Aluminium Company of America, Pittsburgh, 1976.

[46]E. F. Emley and V. Subramanian, Light Metals, 1974, 2, 649.

[47]J. H. Bower, J.Res. NBS, 1944, 33, 199.

[48]E. M. Savinelli and A. P. Black, J.AWWA, 1958, 50, 33.

[49]Gmelin, Handbuch der Anorganischen Chemis, Aluminium, System Number 35, Part B, p.248, Verlag Chemie, Berlin, 1934.

[50]Minerals Yearbook, Vol. 1, U.S. Dept. of Mines, 1977.

[51]K. V. Darragh, "Aluminium Sulfate" in Kirk-Othmer Encyclopaedia of Chemical Technology, 3rd edition, Vol.2, Wiley, New York, 1978.

[52]Aluminium Sulphate, brochure, The Alumina Company Ltd., Widnes.

[53]C. A. Thomas, "Anhydrous Aluminium Chloride in Organic Chemistry", Rheinhold, 1961.

[54]C. M. Marstiller, "Aluminium Chloride" in Kirk-Othmer Encyclopaedia of Chemical Technology, 3rd edition, Vol.2, Wiley, New York, 1978.

[55]W. R. Busler, "Aluminates" in Kirk-Othmer Encyclopaedia of Chemical Technology, 3rd edition, Vol.2, Wiley, New York, 1978.

[56]J. Thery, A. M. Lejus, D. Briancon and R. Collongues, Bull. Soc. Chem. France, 1961, 973.

[57]Canadian Patent No. 964,808 (Nalco Chemical Company) 1975.

[58]U.K. Patent No. 1,311,614 (Arthur D. Little) 1973.

[59]J. W. Mellor, "A Comprehensive Treatise on Inorganic and Theoretical Chemistry", Longmans, Green and Co., London, 1929.

[60]J. R. Van Wazer, "Phosphoric Acids and Phosphates" in Kirk-Othmer Encyclopaedia of Chemical Technology, 2nd edition, Vol.15, Wiley, New York, 1968.

[61]W. D. Kingery, J.Am. Ceram. Soc., 1952, 35, 61.

[62]German Patent No. 2,325,018, G. G. Merkl, 1973.

[63]United States Patent No. 2,957,598, G. G. Merkl, 1976.

[64]R. W. Jones and J. W. Clusky, Cereal Chem., 1963, 40, 589.

[65]Japanese Patent 15, 353 (Takeda Chemical Industries) 1963.

[66]R. C. Mehrattra abd A. K. Rai, J.Indian Chem. Soc., 1962, 39, 1.

[67]Japanese Patent 10,476, N. Shoji and M. Hideo, 1961.

# Silicas and Related Materials

By P. Kleinschmit
DEGUSSA AG., POSTFACH 1345, D-6450 HANAU I, WEST GERMANY

## 1. Introduction and General Survey

With respect to the nature and size of this publication the subject "Silicas and related Materials" requires some limiting definition: This paper will exclude the soluble silicates which have been reviewed before[1] and will be restricted to certain highly dispersed siliceous materials which - although they are in their majority higher priced specialities - mainly are counted as "fillers", a smaller group as "pigments".

Both characteristics, however, are misleading. As the word suggests, "fillers" in the actual sense of the word are cheap materials which are added to higher priced products such as polymers, rubbers, coatings, paints, papers and many other products mainly in order to reduce the price, to "extend" the more expensive ground materials and which, at best, do not cause any loss of quality of the end product. On the other hand, white pigments such as $TiO_2$ have to have refraction indices larger than 1.7, whereas the silica products used in paints and plastics industry as "pigments" - and that is about 15 o/o of the synthetic silica business - have much lower indices in the range of 1.4 - 1.5 and therefore do not correspond to this definition.

So, to choose another approach, some selected classes of synthetic siliceous products will be reviewed that give special and remarkable effects to the formulations they are added to in various, often small quantities and which therefore pay their cost. An impression of the great variety of different applications for synthetic silicas and silicates is given in table 1:

Table 1: Examples for the application of synthetic silicas and silicates

- reinforcement of rubbers and plastics
- thickening and thixotropy of coatings and paints, printing inks, plastics and cosmetics
- matting of lacquers, coatings, paints and plastics
- antiblocking of plastic foils
- free running and free flow of sticky solid or liquid substances
- carrier for pesticides, catalysts
- high temperature insulation
- stabilizing (e. g. of beer, silicon rubbers)
- desiccants
- non-eutrophic water-softening (e. g. "builder" materials in washing powders)

The basic raw material for the production of synthetic silicas and silicates is sand or quartz. There are two main routes to these products: "wet" processes starting from sodium silicate, and "thermal" processes using high temperature reactions in which, in the first step, volatile silicon compounds such as the tetrachloride or the monoxide are prepared as intermediates, the latter being subsequently hydrolysed resp. oxidized, also at elevated temperatures, to give the highly dispersed end products (figure 1).

Despite the apparent simplicity of the chemistry involved in these processes, the properties of the silica products can vary within a large range, and the number of different types sold on the market amounts to well over one hundred. With the exception of the zeolites these products are amorphous colloids, which are characterized and specified mainly by physical-chemical criteria such as specific surface area (e. g. BET-method), average particle size, particle size distribution, average pore diameter, pore diameter distribution, tamped density, moisture content, ignition loss, silanol group density, pH-value in aqueous suspension, sieve residue, dibutylphthalate- or oil absorption number, chemical impurities, whiteness index (e. g. according to HUNTER). However, many of the essential quality

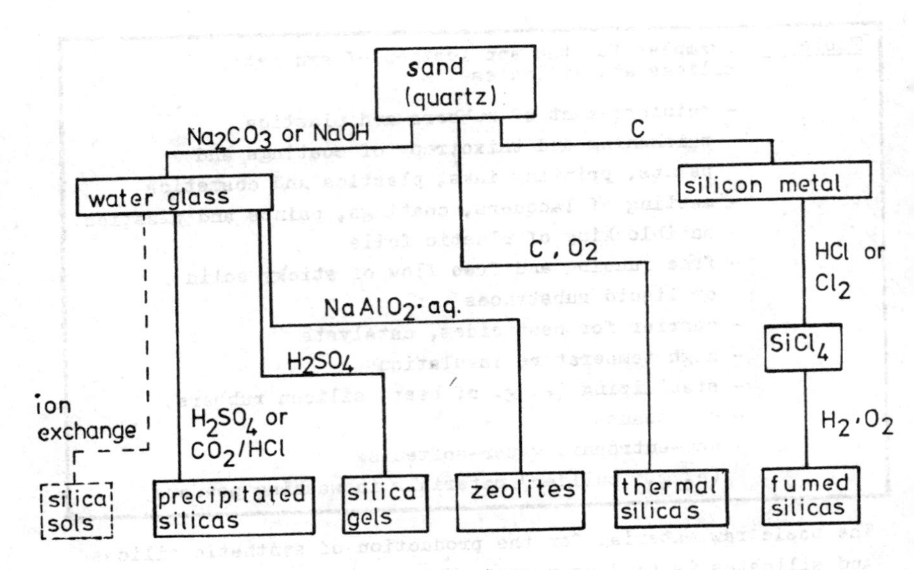


Figure 1: Basic processes for the production of synthetic silicas and silicates

criteria as far as application is concerned cannot be judged by these data. A great number of standardized tests - results of which are also part of the product specifications - have been developed to give information on the performance of such "fillers" or "pigments" in the systems they are added to.

Before a short survey is given on the industrial importance of the main classes of silicas, on the manufacturing processes and on some trends of development, it is useful to mention a class of non-siliceous reinforcement fillers, which historically, applicationwise and with respect to some newer developments have to be seen in close context to the silicas: the carbon blacks.

They are by far the most important group of reinforcement fillers for rubber, since - by a chance discovery made in 1910 - a tyre manufacturer found out that carbon black filled tyres have a much higher abrasion resistance and therefore a longer life-time.[2] In 1980 the production capacity for carbon blacks in the western world exceeded 4.7 mio t/a;[3] more than 90 % of the total amount was used in the rubber industry. Compared to that the synthetic silica business with its 600,000-700,000 t/a production is still a relatively small one.

The fact,however, that the carbon black industry is dependent on petrochemical raw materials has been - at least for countries without domestic oil sources - a problem not only since our recent energy crises and therefore has been a stimulant for researchers to look for adequate substitution products on the basis of better available raw materials.

Some early attempts to find suitable "white carbon" fillers for the rubber industry resulted in the development of highly dispersed calcium carbonates and silicates and special grades of alumina gels. The products are useful for special purposes but their reinforcing power is poor compared to carbon blacks.

## 2. Silicas Produced by Thermal Processes

### 2.1 Fumed Silicas

Another approach to find the "white carbon" was made 1941 by the German chemist Dr. Harry Klöpfer (Degussa)[4] who had developed in the early 1930s a manufacturing process for channel blacks using, instead of natural gas, coal tar oil as feed stock which was vaporized in a stream of hydrogen and partially burnt in diffusion flames. Using a similar kind of equipment and procedure he exchanged the feed stock tar oil by a volatile silicon compound, silicon tetrachloride, which was burnt in an oxy-hydrogen torch to give - by a process of "flame hydrolysis" - highly dispersed silica and hydrogen chloride. A new class of products with remarkable properties had been found. They were called "Aerosils" : "airborn" silicas; in Anglo-Saxon countries they are known as "fumed" silicas.

Considering the different steps involved in the production of fumed silicas it is obvious that these materials are not exactly cheap: At first, sand or quartz is reduced in an electric arc process to silicon metal, then the latter is chlorinated with chlorine or hydrogen chloride to silicontetrachloride which, finally, is burnt in the presence of expensive hydrogen to give $SiO_2$ again.

And indeed, to substitute carbon blacks as reinforcement fillers in rubbers the fumed silicas and other fumed oxides were and are far too expensive. At present, they cost about 2-3 £/kg, whereas common tread blacks are purchased at about 30-35 pence/kg. Moreover, fumed silicas are excellent reinforcement fillers for silicon rubber but not at all for normal rubber, mainly because the unvulcanized raw mixes show a poor processability, for example when being mixed, calendered or extruded: The viscosity of the raw mix is too high, and there is a detrimental reduction in prevulcanizing time.

However, these ultra-pure and ultra-fine fumed silicas found, despite their high price, a continously growing technical application in different industries. Some information on the production capacity and main producers is given in table 2.

Table 2: Production of fumed silicas1980 (estimated)

| | |
|---|---|
| Capacity (world wide): | 60.000 t/a |
| Producers : | DEGUSSA (West Germany, Belgium, USA, Japan) |
| | CABOT ( USA) |
| | WACKER (West Germany) |

Properties and Applications

Fumed silicas are fluffy white powders of amorphous structure. The average diameter of the spherical primary particles can be varied by using different reaction conditions during the flame hydrolysis: In commercial grades within the range of approximately 7-40 nm, corresponding to BET-surface areas between 380 and 50 $m^2/g$ (table 3). Fumed silicas with surface areas $\leq$ 300 $m^2/g$ are non-porous; grades with higher surface area can have a certain amount of micropores. In the case of normal types of fumed silica up to a few hundred primary particles are fused to three-dimensional aggregates (figure 2).

Very important for the outstanding properties of fumed silicas as thickening, thixotropic and reinforcing agents in liquid, plastic and elastic systems are the evenly distributed silanol groups on the silica surface and their ability to form hydrogen bonds. Generally the silanol group density of fumed silicas is in the range of 2-4 per 1 $nm^2$ of surface area. When fumed silica is dispersed in liquids the silanol groups of adjacent particles can interact via hydrogen bonding. In this way a vast three-dimensional net work is formed which is extended throughout the liquid and causes, especially in non-polar liquids, a substantial increase of the viscosity (figure 3). This effect is reversible: By means of shearing forces the net work can be broken down again and the viscosity thus reduced; that means the composition is thixotropic. These relations are demonstrated schematically in figure 4. Technically important systems where fumed silica is used in this way as a thickening agent and thixotrope are e. g. coatings, printing inks, gel-coats, sealants, cosmetics and tooth pastes. For this purpose, normally small amounts (e. g. a few percent) are sufficient to give an optimal effect, also for the application as free-flow and anticaking agents, whereas for the reinforcement of silicon rubber up to 30 weight °/o of fumed silica are used, and in high performance insulation composites the fumed silica is very often the main or sole component.

Table 3: Physical-chemical data of some commercial fumed products

| | | AEROSIL 130 | AEROSIL 200 | AEROSIL 380 | Aluminium-oxid C | Titanium-dioxid P 25 |
|---|---|---|---|---|---|---|
| X-ray structure | | amorphous | amorphous | amorphous | mainly $\gamma Al_2O_3$ | mainly anatase |
| Density | $g/cm^3$ | 2.2 | 2.2 | 2.2 | 2.9 | 3.8 |
| BET-surface area | $m^2/g$ | 130 | 200 | 380 | 100 | 50 |
| Average primary particle size | nm | 16 | 12 | 7 | 20 | 30 |
| Tamped density (standard material) | g/l | ~ 50 | ~ 50 | ~50 | ~60 | ~150 |
| Moisture (2 hours at 105°C) | o/o | < 1.5 | < 1.5 | < 1.5 | < 5 | < 1.5 |
| Ignition loss (2 hours at 1000°C) | o/o | < 1 | < 1 | < 2.5 | < 3 | < 2 |
| pH-value (4 o/o aqueous suspension) | | 3.6–4.3 | 3.6–4.3 | 3.6–4.3 | 4–5 | 3–4 |
| $SiO_2$ (ignited for 2 hours at 1000°C) | o/o | >99.8 | >99.8 | >99.8 | < 0.1 | < 0.2 |
| $Al_2O_3$ | o/o | <0.05 | <0.05 | <0.05 | > 97 | <0.3 |
| $Fe_2O_3$ | o/o | <0.003 | <0.003 | <0.003 | <0.2 | <0.01 |
| $TiO_2$ | o/o | <0.03 | <0.03 | <0.03 | <0.1 | >97 |
| HCl | o/o | <0.025 | <0.025 | <0.025 | <0.5 | <0.3 |
| Sieve residue (by Mocker >45 µm) | o/o | <0.05 | <0.05 | <0.05 | - | - |

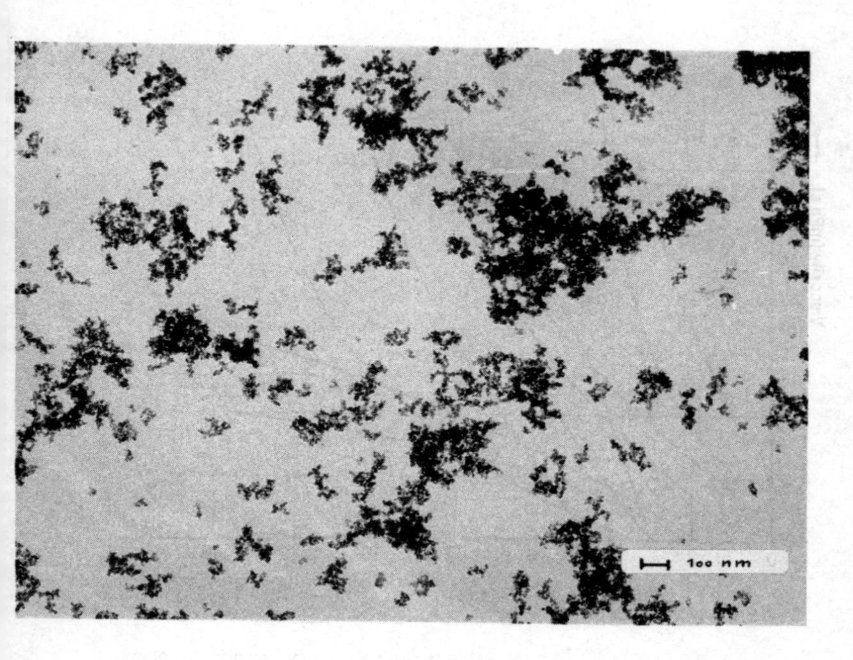


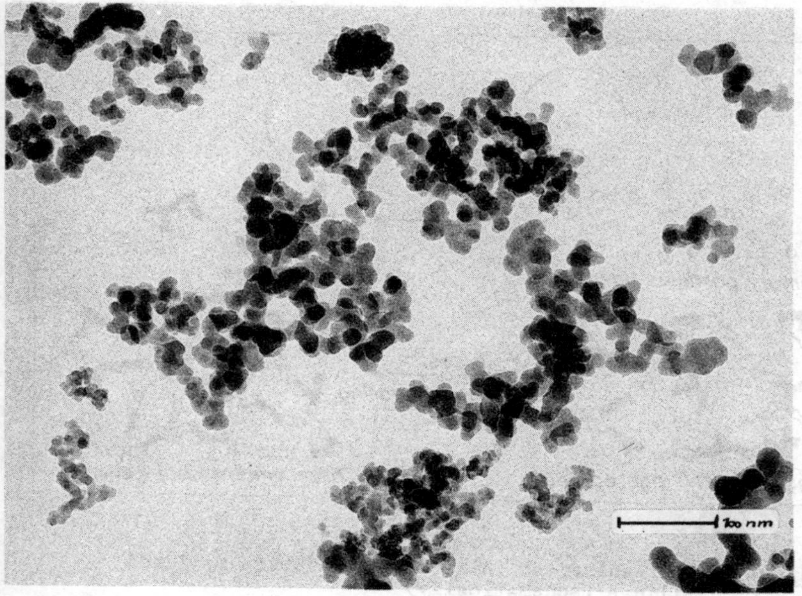


**Figure 2:** Fumed silica Aerosil 130 in two different enlargements

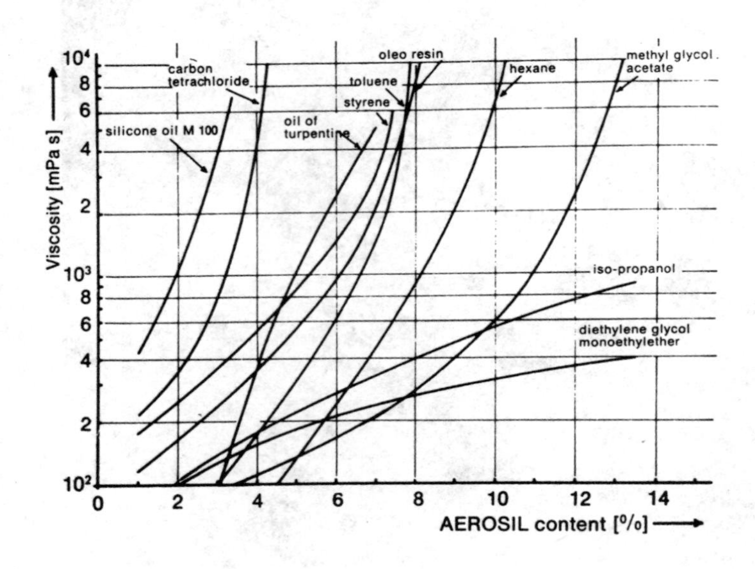


Figure 3: Thickening of various liquids with AEROSIL 200, dispersion by propellor stirrer

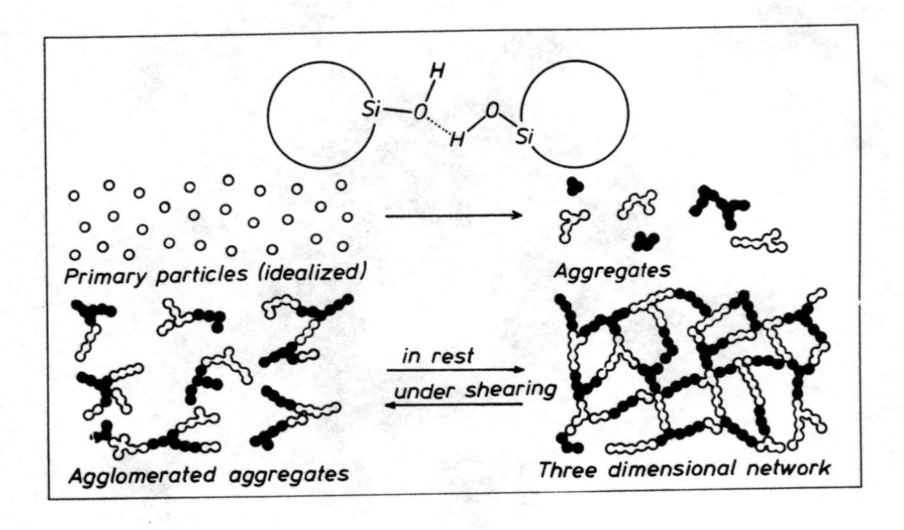


Figure 4: Thickening and thixotropy of liquids by fumed silicas (schematically)

The main applications of fumed silicas are - approximately in the order of their share of the whole business - listed in table 4.

Table 4: Main applications of fumed silicas

| effects | systems |
|---|---|
| reinforcing | silicone rubber |
| thickening, thixotropy, antisettling | plastics (gelcoats, laminating resins, sealing compounds, casting resins, plastisols, adhesives)<br>pharmaceuticals and cosmetics (ointments, creams, lotions, tooth-pastes)<br>coatings, printing inks |
| high temperature insulation | electric storage heaters, electric cooking devices |
| free running, free flow, antistatic | animal feedstuffs, all kinds of hygroscopic and bulking powders |
| flatting | coatings, foils |
| carrier for active substances | pharmaceuticals, catalysts |

Manufacturing Process:

A flow sheet of a modern fumed silica production process is given in figure 5. Whereas Klöpfer initially used a reactor with many small diffusion flames, modern plants are very similar to furnace black plants: the flame hydrolysis takes place in one big flame at temperatures of about 1000°C or more. The ratio of the reactants silicon tetrachloride, hydrogen and oxygen (in form of dry air) which can be varied within a substantial range, determines the properties of the silica, e. g. its particle size and specific surface area. The reaction mixture passes a coagulation and cooling device before the silica particles are separated by means of cyclones or filters. Adjacent HCl is removed, e. g. in a fluid bed reactor, with steam of hot moist air. Bagging is normally combined with a certain kind of compression in order to enlarge the bulk density, which is originally well below 20 g/l.

Economically as well as ecologically it is advantageous to run fumed silica plants in combination with a silicon tetrachloride plant, which produces a good deal of the necessary hydrogen and makes use of the by-product HCl from the silica plant. A simplified flow sheet for such an operation is shown in figure 6. Instead of or in addition to $SiCl_4$ other silanes, e. g. methyltrichlorsilane or other by-products of the silicones manufacture can be used as raw materials for fumed silicas.

Alternative Routes:

In the past a lot of efforts have been made to find alternative - and possibly cheaper - methods for the production of fumed silicas. A promising approach was made by Flemmert[5] who developed the "Fluosil"-process, which uses $SiF_4$ as a raw material:

$$SiF_4 + 2H_2 + O_2 \rightleftarrows SiO_2 + 4HF$$

Due to the fact, however, that the equilibrium of the hydrolysis reaction lies - other than in the case of $SiCl_4$ - only at very high temperatures completely on the right side, the fluosil process is more difficult to run, apart from material problems caused by the handling of HF and $SiF_4$. On the other hand the by-product HF can, theoretically, be recycled, in

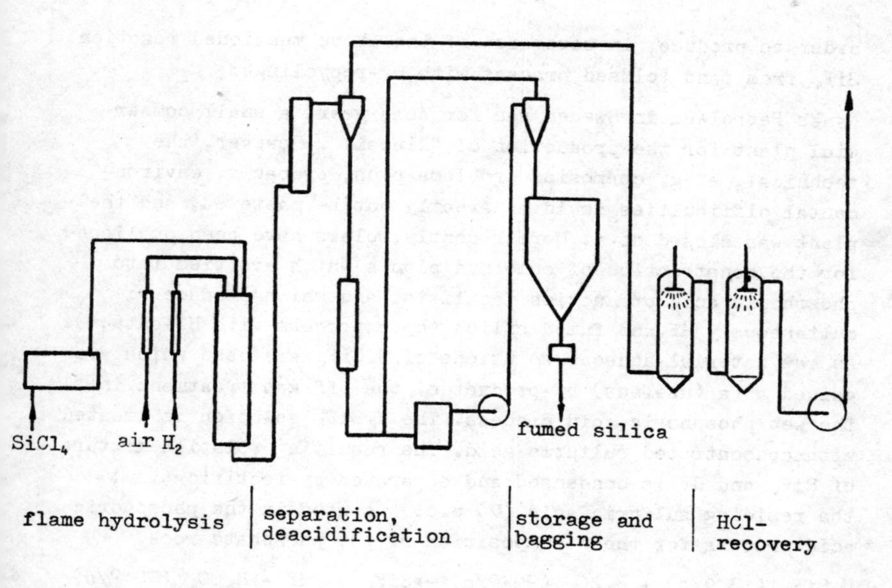


Figure 5: Fumed silica production process

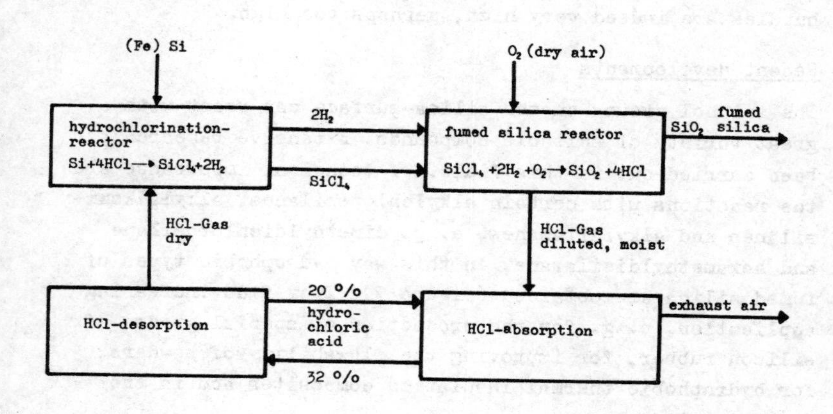


Figure 6: Simplified flow sheet of a combined $SiCl_4$ / fumed silica plant

order to produce, in reversion of the above mentioned reaction, $SiF_4$ from sand (closed process with HF-recycling).

Nynäs Petroleum in Sweden ran for some years a small commercial plant for the production of "Fluosil". However, the technical, e. g. corrosion problems plus, probably, environmental difficulties could apparently not be mastered, and the plant was closed down. More recently, plans have been published for the construction of combined plants which are tied into phosphoric acid production facilities and which produce simultaneously HF and fumed silica (open process with HF-output).[5] As raw material aqueous solutions of $H_2SiF_6$ are used which are gained as a (useless) by-product of the off-gas treatment in the wet phosphoric acid process. The $H_2SiF_6$ solution is treated with concentrated sulfuric acid. The resulting volatile mixture of $SiF_4$ and HF is condensed and separated by rectification; the residing sulfuric acid (70 p.c.) is used in the phosphoric acid process for the decomposition of the phosphate rock.

$$H_2SiF_6\ (13\ \%) + H_2SO_4\ (96\ \%) \longrightarrow SiF_4 + 2HF + H_2SO_4\ (70\ \%)$$

$$SiF_4 + 2H_2 + O_2 \longrightarrow SiO_2\ (\text{fumed}) + 4HF$$

The fact that none of these promising looking plans has yet been realised seems to prove that the technical and ecological hurdles are indeed very high, perhaps too high.

## Recent developments

The silanol groups on the silica-surface can react with a great variety of suitable compounds. Extensive research has been carried out in this field. Of technical importance are the reactions with certain alkylchlorosilanes,[6] alkylalkoxisilanes and alkylsilazanes, e. g. dimethyldichlorosilane and hexamethyldisilazane. In this way hydrophobic types of fumed silica are obtained (figure 7). They find increasing application, e. g. for the production of special grades of silicon rubber, for improving the flowability of powders, for hydrophobic thermal insulation composites and in the cosmetics industry.

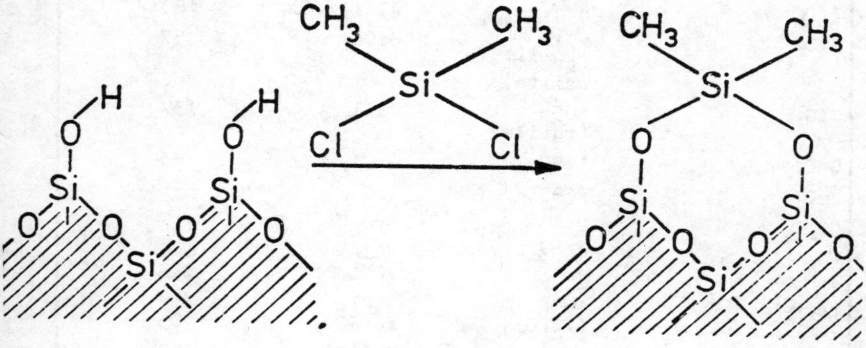

Figure 7: Fumed silica + alkylchlorosilane = hydrophobic silica

It also has to be mentioned that the method of flame hydrolysis for synthesizing highly dispersed materials can be extended to other volatile metal compounds provided that the equilibrium of the hydrolytic reaction is on the side of the oxide and the melting point of the latter is higher than the flame temperature.[4] In table 5 a selection of highly dispersed fumed oxides resp. compounds is listed that have been synthesized by flame hydrolysis, mainly on the laboratory scale. The potential of many of these products, for example as catalysts, has not yet been fully evaluated.[7] Produced on a technical scale and commercially available are, in addition to the silicas, the oxides of aluminium, titanium (see table 3) and iron which are used e.g. as antistatic agents for plastic powders and in the textile industry ($Al_2O_3$), as heat stabilizers in hot curing silicon rubber and as UV-absorbers in sun protection creams ($TiO_2$), and as transparent red pigments for metallic coatings ($Fe_2O_3$).[8]

Table 5: Pyrogenic compounds synthesized in the laboratory

| fumed compound | Fp (°C) | raw material | b.p. °C |
|---|---|---|---|
| $Al_2O_3$ | 2000 | $AlCl_3$ | 180+) |
| $AlBO_3$ | n.n. | $AlCl_3/BCl_3$ | 180+)/12 |
| $AlPO_4$ | n.n. | $AlCl_3/PCl_3$ | 180+)/74 |
| $SiO_2$ | 1713 (cristo-balite) | $SiCl_4$ | 57 |
| $TiO_2$ | 1855 (rutile) | $TiCl_4$ | 137 |
| $GeO_2$ | 1086 | $GeCl_4$ | 83 |
| $ZrO_2$ | 2687 | $ZrCl_4$ | 331+) |
| $Cr_2O_3$ | 2440 | $CrO_2Cl_2$ | 117 |
| $WO_3$ | 1473 | $WCl_6$<br>$WOCl_4$ | 346<br>228 |
| $Fe_2O_3$ | 1600 ($Fe_3O_4$) | $FeCl_3$<br>$Fe(CO)_5$ | 319<br>103 |
| NiO | 1960 | $Ni(CO)_4$ | 42 |

+) sublimation point

## 2.2. Thermal Silicas

In the past, several commercial plants for the production of highly dispersed silicas by electric arc processes have been run.[9] The principle of this method will be described shortly: At temperatures > 2000°C quartz is reduced with coke to volatile SiO which, in a second step, is oxidised with air or steam to silica again ( see figure 8). The energy-intensive procedure and the general performance of these products which are useful as matting agents but, in many respects, cannot compete with the fumed silicas, are responsible for the fact, that today only one commercial plant is left (DEGUSSA).

Other efforts, for example, by a research group of LONZA/ Switzerland,[10] to gain products equivalent to fumed silicas by the reduction of quartz in an alcohol-stabilized plasma torch never materialised in a commercial plant.

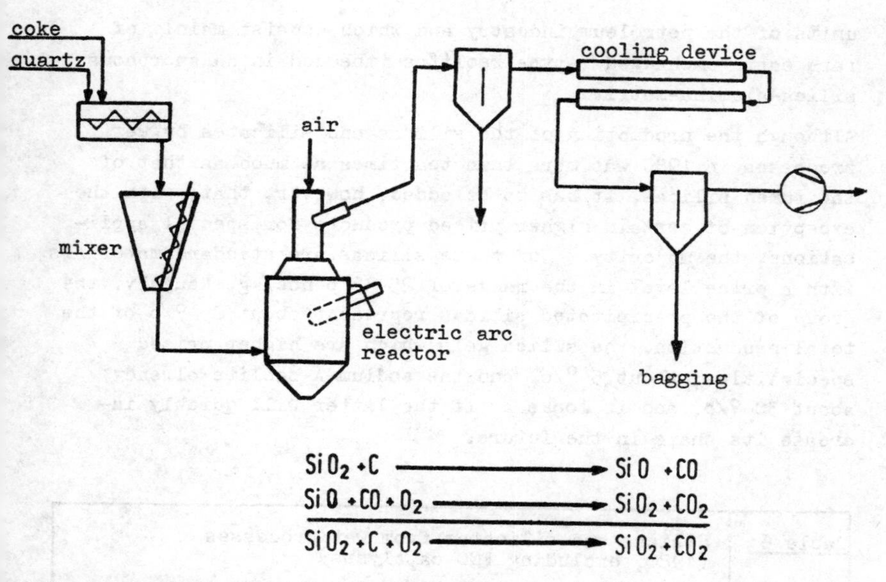


$$SiO_2 + C \longrightarrow SiO + CO$$
$$SiO + CO + O_2 \longrightarrow SiO_2 + CO_2$$
$$SiO_2 + C + O_2 \longrightarrow SiO_2 + CO_2$$

Figure 8: Thermal silica by electric arc process

## 3. Silicas and Silicates by Wet Processes

To this section belong, in accordance with figure 1, the amorphous product groups of the precipitated silicas (resp. silicates) and the silica gels, furthermore the sodium aluminium silicates (zeolites) which are different in many respects, e. g. due to their crystallinity and application. To complete the picture, it has to be mentioned that "silica sols" - colloidal aqueous suspensions, which are obtained mainly by de-ionising dilute sodium silicate solutions by means of cation exchangers followed by concentration (up to 20-50 °/o $SiO_2$) and stabilisation (with small amounts of alkali) - are also produced on a technical scale. They find application as binders, for example in the manufacture of cast mouldings or as granulation aids. For more details see literature.[11]

In 1980, in the western world the estimated production capacity of silicas and silicates gained by wet processes was approximately 900,000 t/a (see table 6), the total market being shared by 20-25 companies. This figure does not include the FCC-catalysts which, up to 200,000 t/a are consumed in the catalytic cracking units of the petroleum industry and which consist mainly of rare earth exchanged Y-type zeolites imbedded in an amorphous silica-alumina-matrix.

Although the production of the silicas and silicates by wet processes in 1980 was more than ten times as much as that of the fumed silicas, it has to be added, however, that, with the exception of certain higher priced products for special applications, the majority of these silicas are standard materials with a price level in the range of 25-35 pence/kg. Roughly, the group of the precipitated silicas represent about 65 °/o of the total production, the silica gels which are higher priced specialities, about 5 °/o, and the sodium-A-zeolite already about 30 °/o, and it looks as if the latter will quickly increase its share in the future.

| Table 6: | Silicas and silicates from wet processes (1980, excluding FCC-catalysts) | | |
|---|---|---|---|
| region | main producers, estimated capacity (t/a) | | |
| | zeolites | silica gels | precipitated silicas |
| USA | 150,000 (ETHYL, HUBER, UCC) | 20,000 (GRACE and others) | 220,000 (PPG, HUBER, PHIL. QUARTZ) |
| South America | - | - | 30.000 |
| Western Europe | 90,000 (DEGUSSA, HENKEL) | 15,000 (GRACE and others) | 280,000 (DEGUSSA, SIFRANCE, CROSFIELD, AKZO, SILQUIMICA and others) |
| Asia | 30,000 (MIZUSAWA, TOYO SODA and others) | (FUJI-DAVISON) | 60,000 |
| Total (western world) | 270,000 | 35,000 | 590,000 |

## 3.1. Zeolites

### Properties and Applications

Zeolites are crystalline alumosilicates with a regular three-dimensional porous lattice structure built up from $SiO_4$- and $AlO_4$-tetrahedrons, in which the negative charges are compensated by mono- or multivalent cations.[12] These cations are exchangeable without the lattice structure being destroyed, thus explaining the zeolites' new and commercially very important application as water-softeners in household detergents, substituting sodium tripolyphosphate (figure 9).[13] There are well over one hundred known types of different zeolites. Some of them are natural minerals, most of them have been synthesized in the laboratory within the last 25 years.

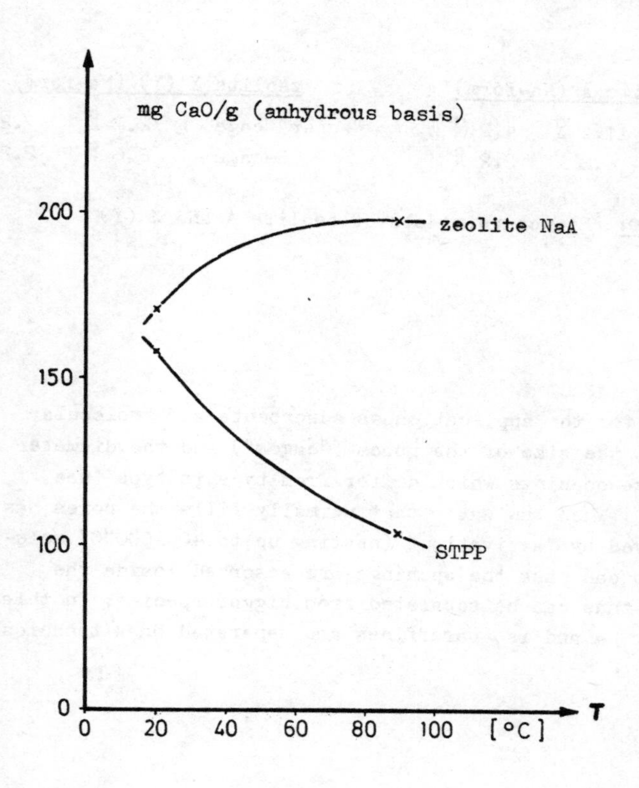


Figure 9: Calcium exchange capacity of sodium A zeolite and STPP

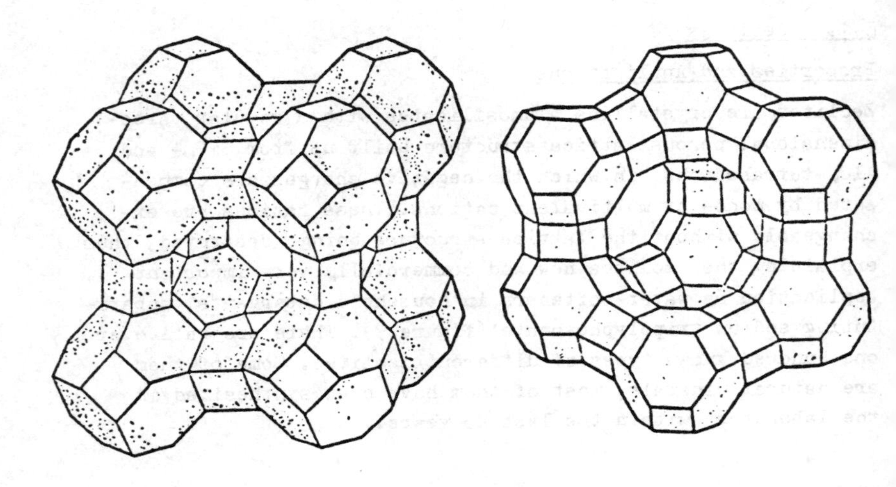

| zeolite A (Na-form) | | | zeolite X (Y) (Na-form) | | |
|---|---|---|---|---|---|
| α-cage | 11.4 Å | 4.2 Å | "super" cage | 12.7 Å | 7.2 Å |
| ß-cage | 6.6 Å | 2.2 Å | ß-cage | 6.6 Å | 2.2 Å |

Figure 10: Structure models of zeolite A and X (Y)

Important for the application as adsorbents and "molecular sieves" is the size of the pores ("cages") and the diameter of the cage openings which differ from type to type (see figure 10): When the water that normally fills the pores has been removed by "activation" (heating up to 400-500$^{\circ}$C) molecules that can pass the openings are adsorbed inside the cages and thus can be separated from bigger species. In this way, e. g. n- and iso-paraffines are separated on a technical scale.

The zeolites are the "newcomers" in the synthetic silicas/ silicates business. They left laboratory scale only in the late 1950s. Used at first as ion exchangers and adsorbents, later on their out-standing catalytic properties were discovered which resulted in a fast growing market for special grades, such as the already mentioned rare earth exchanged X- and Y-types as ingredients of cracking catalysts.[14] Research on zeolite-catalysts is still going on world-wide and ZSM-5, the MOBIL catalyst for the methanol conversion to gasoline, will not be the last word.

As far as tonnages are concerned, however, a new dimension has been added to the synthetic zeolites business in the USA, W. Germany and Japan when the sodium-A-type turned out to be - at least partially - a suitable non-eutrophic and comparatively cheap substitution product for STPP as detergent builder (table 7). In 1980, in the western world, more than 270,000 t/a of zeolite 4 A will be produced for that purpose, thus making this product a pseudo-commodity within the field of synthetic silicas and silicates.[15]

Table 7: Some properties of a commercial detergent zeolite (NaA-type)

| | | |
|---|---|---|
| medium particle size | [nm] | 3-4 |
| maximum particle size | [nm] | 25 |
| sieve residue > 45 µm | | < 0.1 % |
| tamped density | [g/l] | ~450 |
| chemical analysis | | |
| $Na_2O$ | [%] | 17 |
| $Al_2O_3$ | [%] | 28 |
| $SiO_2$ | [%] | 33 |
| $H_2O$ | [%] | 22 |
| $Ca^{2+}$ exchange capacity | [mg CaO/g+)] | >160 |

+) anhydrous basis

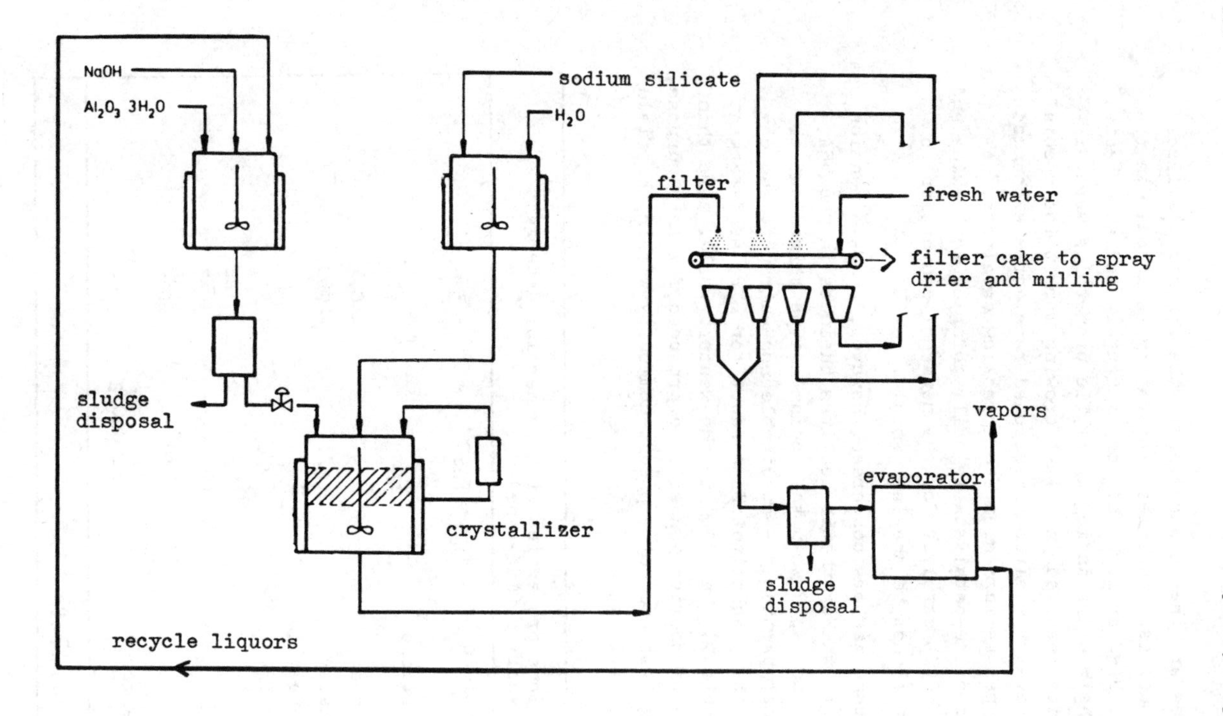


<u>Figure 11:</u> Flow sheet of zeolite NaA-production process[16]

Manufacturing Process

Commonly used raw materials for the industrial production of sodium-A-zeolite,

$$Na_{12}(AlO_2)_{12}(SiO_2)_{12} \cdot 27H_2O,$$

are alumina trihydrate which is reacted with caustic soda to give an aqueous solution of sodium aluminate. In the next step this is - continuously or in a batch process - intensively mixed with a sodium silicate solution. Then the mixture is heated up for crystallisation. As in the system $Na_2O/Al_2O_3/SiO_2/H_2O$ a variety of zeolites with different crystal structures and pore sizes can be synthesized; a careful control of the reaction parameters such as the molecular ratio of the reactants, concentration, temperature and reaction time is essential. When the crystallisation is finished the suspension is filtered. The filter cake is washed and reliquified, and then either spray-dried or - after stabilisation - purchased as a pumpable master-batch.[13] As an example, a flow sheet of the HUBER-process is given in figure 11.[16]

## 3.2 Silica Gels

Unlike the zeolites the silica gels (in addition to the fumed and the precipitated silicas) are amorphous products. They are the oldest group of synthetic silicas. Commercial production was started about 1920 by DAVISON,[11] now belonging to GRACE, which is still the biggest producer of these materials in the world. They are mainly used as adsorbents, for example desiccants, but also matting, free flow and anti-blocking agents.

As the silica gels were reviewed on the occasion of a former symposium,[1] only a few concise remarks on the production process and their general properties will be made.

In synthesizing a silica gel, a hydrosol is produced as an intermediate by mixing sodium silicate with ( preferably) $H_2SO_4$ at an acid pH. The mixture is allowed to set in vessels or on conveyer belts. The resulting transparent rigid mass is then broken and washed salt-free with water, for example in counter-current reactors, dried and classified. For special purposes a milling step can be added (see figure 12).

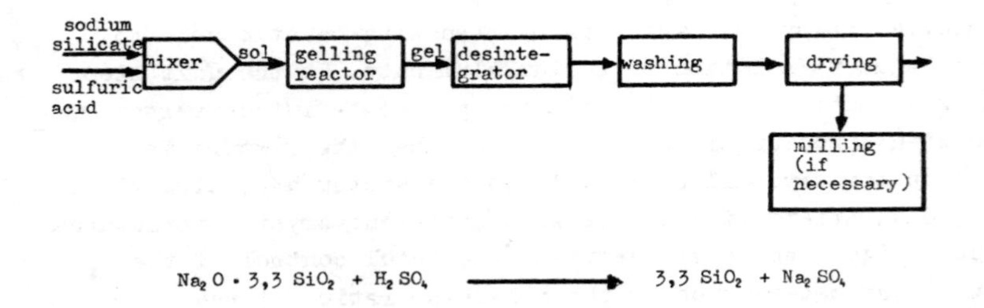


$$Na_2O \cdot 3{,}3\ SiO_2 + H_2SO_4 \longrightarrow 3{,}3\ SiO_2 + Na_2SO_4$$

Figure 12: Flow sheet of silica gel production

Characteristic for silica gels is their pore structure; the pores are originally very narrow (within the range of 2-3nm), the pore size distribution curve shows a distinctive maximum, and the surface area can go up to more than 800 $m^2/g$.

By modifying the washing and aging process - using higher temperature and/or pH-value - the average pore diameter and the pore volume increase; this is combined with a reduction of surface area. In this way a great variety of gel types can be prepared. Commercial products have surface areas within the range of 800-100 $m^2/g$, corresponding to average pore diameters of 2-20 nm and pore volume of 0.4-2.0 ml/g.[11]

## 3.3 Precipitated Silicas and Silicates

The precipitated silicas which are, quantity-wise, the most important group of synthetic silica products represent, historically, the third generation of white reinforcement fillers for rubber. They were developed in the early 1940s, following and gradually replacing their predecessors calcium carbonates and alumina gels resp. calciumsilicates and aluminiumsilicates. (Nevertheless the latter products are still produced and used, for example as less active fillers in rubbers, as extenders for titania in paints, in the paper industry and as flame retardants in plastics.)

### Properties and Applications

Unlike the fumed silicas and in common with the silica gels the precipitated silicas are porous. The average pore diameter is, in general, bigger than that of the gels ($>$30 nm), and the size of the pores varies within a large range. The silanol group density exceeds that of the fumed silicas (5-6 SiOH/$nm^2$ instead of 2-4 SiOH/$nm^2$), and also the moisture content (2 h, 105°C) normally is higher (3-7 °/o). The specific surface area can vary - depending on the reaction conditions - between approximately 50-800 $m^2$/g. The thickening effect in liquids, as in the case of the silica gels, cannot compete with that of the fumed silicas.

Besides the pure silicas a certain amount of calcium - resp. aluminium containing silicas are still produced on a technical scale. These silica products which are non-stoichiometric coprecipitates of silica and silicates are mainly used for special applications in the rubber industry and as fillers and extenders, for example in the paper industry.

A small amount of the commercially available precipitated silicas are hydrophobic types; they are obtained by chemical reaction of the silanol groups with certain silanes (e. g. as described in the chapter "Fumed Silicas", see above). Others are coated, e. g. with waxes, in order to improve the dispersion in certain organic systems.

Table 8, where the main applications of precipitated silicas are listed, shows that the majority of these products go, as reinforcement fillers, into the rubber industry. But the total amount of silicas used today for this purpose is only about 10 % of the carbon blacks consumption of the same industry.

Table 8: Main applications of precipitated silicas

| | |
|---|---|
| reinforcement of rubber | ~ 70 % |
| carriers for active substances (e.g. pesticides, catalysts, anti-foaming agents)<br>extender fillers (papers, paints)<br>matting agents (coatings, paints, plastics)<br>anti-blocking agents (plastic foils)<br>free running agents (animal feedstuffs, fire extinguishing powders)<br>abrasive fillers (tooth pastes)<br>beer clarifiers | ~ 30 % |

And the fact, that about 60 % of these "rubber silicas" are used for the fabrication of shoe soles, demonstrates that the above mentioned ambitious target of the silica researchers to find a fully competitive substitute product for carbon blacks obviously has not been reached yet.

Manufacturing Process

The production process for precipitated silicas (and "calcium"- resp. "aluminium-silicates") is, in the main step, the precipitation, a discontinuous one (see figure 13).

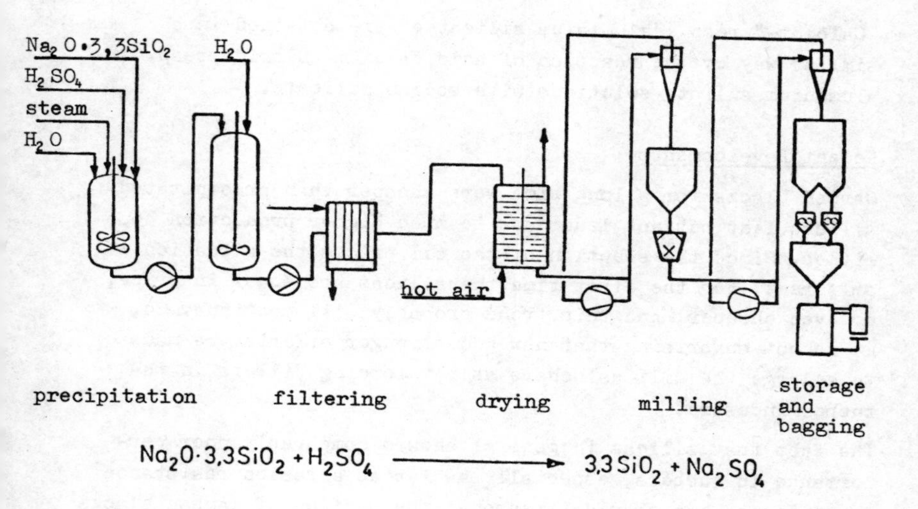


$$Na_2O\cdot 3{,}3SiO_2 + H_2SO_4 \longrightarrow 3{,}3\,SiO_2 + Na_2SO_4$$

Figure 13: Flow sheet of precipitated silica production

Sodium silicate and sulfuric acid - instead of sulfuric acid a mixture of $CO_2$ and HCl can also be used - are reacted at higher temperatures and, normally, at alkaline pH-values, while the reaction mixture is vividly stirred.

Thus, the gelation state is avoided and discrete silica particles are formed which grow by further precipitation of silica on the surface of the primarily formed seed particles.

The silica is separated and washed, using filter presses or filter drums. Standard types are dried in disc or drum driers, milled, classified and bagged or shipped as bulk materials. By modification of the precipitation parameters, such as reaction time, temperature, concentration, pH-value and shearing forces, as well as by using specific after-treatment steps, for example, spray-drying of the reliquified cake, finemilling of the dried products in e. g. jet-mills, a great variety of precipitated silicas are produced on a technical scale.

"Calcium-" resp. "aluminium silicates" are obtained in a similar way by the reaction of acid calcium chloride resp. aluminium sulfate solutions with sodium silicate.

## Recent Developments

Carbon blacks for a long time were cheaper than precipitated silicas,last but not least due to much higher production capacities. Since the eruption of the oil prices the situation is different: For the first time the silicas are equal in price, or even cheaper, and this trend probably will continue. So, it is not surprising that new and stronger efforts are made to enlarge the silicas' share as reinforcing fillers in the rubber industry.

The fact that silicas in general show a comparably poor performance in rubbers, especially as far as abrasion resistance is concerned has chemical reasons: The surface of carbon blacks is hydrophobic and has functional groups, for example, olefinic or conjugated double bonds which can react or at least physically interact with the double bonds of the rubber matrix, thus giving good reinforcement. The hydrophobic silanol groups, on the other hand,disturb the vulcanisation process, and the silicas give a certain reinforcement to the vulcanized rubber due only to their small particle size.

A promising approach to improve the performance of silicas in rubbers substantially - both the processability of the unvulcanized raw mixes and the mechanical properties of the cured material - has been made by the development of certain bifunctional organosilanes, which are used as "reinforcement agents" in silica containing rubber mixtures.[17] A very effective compound is e. g. the bis -(3-[triethoxysilyl]-propyl)-tetrasulfane ("Si 69", DEGUSSA),

$$(H_5C_2O)_3-Si-(CH_2)_3-S_4-(CH_2)_3-Si(OC_2H_5)_3,$$

which can, on one side, react with the silanol groups at the silica surface, forming stable siloxane bonds, and, on the other side, with the polymer network via the tetrasulfane groups, thus enabling a chemical fixation of the filler particle within the polymer (see figures 14 + 15).

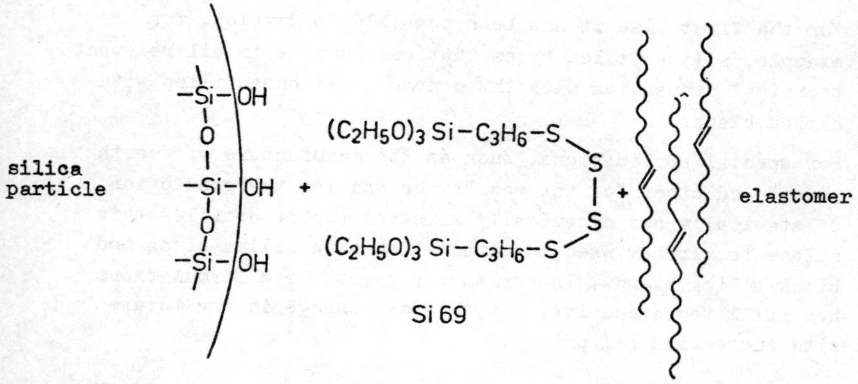

silica particle

$-Si-OH$

$(C_2H_5O)_3Si-C_3H_6-S-S-S-S-C_3H_6-Si(OC_2H_5)_3$

Si 69

elastomer

Figure 14: Coupling agent "Si 69" before vulcanisation

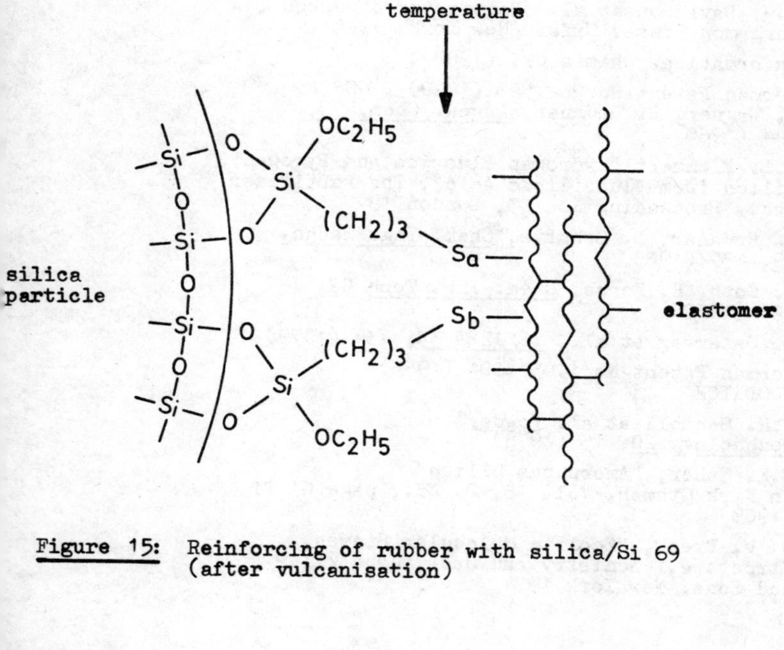

temperature

silica particle

$OC_2H_5$

$(CH_2)_3$

$S_a$

$S_b$

elastomer

Figure 15: Reinforcing of rubber with silica/Si 69 (after vulcanisation)

For the first time it has been possible to develop, for example, silica-filled tyres that can compete in all relevant technical properties with the conventional ones filled with carbon blacks.[17]

For special applications, such as the manufacture of "earth mover" and other "off the road" tyres and for the production of statically and dynamically stressed rubber articles this silane is already used in combination with silica or carbon black/silica blends. In normal car tyres these formulations are still too expensive, but this may change in the future with increasing oil prices.

## References

1. D. Barby et al. in R. Thompson "The Modern Inorganic Chemicals Industry", The Chem. Soc., Special Publ. No. 31, 1977, page 320 ff.
2. H.W. Davidson et al. "Manufactured Carbon", Pergamon Press, Oxford-New York, 1968
3. Informations Chimie G/7 (1980).
4. German Patent No 762 723 (1942) DEGUSSA E. Wagner, H. Brünner, Angew. Chem. 72, 744 (1960).
5. G.L. Flemmert "Hydrogen Fluoride and Pyrogenic Silica from Fluosilicic Acid", The Fertilizer Soc., Proceeding No 163, London 1977
6. H. Brünner, D. Schutte, Chem.-Ing. Techn. 89, 437 (1965).
7. D. Koth, H. Ferch, Chem.-Ing. Tech 52, 628 (1980)
8. W. Ostertag et al., DEFAZET 33, 434 (1979)
9. German Patent No 1.034.601 (1955) GOODRICH
10. E.R. Schnell et al. Powder Technology 20, 15 (1978)
11. P.K. Maher, "Amorphous Silica" in Kirk/Othmer, Vol. 18, 2. Ed., page 61 ff (1969)
12. D. W. Breck, "Zeolite Molecular Sieves, Structure, Chemistry and Use", John Wiley and Sons, New York 1974

13. H. Strack, P. Kleinschmit, Chem. Ztg 104, 1 (1980)

14. P.B. Venuto, E.T. Habib, "Fluid Catalytic Cracking with Zeolite Catalysts", Marcel Dekker, New York, 1979.

15. Chemical Week, July 23, 1980, p. 44 ff.

16. German Patent OS No 26 33.304 (1976)

17. F. Thurn und S. Wolff, Kautschuk und Gummi, Kunststoffe 28 (1975), S. 733-739

S. Wolff, Kautschuk und Gummi, Kunststoffe 30 (1977), S. 516-523

S. Wolff, "Effects of Bis-(3-Triethoxisilyl-propyl)-tetrasulfide-modified Silicas in NR", ACS Meeting 1979, Washington DC/USA

# The Preparation and Uses of Inorganic Titanium Compounds

By G. F. Eveson
LAPORTE INDUSTRIES LTD., P.O. BOX 26, GRIMSBY, SOUTH HUMBERSIDE DN37 8DP, U.K.

## Introduction

No single unifying theme is put forward to support the selection of the titanium compounds discussed in this article although each has industrial applications, or potential applications, which are of considerable importance in most industrialised communities. Detailed information concerning the annual production, consumption, import and export of these compounds is not available from the Departments of Trade and of Industry. Such information for titanium compounds is classified under an NES ("not elsewhere specified") heading, which includes derivatives of other elements including boron, molybdenum and tantalum. Companies which manufacture or use these compounds generally are unwilling to disclose production and consumption statistics on the grounds that this information may be helpful to a competitor. Most of the selected compounds are difficult to prepare as pure, stoichiometric materials. Consequently, considerably discrepancies are often found in reported values for physical and thermodynamic properties. For this reason, detailed reference to precise values for such properties has been omitted from most sections of this article.

## Titanium Tetrachloride

Very substantial quantities of purified titanium tetrachloride are produced each year, mainly for use in the production of titanium dioxide pigments and titanium metal. Several titanium-bearing materials are being used in commercial-scale plants, including naturally occurring mineral rutiles and ilmenites and beneficiated ilmenite products obtained by chemical or pyrometallurgical processes (Table 1).

Rutile is the preferred feedstock since, as a consequence of its low impurities content, its use introduces few serious technical problems. Unfortunately, there are limited quantities in the proved economic reserves of this mineral. Consequently, producers of titanium tetrachloride prefer

Table 1. Chemical Analyses of Typical Titanium-Bearing Feedstocks

| Component (Wt. %) | Rutile | W. Aust. Ilmenite | Leucoxene | Benelite | Richard's Bay Slag |
|---|---|---|---|---|---|
| $TiO_2$ | 95 - 96 | 54.5 | 88 - 90 | 91 | 86 |
| $Fe_2O_3$ | 0.4 | 17.5 | 5 | 4 | |
| FeO | 0.2 | 23 | 1 | 0.4 | 9.8 |
| $Al_2O_3$ | 0.3 - 0.6 | 0.6 | 1.3 | 0.6 | 1.2 |
| CaO | 0.01 | 0.02 | 0.06 | 0.01 | 0.18 |
| $Cr_2O_3$ | 0.14 - 0.20 | 0.04 - 0.08 | 0.13 | 0.07 | 0.24 |
| MgO | 0.01 | 0.2 | 0.06 | 0.05 | 1.1 |
| MnO | 0.01 | 1.5 - 2.0 | 0.12 | 0.04 | 1.9 |
| $Nb_2O_5$ | 0.35 - 0.45 | 0.24 | 0.4 | 0.3 | 0.1 |
| $P_2O_5$ | 0.01 | 0.03 - 0.06 | 0.04 | 0.06 | |
| $SiO_2$ | 0.4 - 1.0 | 0.5 | 0.3 | 1.7 | 2.6 |
| $SnO_2$ | 0.02 - 0.05 | 0.003 | 0.02 | 0.002 | <0.01 |
| $V_2O_5$ | 0.6 | 0.14 - 0.20 | 0.3 | 0.11 | 0.48 |
| $ZrO_2$ | 0.5 - 1.0 | 0.5 - 1.5 | 0.6 | 0.12 | 0.27 |

to operate equipment which can process a number of feedstocks, including blends of materials. This enables maximum advantage to be taken of the fairly frequent fluctuations in the availability and price of raw materials.

The chlorination of rutile is an endothermic reaction and carbon is added to the rutile to supply the heat requirement. The overall reactions for the process may be written:-

$$TiO_2\ (impure) + C + 2Cl_2 \rightleftharpoons TiCl_4 + CO_2$$

and/or

$$TiO_2\ (impure) + 2C + 2Cl_2 \rightleftharpoons TiCl_4 + 2CO$$

Chlorination can be carried out in a shaft kiln, using briquetted or pelletised rutile-carbon feedstock, or in a fluidised bed reactor. The latter is used in modern large-scale plants since it lends itself to continuous operation with reasonable control of bed temperature. Even so, a vertical temperature gradient exists within the fluidised bed. The bed temperature may be raised to any desired level by the addition of excess carbon and oxygen. Generally, it is maintained in the range 900-1000°C. Chlorine slip increases at temperatures below about 900°C, resulting in the presence of unreacted chlorine in the exit gas. At temperatures above 1100°C, the sintering temperature of titanium dioxide is approached and the rate of corrosion of the refractory lining of the fluidised bed reactor increases rapidly.

Petroleum coke, containing 1-2 per cent. sulphur and 0.2-0.5 per cent. ash, is frequently used as the source of carbon. It is possible to use coke or char derived from coal, but this can introduce additional operating problems arising from the relatively high ash content of such materials.

The coke and rutile fed to the chlorinator must be dry. The size range of each material is selected to minimise segregation of coke and rutile in the fluidised bed. The bulk of the chlorine required by chlorinators incorporated in a titanium dioxide pigments production plant is derived from the oxidation section of the process where purified titanium tetrachloride is reacted with oxygen. The chlorine re-cycled from the oxidation stage, containing some impurities, is mixed with make-up chlorine before it is fed to the chlorinator. There is no re-cycle gas stream in plants which are concerned solely with the production of purified titanium tetrachloride.

The chlorination stage is basically a straightforward operation which can be complicated by the influence of certain impurities present in the raw materials. The chlorides of certain elements - principally iron, manganese, magnesium and calcium - which have high melting points, also have low vapour pressures under the normal operating conditions for a chlorinator. Such chlorides might accumulate in the fluidised bed if the vapour pressure of the chloride at the localised bed temperature is less than its partial pressure in the reactor. This would promote the formation of sticky agglomerates of particles, impairing the efficient operation of the fluidised bed.

Chloride impurities present in the chlorinator off-gases will condense as a solid if the dew point of the gas stream for each chloride is below its melting point. Otherwise, the chloride will condense as a liquid, giving rise to the build-up of sticky deposits in ducting.

Careful attention must be given to the chemical compositions of all of the materials fed to a chlorinator. The chlorination of low-grade feedstocks (e.g. ilmenite) involves a substantial increase in the chlorine consumption for the production of a given quantity of titanium tetrachloride. In addition, the condensation section of the plant must be increased in size to deal with the increased quantity of waste product, disposal of which creates a serious effluent problem.

It is possible to operate a fluidised bed chlorinator to convert the iron content of the feed to ferrous chloride or to ferric chloride. The $FeCl_2/FeCl_3$ ratio is influenced by the concentrations of iron and chlorine in the reaction zone. Relatively deep beds and/or high iron contents promote the formation of ferrous chloride. Ferric chloride is obtained when using relatively shallow beds and/or low iron contents.

The formation of ferrous chloride offers the advantage of reduced chlorine consumption, coupled with the disadvantage of higher coke consumption, in order to achieve autothermal operation and because of the need to operate at higher temperatures to avoid the condensation of ferrous chloride. Technically, it is possible to oxidise ferric chloride, liberating chlorine which can be re-cycled to the chlorinator. There is no extensive demand for the iron oxide by-product of this oxidation step.

Gases leaving the chlorinator entrain finely-divided particles of coke and rutile (or other titanium-bearing feedstock) which, along with other impurities, have to be removed in the purification section of the plant.

The chlorinator off-gases are cooled to a temperature of about 200°C. A large proportion of the chlorides are condensed out of the titanium tetrachloride vapour and, together with the entrained particles of unreacted feed materials, are removed in a cyclone separator. Some contaminants remain in the gas phase and are condensed with the titanium tetrachloride in which they either dissolve or form a suspension. The final traces of titanium tetrachloride are removed from the chlorinator off-gases in a refrigerated condenser. The off-gases pass to the gaseous effluent scrubbing system.

The crude liquid titanium tetrachloride has a yellowish colour due mainly to the presence of chlorides of vanadium, iron and chromium. It also contains the colourless chlorides of tin and silicon. It is essential to remove all of those chlorides, which give rise to coloured oxides, from titanium tetrachloride which is used for pigment manufacture. Equally stringent purity specifications are demanded for titanium tetrachloride used for titanium production.

Impure tetrachloride, in the liquid or vapour phase, may be reacted with anhydrous hydrogen sulphide, precipitating any vanadium,iron and molybdenum present in the liquid or vapour. The temperature should be kept below about 200°C to minimise the production of titanium sulphides.

Alternatively, the crude tetrachloride, maintained at a temperature just below its boiling point, is mixed with a small quantity of a mineral or vegetable oil having a high iodine number. The oil is charred, forming insoluble, non-volatile complexes with the heavy metal impurities. Purified titanium tetrachloride is recovered from the sludge by distillation. The tin and silicon contents of the purified product are controlled by adjusting the reflux ratio in a final distillation step.

The conversion of titanium tetrachloride into titanium dioxide pigments by the chloride process has been described in a recent publication[1]. It is estimated that about 130,000 Mg of titanium tetrachloride are produced each year in the U.K. for this purpose.

Commercial-scale production of titanium metal began in the late 1940's. Two processes (Kroll; Hunter) have remained dominant in the few countries where plants to produce titanium metal have been built. Each process utilises titanium tetrachloride which has been up-graded by distillation in argon or nitrogen to achieve a purity of about 99.8 per cent.

In the Kroll process[2,3,4], a weighed amount of clean magnesium is placed in a cylindrically-shaped steel reaction vessel. A gas-tight seal must be obtained between the vessel and its lid. The atmosphere in the vessel is replaced by argon or helium. The vessel is placed in a furnace and heated until the magnesium melts. Titanium tetrachloride is added at a controlled rate, so that the highly exothermic reaction maintains the temperature within the range 750°-800°C. Magnesium chloride is removed as a liquid by tapping several times during the batch reduction operation. Sufficient titanium tetrachloride is added to react with 85-90 per cent. of the magnesium.

A titanium metal sponge is formed in the reaction vessel. Magnesium chloride fills the pores and surrounds individual pieces of the metal. The reaction mass also contains magnesium, which is used in excess of the stoichiometric quantity in order to ensure complete reduction of the titanium tetrachloride. The reaction vessel is opened in a temperature-controlled, dry room. This prevents absorbtion of moisture by the hygroscopic magnesium chloride[3]. The vessel contents are removed by machining on a large lathe. Care is taken not to remove with the sponge a thin iron-contaminated layer of material which forms adjacent to the wall of the vessel.

The sponge turnings are heated to a temperature of about 900°C in a vacuum furnace for 30-40 hours. The magnesium and magnesium chloride impurities are volatilised and collected in a water-cooled condenser. Some operators of the Kroll process purify the sponge turnings by leaching them with dilute hydrochloric acid. This technique consumes far less energy than vacuum distillation but it has been claimed that leaching produces a 'dirty' sponge, containing more magnesium and magnesium chloride than does sponge purified by vacuum distillation[2].

In the OREMET variation of the Kroll process, the impurities are removed from the sponge at the end of the reduction process by passing a stream of helium through the reaction vessel, maintained at a temperature of about 1000°C[4]. There is no leaching or vacuum distillation step.

In production units where practically all of the magnesium chloride is recovered by tapping and vacuum distillation, the chloride can be returned to electrolytic cells which produce magnesium for re-use in the reduction process, and chlorine which can be used in the manufacture of titanium tetrachloride. Where leaching of the sponge is practised, the disposal of an aqueous solution of magnesium chloride must be considered to constitute an environmental and economic problem[4].

The Hunter process, in which metallic sodium is used to reduce titanium tetrachloride, was developed to production scale by I.C.I. Ltd. in the period 1951-53. It is claimed that sodium reduction is more efficient than the Kroll process. A higher purity titanium metal can be prepared from the impure sponge by leaching, eliminating the need for vacuum distillation[2].

The original Hunter process, like the Kroll process, was carried out in a single step. A variation of the process, introduced by Reactive Metals Inc., has two stages[4]. Titanium tetrachloride is reacted with metallic sodium in a continuous reactor in which an argon atmosphere is maintained. The reaction product, a mixture of titanium dichloride and sodium chloride, is charged into sinter pots with more sodium. The atmosphere in each pot is replaced with argon. On heating to a temperature of approximately 1000°C, the titanium dichloride is reduced to titanium sponge and sodium chloride. The pot is cooled, opened and the mixture of titanium sponge and sodium chloride chipped out with a remote-controlled pneumatic hammer. The chippings are crushed, leached with dilute acid, washed with clean water and dried in a continuous vacuum drying unit.

The Titanium Metals Corporation of America has operated a semi-commercial scale electrolysis unit consisting of five cells, each capable of producing about 190 lb titanium metal per day[4]. Titanium tetrachloride is fed into the interior of a perforated metal basket-type cathode which is surrounded by a number of graphite anodes. The electrolyte is composed of various alkali or alkaline earth chlorides. The operating temperature depends upon the nature of the electrolyte; for sodium chloride, it is approximately 850°C. Titanium tetrachloride is only sparingly soluble in molten alkali chlorides but titanium tri- and di-chlorides are very soluble. Titanium metal deposits on the wall of the basket cathode. It forms a porous structure which acts as a diaphragm, dividing the electrolyte into anolyte and catholyte portions.

The cathode, containing titanium metal and entrained electrolyte, is removed from the cell and cooled. The electrolyte is removed by leaching with dilute acid. The titanium metal is of high purity, having a low iron content.

This single-stage reduction process is reported to have the potential to reduce costs but major technological improvements are needed.

The processes whereby titanium sponge is converted to titanium metal and alloys in plate and tube forms are outside the scope of this article but they have been discussed in recently published papers[2,3]. Installed annual plant capacity in non-Communist countries for producing titanium sponge, in 1979, is reported[5] as:- U.K. 3600 Mg; U.S.A. 20,400 Mg; Japan 18,500 Mg.

## Potassium Fluotitanate

Anhydrous potassium fluotitanate ($K_2TiF_6$) is prepared on a commercial scale by dissolving titanium dioxide in dilute hydrofluoric acid. The exothermic reaction is carried out in an agitated, carbon tiled, rubber lined mild steel vessel fitted with a Monel metal cooling coil. Both the anatase and rutile forms of pigmentary titanium dioxide can be used. Anatase is soluble in cold hydrofluoric acid but it is necessary to warm the acid to initiate the reaction with rutile.

A clear solution of hydrofluotitanic acid ($H_2TiF_6$) is formed which is neutralised with concentrated potassium hydroxide solution. This is a more exothermic reaction resulting in the formation of small, lustrous white particles of anhydrous potassium fluotitanate. The contents of the reaction vessel are dropped into a perforated basket centrifuge, de-watered and washed with cold raw water to remove soluble impurities (mainly potassium fluoride). Fluotitanates are sparingly soluble in cold water. The washed product is dried, milled if necessary and packed.

There is a small annual consumption of finely-ground potassium fluotitanate in the preparation of dental fillings. A relatively small quantity is used as a cleaning flux in metal casting processes, principally in the non-ferrous industries. The flux reduces the melting point of the associated slag, enabling it to be removed more readily from the molten metal.

The major use of potassium fluotitanate is in the grain refining of aluminium. Grain refining agents are primarily used in the casting of wrought aluminium and its alloys by Direct Chill (DC) and continuous casting processes. They promote the production of a small grain size as the molten alloy cools and solidifies. The mechanical properties and surface finish of the wrought alloy products are improved as the grain size decreases. In addition, efficient grain refining leads to faster casting speeds[6].

Originally, mixtures of potassium fluotitanate and potassium fluoborate ($KBF_4$) or borax were added to the molten aluminium, forming titanium and boron compounds dispersed in an unreacted aluminium matrix. This technique had a number of disadvantages. A high addition rate was required to obtain an acceptable level of grain refinement. This increased the variable costs of the casting process and, in plant where scrap material was re-cycled, there was an undesirable increase in the titanium and boron contents of the cast alloys.

Salt mixtures have largely been replaced by a wide range of aluminium-based master alloys containing either titanium or titanium and boron. The master alloys are produced in the form of waffle plate, which is added to the aluminium holding furnace, or of rod, which is fed into the casting trough countercurrent to the flow of molten metal[7]. There is a substantial annual consumption of potassium fluotitanate in the manufacture of these alloys. Binary titanium-aluminium alloys usually contain 6-10 per cent. (w/w) of titanium. The ternary alloys provide Ti:B ratios in the range 5:1-50:1 and generally contain 5 per cent. of titanium[7]. The microstructure of the binary master alloys consists of titanium aluminide ($TiAl_3$) particles in a matrix of aluminium whereas, in the ternary alloys, particles of the aluminide and of the refractory compound titanium diboride ($TiB_2$) are dispersed in an aluminium matrix[8]. It is reported[6] that when a ternary master alloy is stirred into molten aluminium, the titanium aluminide particles dissolve rapidly whilst the titanium diboride particles, which have an average size in the range 0.4-1μm, are dispersed throughout the melt. The experimental evidence indicates that these particles are a most efficient nucleant for aluminium under the optimum grain refining conditions but the precise mechanism of nucleation has not yet been established.

## Titanium Disulphide

Titanium disulphide is the most interesting of the sulphides of titanium in so far as considerable efforts have been made during the last 10 years to develop a commercial use for this, the lightest of all the layered dichalcogenides. It has several properties which make it attractive as a reversible cathode material in secondary batteries, where it is used in conjunction with a lithium anode[9]. For example:-

a) it has a high electrical conductivity[10];
b) it has a low solubility in common organic electrolytes;
c) when coupled with a lithium anode, it produces a cell with a high theoretical energy density (480 Wh/kg) based solely on the weights of lithium and titanium disulphide;
d) the charge-discharge mechanism of the cell involves only the intercalation of lithium ions between adjacent sulphur layers in the titanium disulphide lattice, allowing the crystalline structure to be retained during the reaction;
e) it is highly reactive only to elements or compounds which can inter-

calate the lattice structure and, therefore, it is relatively non-corrosive to materials used for the battery casing.

Generation of electrochemical energy by intercalation, involving the production of no new phase, is the key to the potential value of solid solution electrodes, such as titanium disulphide. The intercalation process has been studied extensively[11,12,13,14,15]. Two methods have been used, each operating at ambient temperature and atmospheric pressure.

i) Titanium disulphide is immersed in n-butyl-lithium dissolved in an aprotic solvent such as benzene or toluene. The intercalation process must be carried out in an inert atmosphere and requires many hours to proceed to completion.

ii) An electrochemical cell is constructed with a lithium anode, a titanium disulphide cathode and an electrolyte consisting of a lithium salt (e.g. the iodide) in an organic solvent (e.g. propylene carbonate). The reaction is initiated by shorting the cell through an external load. The degreee and rate of intercalation can be controlled by adjusting the cell e.m.f. and/or current flow.

The c axis spacing of the titanium disulphide expands in a continuous manner from 5.7Å to 6.2Å as the intercalation proceeds to completion, corresponding to the formation of $LiTiS_2$ (Fig. 1). The cell is then in a fully discharged condition.

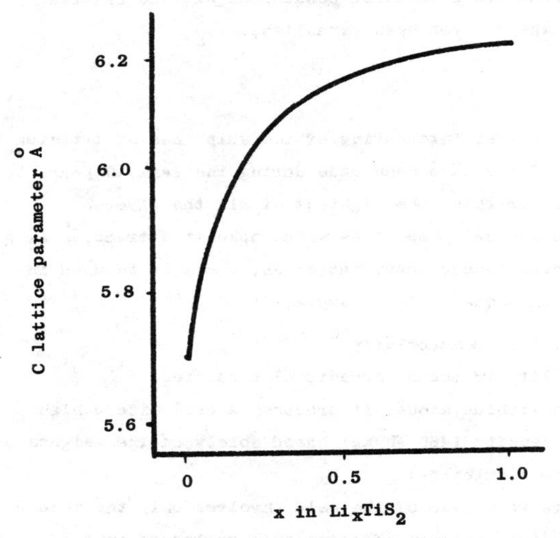


Fig. 1 Lattice Expansion of $TiS_2$ on Intercalation of Lithium. (Ref. 11).

For crystals smaller than 100μm with perfect hexagonal morphology, the crystal structure remains essentially unchanged, free from cracking, during intercalation and de-intercalation. Consequently, these processes proceed rapidly[13]. Larger crystals and crystals with imperfections frequently show extensive cracking during intercalation.

Many of the transition metal dichalcogenides exhibit a range of non-stoichiometry. The existence of excess titanium ions in the van der Waals layers of titanium disulphide restricts expansion of the lattice during intercalation. Diffusion of lithium ions into the lattice is impeded, corresponding to a reduction in the maximum power which may be attained in a battery.

Stoichiometric titanium disulphide used for laboratory-scale electrochemical investigations has generally been synthesised by direct reaction between the metal and sulphur vapour[14,15,16,17,18]. Titanium foil, or finely-divided powder, and sulphur are sealed in opposing ends of an evacuated ($10^{-4}$ torr) quartz tube which is heated slowly in a tube furnace to about 600°C for several days. Large single crystals (e.g. 5mm x 5mm x 20-100μm thickness) of titanium disulphide have been prepared from the powder using a chemical vapour transport technique which proceeds via the reversible reaction

$$TiS_2 + 2I_2 \rightleftharpoons TiI_4 + S_2$$

Preparative methods of this type do not provide a basis for an economic, commercial-scale, continuous production process. Titanium disulphide for use in battery cathode manufacture should have the following physical and chemical characteristics.

i) The size range of the particles should be fairly small (e.g. 1-10μm), facilitating control over the porosity of the final, manufactured cathode structure. The reactor design must be capable of consistently producing a large proportion of the titanium disulphide product in the required size range, since the lubricating properties of the material precludes comminution as an efficient method for controlling particle size.

ii) The particles should have a small aspect ratio (i.e. ratio of basal plane axis to vertical axis) or be built up from crystallites having a small aspect ratio. During intercalation, lithium atoms diffuse, parallel to the basal plane, inwards from the crystallite edge. The diffusion distance, corresponding to complete intercalation, decreases with decrease in the length of the basal plane axis. This results in higher battery discharge rates.

iii) The chlorine content should be low since, in a battery, this element reacts with the lithium anode, reducing the reversibility of the system.

iv) The free sulphur content should be low. Elemental sulphur is an inert constituent in the battery system, limiting the energy density which can be achieved.

v) The chemical composition should correspond to $Ti_xS_2$, where $1 < x < 1.05$.

There are several predominantly physical measurements which can be used to assess the chemical composition of titanium disulphide. The most rapid of these methods is to measure the thermoelectric e.m.f. (Seebeck coefficient). A pellet of compressed particles is subjected to a temperature differential of 10-15°C across one of its faces and the e.m.f. is measured using a high impedance potentiometer. The relationship between stoichiometry and thermoelectric e.m.f. is shown in Fig. 2. Material suitable for use in battery cathodes has a Seebeck coefficient in the range 130-150µV/K. Values in excess of 250µV/K indicate the probable presence of some titanium trisulphide, which has a coefficient of 510µV/K.

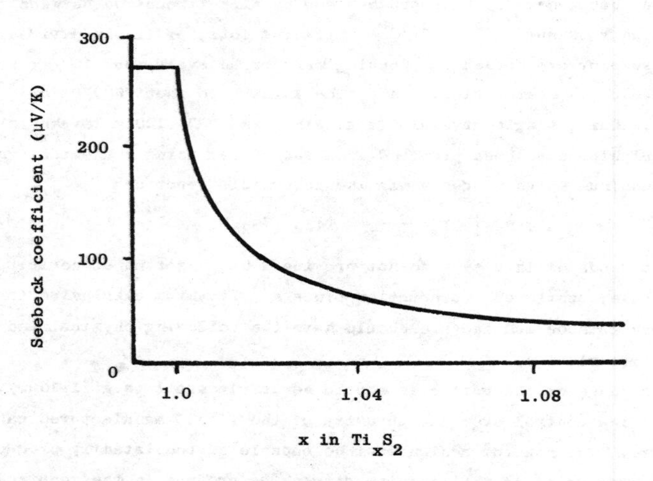


Fig. 2 Relationship between Seebeck Coefficient and Composition for $Ti_xS_2$. (Ref. 16).

The rate of intercalation of anhydrous, liquefied ammonia at 20°C into titanium disulphide decreases rapidly as the amount of interlayer titanium in the crystal lattice increases[16]. A more precise, but allied, measurement of stoichiometry is to use X-ray diffraction techniques to determine the extent to which a sample of titanium disulphide has been intercalated by pyridine at a temperature of 150°C. A detectable degree of intercalation by pyridine occurs only over the stoichiometry range $TiS_2$ - $Ti_{1.005}S_2$.

A continuous process[19,20] has been developed and operated at pilot scale, producing a high yield of titanium disulphide which is satisfactory for use in electric batteries. Titanium tetrachloride vapour and dry, oxygen-free hydrogen sulphide are separately pre-heated before being fed to a tubular reactor maintained at a temperature of approximately 500°C. Titanium disulphide particles are separated from the gaseous reaction products, cooled under nitrogen and packed. The particles (Fig. 3) comprise orthogonally intersecting hexagonal plates, or plate segments, with a relatively high surface area ($\geqslant 4\ m^2g^{-1}$). The morphology is quite different from that of the flat platelets produced from the reaction between titanium metal and sulphur vapour. The material has a low bulk density (70-100 kg $m^{-3}$) and can be stored for prolonged periods, without decomposition, under dry air or nitrogen at temperatures below 20°C.

The current density of a Li/$TiS_2$ cell ($\approx$10mA $cm^{-2}$) is appreciably lower than that reported for some high-temperature cells (50-250mA $cm^{-2}$). A cell construction involving a multiplicity of thin plates of anode and cathode materials will probably be necessary to achieve a satisfactory energy density (say 100 Wh $kg^{-1}$) required in batteries for large-scale applications such as electric vehicle propulsion and load levelling for electricity generating plants. However, the most serious problem to be solved in developing this type of battery is the selection of a suitable non-aqueous electrolyte. The requirement is for a non-hazardous, solid organic electrolyte having a relatively high conductivity for alkali ions at ambient temperature[21]. In the meantime, small 'button' cells are available, incorporating non-hazardous electrolytes (e.g. lithium perchlorate in propylene carbonate) and are being used quite extensively in watches and pocket calculators.

Titanium disulphide gives a low coefficient of friction ($\mu$ = 0.31) when present as a thin film between steel surfaces and a lower value ($\mu$ = 0.26) when sliding as a solid substance on itself[22]. Corresponding values for molybdenum disulphide are 0.2 and 0.22 respectively. However, titanium disulphide has not been used extensively as a lubricant because it does not adhere strongly to metal surfaces. It is not as effective as molybdenum sulphide in replacing metal/metal junctions by sulphide/sulphide junctions of approximately equal area.

The compounds discussed in the remainder of this article are ceramics in the sense that they are inorganic, non-metallic materials having melting or sublimation points in excess of 1500°C. Some ceramics (e.g. carbides; borides) have electrical properties similar to those of metals. At high temperatures, ceramics have important advantages over metals (e.g. greater chemical stability; greater abrasion resistance) but, on the other hand, they are less resistant to thermal and mechanical shock. Inevitably, attempts have

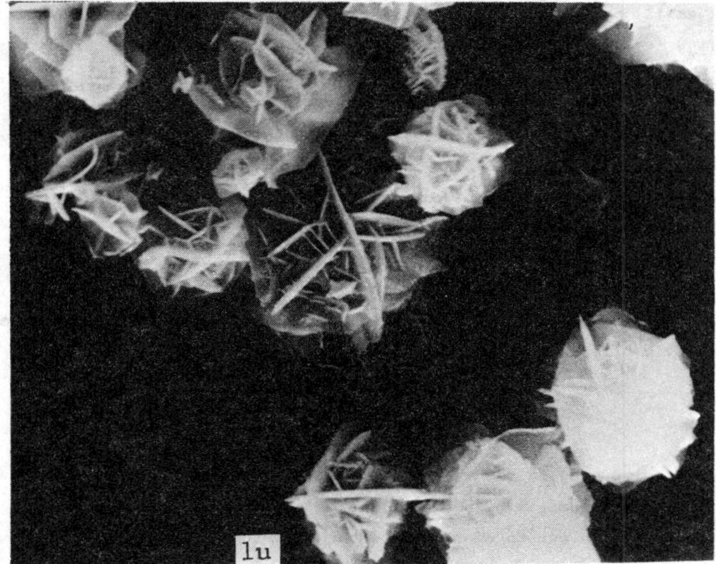


**Fig. 3 Scanning Electron Micrograph of Titanium Disulphide Particles. (Courtesy of R.W. Francis and J.R. Alonzo, Exxon Research and Engineering Co.).**

been made to combine the more desirable characteristics of ceramics and metals in one product. This work led to the development of cermets, a class of materials in which the proportion of ceramic to metal is varied considerably, in a controlled manner, in order to obtain the required properties. The presence of inhomogeneities (e.g. large pores; large areas of different phases) can seriously modify the properties of ceramics. Undoubtedly, the discrepancies which exist in reported values for a physical property of a particular ceramic material frequently arise because of unidentified departures from perfect homogeneity or stoichiometry.

## Barium Titanate

The permittivity of the rutile form of titanium dioxide is much higher than that of most other insulating materials. Single crystals exhibit a marked degree of anisotropy but, in the polycrystalline form, the average value of permittivity is 100-110[23]. Rutile was used in the manufacture of small, thin-walled, tubular, ceramic capacitors during the 1930's. Development work showed that the dielectric properties of rutile could be improved by the incorporation of other oxides, expecially barium oxide, leading to the discovery of the remarkable properties of barium titanate ($BaTiO_3$). This work gained additional impetus as a result of the exacting requirements created during World War II. It was demonstrated that many inorganic compounds were capable of modifying barium titanate to provide a wide range of electrical, temperature and ageing characteristics. Barium titanate derivatives have remained the subject of intensive investigation.

Barium titanate, at temperatures below 1460°C, has a cubic crystal structure which is similar to that of the mineral perovskite ($CaTiO_3$)[24].

It is a ferroelectric material. When such a material is incorporated into a capacitor, the charge on the condenser increases non-linearly with increase in applied voltage. The charge does not fall to zero when the voltage is removed[25]. There is some remanent polarisation due to alignment of electric dipoles within each of the multiplicity of small domains within the material. This does not occur with non-ferroelectrics.

At a frequency of 1 kHz, the dielectric constant is about 2,000 at room temperature. It decreases slightly as the temperature is increased to about 70°C. Permittivity then increases rapidly with increase in temperature, reaching a value of about 7000 at 120°C. Above this temperature of maximum permittivity (Curie point), barium titanate is non-ferroelectric and the permittivity decreases rapidly with increase in temperature[25]. The value of dielectric constant shows a marked dependence on frequency; e.g. for a temperature below the Curie point, at a frequency of 3 GHz, the value is less than half of that at a frequency of 1 kHz[24].

The main effect of applying a strong D.C. field to barium titanate is to 'pole' the material[24]. This polarisation imparts permanent piezoelectric properties. When a piezoelectric material is subjected to a mechanical strain, a change in polarisation occurs and electric charges appear on opposite faces of the specimen. The polarity of the charge depends on the direction of the mechanical deformation. The inverse piezoelectric effect can also be observed, a mechanical strain being set up in a specimen subjected to an electric field. The piezoelectric properties are degraded when the ceramic is heated to just below the Curie point and disappear at temperatures above the Curie point. Barium titanate was the first ceramic piezoelectric material to be produced.

The use of barium titanate in electronic devices has been limited for several reasons. There is an extensive list of Patents and scientific papers relating to modifications of the chemical composition of barium titanate ceramic which produce improved electrical properties[24]. The rather high power factor of barium titanate makes the material unsuitable for use in tuned circuits. The power factor can be reduced by incorporating compounds which promote the formation of very small crystallites during the calcination stage of manufacture or which reduce the freedom of movement of the domains under the influence of an applied electrical field[23].

The rapid change of permittivity near the Curie point would normally preclude the use of barium titanate capacitors at a temperature above about $100^{o}C$. The temperature dependence of permittivity has been reduced by adding small amounts of compounds (e.g. tantalum oxide; bismuth titanate; zirconates) and by reducing crystallite size[26].

Other additives are employed primarily to change the Curie point, although the maximum value of permittivity and the variation of permittivity with temperature may also be modified. Strontium titanate reduces the Curie point whereas lead titanate increases it. Each of these compounds has a perovskite structure and forms solid solutions with barium titanate[23].

Barium titanate of high purity and the correct stoichiometry can be prepared by a variety of co-precipitation methods using solutions of pure chemicals in precisely controlled amounts. Hydrolysis of organic tetravalent titanium esters in aqueous, alkaline solutions in the presence of barium ions can produce barium titanate powder of almost perfect stoichiometry and small crystallite size but it is difficult to control both stoichiometry and size[27]. Barium titanate, with a $BaO:TiO_2$ mole ratio of 1.02:1, has been prepared by heating barium titanyl oxalate in air, using a carefully controlled temperature-time cycle to ensure complete decomposition of each of the chemical intermediates which are formed during the pyrolysis process[27]. Most of the processes based on co-precipitation involve a final step of calcining a precipitate and, frequently, problems associated with sintering of the product are encountered. Barium titanate made by these processes is more costly than that prepared by the commercial process involving the reaction between titanium dioxide and barium carbonate.

The large number of variables associated with the commercial process accounts for much of the difficulty experienced in obtaining reproducible properties in barium titanate-based ceramics. Carefully controlled quantities of commercially pure barium carbonate and anatase, plus selected additives, are intimately mixed. This is a key step in the process[24]. The electrical properties of the final ceramic product are adversely affected by lack of homogeneity. Milling and mixing operations carried out in later stages of the process generally are unable completely to compensate for inadequate mixing of the raw materials.

The mixture is calcined at a temperature of approximately 1350°C. The heterogeneous reaction between titanium dioxide and barium carbonate to produce barium titanate is complex, involving the formation of several intermediate compounds, including $Ba_2TiO_4$ and $BaTi_3O_7$. It is essential that the chemical reaction proceeds to completion, eliminating such intermediates. In particular, the presence of $Ba_2TiO_4$ in the calcined product must be avoided, since it is hygroscopic and decomposes, with swelling, in moist air[24]. This would produce a large decrease in permittivity, followed by crazing or cracking of the ceramic.

Calcination temperatures should be kept below 1460°C. A stable hexagonal form of barium titanate, which has no piezoelectric properties, exists at higher temperatures and this crystalline form can persist, metastably, at room temperature, although the transformation from hexagonal to cubic structure generally occurs as barium titanate is cooled below 1460°C[24]. The hexagonal structure can also be formed if the material is calcined in the presence of reducing agents or in an insufficient supply of oxygen. Furthermore, under the latter condition there is a tendency for the undesirable reduction of $Ti^{4+}$ to occur.

The presence of excess barium ions in the calcined product leads to a reduction in grain size. The presence of excess titanium ions promotes the formation of larger grains and a product which is more readily sintered[24]. The calcined material is milled to yield a product, sized 1-10μm, which is formed into ceramic shapes by a variety of techniques including extrusion, isostatic pressing and centrifugal casting[23]. Organic compounds are frequently used as binding agents. They are burned out during the subsequent firing stage and do not influence the chemical composition of the ceramic.

The firing is carried out at a temperature of 1350°-1450°C. It is important that the furnace atmosphere should be controlled to ensure the elimination of sulphur, which can cause low density in the fired ceramic body. The product should have a density of at least 95 per cent. of the theoretical value ($6.02\ g/cm^3$). Generally, barium titanate produced on an industrial scale has a purity of about 98 per cent.[24].

A future trend in the field of electro-ceramics will be the development of materials which can be fired at temperatures below 1000°C, thereby reducing fuel consumption.

The original barium titanate-based capacitors were tubular. During and immediately following World War II, processes were developed for producing a thin sheet of green ceramic, which was required for disc and multilayer capacitors[28]. Continuous refinement of tape casting techniques reduced the thickness of the fired ceramic product but it also made more stringent the specifications for the raw materials used in the production of titanate capacitors.

The major impetus to the growth of barium titanate-based materials was the use of high permittivity disc capacitors for applications where a rather poor power factor can be tolerated; e.g. as decoupling and smoothing capacitors for the H.T. supplies of radio and television sets and for suppressing radio interference from electrical appliances[23]. Colour television, with three major picture circuits rather than one, has been a particularly important application. Millions of capacitors have been required for computers, electronic calculators and for control and monitoring devices developed for industrial use[28].

The most important piezoelectric ceramics are lead zirconate titanate and barium titanate in which a small amount of lead titanate has been incorporated. Piezoelectric devices are used in the field of acoustics (preferably ultra-sonics). They are not suitable for low frequency (e.g. 50 Hz) applications[24]. The more important applications include record player pick-up elements; accelerometers; smoke detectors; seat belt warning devices; delay line transducers; ultrasonic cleaning transducers; underwater detection devices (hydrophones, sonar and echo sounding systems).

## Titanium Carbide

Commercial-scale production of titanium carbide began in 1938[29]. Briquettes, made from a mixture of titanium dioxide, coke and partially reacted material from previous production cycles, were fed into an arc furnace. At the end of a 30-35 hour cycle, the furnace was allowed to cool. The outer layer of partially reacted material was removed for subsequent incorporation in the briquettes. A central core of titanium carbide, containing about 6 per cent. of free carbon, was re-processed yielding a product containing less than 0.5 per cent. free carbon.

In a development of this process, titanium dioxide and excess carbon are heated with metallic iron in an arc furnace to a temperature of approximately 3000°C[29]. After cooling, crystalline titanium carbide is removed from the solidified iron by mechanical or chemical processes. The titanium carbide is of high purity and stoichiometry, containing little iron and free carbon.

Small deviations from stoichiometry and structural defects are frequently encountered in commercially-produced material. These deviations have a considerable influence on the chemical, physical and thermodynamic properties of the product[30]. Titanium carbide has a high melting point (reported values in the range 3050°-3150°C). It is chemically unreactive at room temperature, being slowly attacked only by very concentrated acid solutions. It oxidises readily in air at high temperatures, forming titanium dioxide[30]. Titanium carbide has good thermal conductivity but its resistance to thermal shock is relatively poor. The thermal shock resistance can be improved by reacting the carbide with graphite in an arc fusion process, followed by casting the melt to form the required shape[31].

The extreme hardness of titanium carbide at room temperature can not be fully utilised because of the brittleness of the material. Surface flaws and internal pores act as crack initiation sites. Insufficient plastic flow occurs at the tip of a crack to absorb enough energy to halt the propagation of the crack. Titanium carbide deforms plastically at temperatures above about 1000°C. Self-bonded carbides have great strengths up to about 1800°C and can be used as high temperature structural materials[30].

Titanium carbide forms the basis of a series of cermets, which have particularly high impact strength and oxidation resistance. These materials are used as cutting tools; wear resistant parts in wire drawing; extrusion and pressing dies; drilling tools used in mining operations; spikes for snow tyres[30].

Cutting tools are made from a mixture of tungsten and titanium carbides bonded in a metal matrix, usually cobalt. This technique circumvents the inability of carbides to self-sinter except at very high temperatures. In addition, the metal increases the toughness of the cutting tool without producing a great reduction in hardness. Cobalt wets the surfaces of the carbide particles sufficiently to ensure a good bonding action but its solubility in the carbides is so low that they remain relatively pure after sintering. Tungsten carbide-cobalt cermet tools wear rapidly as a result of localised welding of the tool with the metal object being cut. Tungsten carbide forms a volatile oxide under such conditions. The surfaces of titanium carbide particles are oxidised to titanium dioxide, which has a low vapour pressure at the temperatures attained in cutting operations and hence protects the tool from rapid wear[30].

Attempts have been made in the United States to develop titanium carbide cermets which will withstand the high centrifugal forces, thermal shock and abrasion to which turbine blades are subjected[23].

## Titanium Diboride

The principal method for making technical grade material (purity >98 per cent.) is based on the carbothermic reduction of titanium dioxide in the presence of boron or boron carbide. The chief impurities in the product are residual carbon and oxides. A second heat treatment is required to reduce the oxygen content of the product to a low level[31].

The electrolysis of titanium dioxide, or of mineral rutile concentrates, dissolved in a mixed electrolyte at $1050^{o}C$, using carbon electrodes or a carbon anode in conjunction with a titanium cathode, produces relatively impure (96-98 per cent.) titanium diboride[32]. The composition of the electrolyte was: $TiO_2$ 3 mole per cent; $Na_2CO_3$ 3; $Na_3AlF_6$ 9; NaCl 72; $Na_2B_4O_7$ 13. The sodium chloride prevented the co-deposition of $Ti_2O_3$ and a mixture of TiC and TiO.

High purity material for research purposes has been made in small quantities by several methods.

A stoichiometric mixture of titanium dioxide and boron was pressed into pellets, placed in fused alumina crucibles and heated in a vacuum furnace at a pressure of $10^{-4}$mm mercury. Complete conversion of titanium dioxide to the diboride was obtained at a temperature of $1700^{o}C$[33]. The oxygen is eliminated in the form of volatile boron monoxide. Although titanium diboride having a composition close to stoichiometric can be produced by this method, it is difficult to avoid an excess of free boron in the product[31].

Other methods include the vapour deposition of titanium diboride from a mixture of boron and titanium tetrachloride in the presence of hydrogen and the direct combination of the elements in induction-heated crucibles or in vacuum arc furnaces[31].

Fabrication of titanium diboride products has been carried out by hot pressing the powder at a temperature of $1400^{o}$-$1650^{o}$, or by cold pressing followed by sintering in an inert or reducing atmosphere[23]. Titanium diboride has also been flame sprayed using a plasma jet. Vacuum arc casting has been used to produce an essentially 100 per cent. dense, single phase titanium diboride. The molten material readily wets the graphite hearths and moulds used in the casting process[31].

Titanium diboride is a very hard, brittle material at room temperature. It has high electrical and thermal conductivities. These properties, coupled with the fact that it is readily wetted by, and only slightly soluble in, molten aluminium, has led investigators to give serious consideration to the use of titanium diboride as electrodes in aluminium reduction cells[31].

Although titanium diboride is chemically inert in many severe environments, it is attacked by molten ferrous metals and alloys and, above a temperature of about 1000°C, its rate of oxidation in air increases rapidly. It is outstanding among high temperature materials for its high strength to weight ratio over the temperature range 1600°-2000°C[34].

The small annual production of titanium diboride is used for the manufacture of vessels for molten beryllium and other metals; for thermowells used in high temperature environments[35]. A considerable amount of work was devoted to its use in the manufacture of rocket nozzles. A titanium diboride cermet with iron powder and boron carbide has been used for surfacing machine parts operating in abrasive conditions. A Russian claim has been reported that a cermet containing titanium diboride, tungsten and cobalt has been used successfully to protect missiles during re-entry into the earth's atmosphere.

## References


1. R. Thompson, "The Modern Inorganic Chemicals Industry", Special Publication No. 31, The Chemical Society, 1977.
2. T.W. Farthing, Proc. Inst. Mech. Eng., 1977, 191, (9), 59.
3. R.J. Dowsing, Metals and Materials, 1979, July/August, 27; October, 31; December, 43.
4. National Materials Advisory Board, 1974, Report No. NMAB-304, PB 232 545.
5. W.W. Minkler, Fourth International Conference on Titanium, Japan Institute of Metals, 1980.
6. G.P. Jones and J. Pearson, Metallurgical Transactions B, 1976, 7B, 223.
7. J. Pearson and M.E.J. Birch, J. Metals, 1979, 31, (11), 27.
8. J. Pearson, Proc. International Symposium Puerto Madryn, Chubuit, Argentina, 21-25 August, 1978, 389.

9. L.H. Gaines, R.W. Francis, G.H. Newman and B.M. Rao, Eleventh Intersociety Energy Conversion Engineering Conference, Statline Nevada, 1976.

10. A.H. Thompson, Phys. Rev. Lett., 1975, 35, 1786.

11. M.S. Whittingham, Chem. Tech., December 1979, 766.

12. J.V. McCanny, J. Phys. C.; Solid State Phys.,1979, 12, 3263.

13. R.R. Chianelli, J. Crystal Growth, 1976, 34, 239.

14. D.A. Winn and B.C.H. Steele, Mat. Res. Bull., 1976, 11, 55.

15. D.A. Winn, J.M. Shemilt and B.C.H. Steele, Mat. Res. Bull., 1976, 11, 559.

16. A.H. Thompson, F.R. Gamble and C.R. Symon, Mat Res. Bull., 1975, 10, 915

17. U.S. Patent 4,007,055; 1977, assigned to Exxon Research and Engineering Co.

18. U.S. Patent 4,069,301; 1978, A.H. Thompson, assigned to Exxon Research and Engineering Co.

19. British Patent Application 50391/76, assigned to Laporte Industries Ltd.

20. British Patent Application 53781/77, assigned to Laporte Industries Ltd.

21. B.C.H. Steele, New Scientist, 7th September, 1978, 705.

22. F.K. McTaggart and A. Moore, Aust. J. of Chem., 1958, 11, (3), 481.

23. B.E. Waye, "Introduction to Technical Ceramics", Maclaren and Sons Ltd., London, 1967.

24. B. Jaffe, W.R. Cook and H. Jaffe, "Piezoelectric Ceramics", Academic Press, London and New York, 1971.

25. M. McQuarrie, Am. Ceram. Bull., 1955, 34, (6), 169.

26. R.J. Lockhart and J. Magder, J. Am. Ceram. Soc., 1966, 49, (6), 299.

27. K. Kiss, J. Magder, M.S. Vukasovich and R.J. Lockhart, J. Am. Ceram. Soc., 1966, 49, (6), 291.

28. D.D. Wheeler, Ceram. Ind. Mag., 1974, 103, (1), 22.

29. J.R. Tinklepaugh and W.B. Crandall, "Cermets", Rheinhold Publishing Corporation, New York, 1960.

30. L.E. Toth, "Transition Metal Carbides and Nitrides", Academic Press, New York and London, 1971.

31. J.E. Hove and W.C. Riley, "Modern Ceramics, Some Principles and Concepts" John Wiley and Sons Inc.,New York, London and Sydney, 1965.

32. J.M. Gomes, K. Uchida and M.M. Wong, U.S.B.M., R.I. 8053, 1975.

33. P. Peshov and G. Bliznakov, J. Less Common Metals, 1968, 14, (1), 23.

34. V. Mandorf, J. Hartwig and E.J. Seldin, Met. Soc. Conf., 1961 (published 1963), 18, 455.

35. A.K. Ganesan, M. Sundaram and V. Aravamuthan, High Temp. Mat., Proc. Symp. Mater. Sci. Res., 1972, 281.

# The Production, Properties, and Uses of Zirconium Chemicals

By F. Farnworth, S. L. Jones, and I. McAlpine
MAGNESIUM ELEKTRON LTD., P.O. BOX 6, SWINTON, MANCHESTER M27 2LS, U.K.

## 1. Introduction

Zirconium chemicals are considered by many people to have few industrial applications and these to be confined in the main to the ceramics industry. In fact zirconium chemicals are used today in a wide range of industries and with advances in modern technology the diversity of their applications is constantly increasing. Far from being laboratory curiosities, zirconium chemicals have been manufactured in commercial quantities for upwards of 30 years and with developments in processing techniques the volume and range of products continues to grow. In terms of raw material supply, zirconium is a fairly common element, ranking twentieth in order of abundance in the earth's crust[1], and is widely distributed, especially as its most abundant mineral, zircon.

## 2. Historical

The mineral zircon has been known to man from ancient times. It is mentioned in both the Old and New Testaments usually under the name hyacinth or jargon. In the eighteenth century colourless zircons were regarded as inferior or imperfect diamonds and were known as Matara Diamonds.

It was not until 1789 that it was recognised that zircon contains a distinguishing oxide. M.H. Klaproth fused zircons from Ceylon with sodium hydroxide, extracted the reaction product with hydrochloric acid, and found the solution to contain an element of novel behaviour[2]. When precipitated by a base, the precipitate would not dissolve in an excess of sodium hydroxide. Klaproth proposed the name 'zirkonerde' for the new oxide.

Some thirty five years elapsed before the element zirconium was extracted from its compounds. In 1824 J.J. Berzelius heated a mixture of potassium metal and potassium fluorozirconate in a closed vessel and obtained a black powder which he regarded as elementary zirconium.[3] This elementary zirconium was very crude, since zirconium is extremely reactive chemically, picking up oxygen and nitrogen avidly. Because of this reactivity it was very nearly 100 years before pure zirconium metal was obtained.[4]

It was not recognised until 1923, when Von Hevesy discovered hafnium by X-ray studies of zircon,[5] that all zirconium compounds found in nature contain a small proportion of hafnium. Hafnium is chemically almost identical to zirconium and for all practical applications of zirconium, other than uses in atomic energy, the hafnium is not separated from the zirconium. Its presence, commonly at a level of about 2% of the zirconium, has the effect of a heavy isotope of zirconium.

The need to produce pure zirconium metal provided the stimulus for most of the developments in the extraction technology of zirconium. Two major outlets which reinforced the need for this work were the incorporation of zirconium in magnesium alloys[6] and the use of hafnium-free zirconium alloy fuel cans for nuclear reactors.[7] The extraction technology developed made possible the production of a range of pure zirconium chemicals at reasonable cost, thereby promoting the use of zirconium chemicals in industry.

The first industrial applications for zirconium chemicals were the uses of zircon ($ZrO_2.SiO_2$) and zirconia ($ZrO_2$) in refractory furnace linings, to withstand the corrosive action of salt vapours, and in the production of opaque vitreous enamels and ceramic glazes. However, these early uses were largely based on the physically beneficiated minerals themselves or crude zirconium oxide. The first major application requiring a pure zirconium oxide, and the one that did most to promote the production of zirconium chemicals, was the discovery in 1948 of the first of the temperature stable glaze resistant zircon ceramic pigments.[8] These zirconium silicate pigments, synthesised from zirconium oxide and silica in the presence of colouring metallic oxides, are widely used today by the ceramic industry.

The first industrial application of a zirconium chemical which made use of its reactivity in an aqueous environment was the introduction, in 1937, of wax emulsions containing dissolved zirconium oxychloride for the waterproofing of textiles.[9] This was the precursor of a range of applications which take advantage of the strong affinity of zirconium for oxygen containing species. Invariably organic polymers, both natural and synthetic, have been involved and have resulted in important applications especially in the paint, paper and leather industries.

Increasing knowledge of the aqueous chemistry of zirconium has been the main reason for the large diversity of applications developed for zirconium chemicals during the past 30 years. New chemicals have been produced with the special properties required for these applications, including most recently special zirconias for the newer developments in ceramics. In this latter context it is interesting how history repeats itself. The earliest use for a zirconium chemical was zircon as a gemstone; today very pure zirconium oxide is being used for the production of synthetic gemstones.

## 3. Raw Materials

Zirconium is found in crystalline rocks (especially in granular limestone) in chlorites and schists, in gneiss, syenite, granite, iron ore beds, pegmatite, sandstone and ferruginous sands. It does not occur as the element, but as the oxide baddeleyite, the double oxide zircon and other more complex oxides.

Zircon ($ZrO_2.SiO_2$) is the most abundant and widely distributed zirconium mineral. It is found as an accessory mineral in igneous rocks. Because of its high specific gravity and resistance to chemical decomposition and erosion, it has been concentrated with other heavy minerals such as rutile, ilmenite, monazite and garnet in river and beach sands. Commercially important deposits are found in Australia (East and West coast), the USA (Florida), South Africa (Richards Bay), India, Sri Lanka, Malaya, Brazil, the U.S.S.R. and China.

The zircon is mined as a co-product with rutile, ilmenite and monazite and is separated by electrostatic, magnetic and gravitational techniques. The degree of purity depends on the efficiency of ore dressing but is usually 99% $ZrO_2.SiO_2$ with the main contaminants being titanium and iron. The main zircon deposits and production estimates[10] for 1978 are shown in Table 1.

Table 1. Zircon Production in 1978 ('000 tonnes)

| | |
|---|---|
| Australia | 395 |
| U.S.A. | 85 |
| South Africa | 65 |
| India | 15 |
| Sri Lanka | 8 |
| Others | 2 |
| Total | 570 |

The largest proportion of this zircon production goes directly into foundry applications, such as refractory bricks and crucibles (Table 2). Zircon is, however, the most important ore for production of zirconium chemicals, because of its availability and price.

Table 2. Consumption of Zircon by Industry 1978 ('000 tonnes)

| | |
|---|---|
| Foundry | 175 |
| Refractory | 160 |
| Ceramic | 80 |
| Abrasive | 25 |
| Zirconium Metal, Chemicals and Others | 40 |

Baddeleyite ($ZrO_2$) is the second most important zirconium mineral. It is found in both primary and residual deposits in areas such as Brazil (Sao Paulo and Minas Geraes) and South Africa (Phalabora). The ore in Brazil, which was the main source of baddeleyite for many years, is in fact an impure baddeleyite, known as zirkite, contaminated with silica. Within the last decade South Africa has become the main source of baddeleyite with production figures of 11,000 tonnes in 1976. At Phalabora virtually silica free baddeleyite is obtained as a by-product, along with uranium, from copper and phosphate extraction. Mineral dressing and chemical treatment yield baddeleyite of 97% to greater than 99% purity.

Baddeleyite is used widely as a direct source of zirconium oxide, but its properties are limited to those of the naturally occurring mineral. If variations in properties are required chemically precipitated forms of zirconium oxide need to be produced for which baddeleyite can be used as a raw material.

## 4. Extraction of Zirconium from its Ores

There are a number of processes whereby zirconium can be extracted from its ores.[11] However, the choice of process is very much dependent upon (a) the purity, properties and range of chemicals required (b) the waste products and their method of disposal and (c) the cost.

In discussing the extraction of zirconium from its ores we need to consider firstly the decomposition of the ore, which can be by thermal or chemical means, secondly the treatment of the decomposition products and finally the isolation of substantially pure zirconium compounds. The ore most commonly used is zircon, where the main concern is removal of the silica, but routes via baddeleyite are also discussed. Emphasis is placed on those routes which are used commercially.

4.1. Thermal Decomposition of Zircon. Zircon can be completely dissociated by heating at temperatures in excess of 1800°C. If the products are slowly cooled reassociation occurs to form zircon and so rapid cooling is required if the zircon is to remain dissociated. Processes based on conventional furnace techniques, including the addition of small amounts of $Na_2O$, have been described,[12] but only since the advent of the plasma arc furnace has the thermal dissociation of zircon been a viable commercial process. In the plasma arc process[13] zircon sand is injected into the plasma arc where it is melted and dissociated into zirconia and silica. On leaving the plasma flame the molten particles assume a spherical shape and are cooled rapidly as they pass down the furnace chamber. Solidification occurs with the formation of extremely small zirconia crystallites in an amorphous silica matrix.[14]

Zircon can also be decomposed by heating in an electric arc furnace with carbon at above 2000°C.[15] Silicon monoxide is volatilized and reoxidised to silica at the furnace mouth. The zircon is converted to 'zirconium cyanonitride' which forms a golden yellow metallic plug in the furnace shell. This process was one of the earliest commercial decomposition routes leading to the production of zirconium metal via zirconium tetrachloride.

4.2. Chemical Decomposition of Zircon and Baddeleyite. Zircon can be chlorinated in the presence of carbon to produce zirconium tetrachloride and silicon tetrachloride.

$$ZrO_2.SiO_2 + 4Cl_2 + 4C \longrightarrow ZrCl_4 + SiCl_4 + 4CO$$

The zircon sand is milled, pelletized with coke and chlorinated in a fluidized bed or shaft furnace at 800-1200°C. $ZrCl_4$ is collected in the primary condensers at 150-180°C and the $SiCl_4$ in the secondary condensers at -10°C.[16]

Sintering with alkali or alkaline earth oxides has been employed widely for the decomposition of zircon and baddeleyite. Sodium hydroxide reacts with zircon at temperatures of around 600°C to produce a mixture of sodium zirconate, sodium silicate and sodium zirconium silicate.[17] Careful control of the zircon:NaOH ratio and reaction conditions can produce almost complete conversion to sodium zirconate and sodium silicate.[18]

$$ZrO_2.SiO_2 + 6NaOH \longrightarrow Na_2ZrO_3 + Na_4SiO_4 + 3H_2O$$

Sodium carbonate requires a higher reaction temperature, about 1000°C, in order to decompose zircon. At mole ratios of 1:1 sodium zirconium silicate is formed

$$ZrO_2.SiO_2 + Na_2CO_3 \longrightarrow Na_2ZrSiO_5 + CO_2$$

whereas at higher ratios sodium zirconate and sodium silicate are also produced.[19]

$$Na_2ZrSiO_5 + Na_2CO_3 \longrightarrow Na_2ZrO_3 + Na_2SiO_3 + CO_2$$

Lime or dolomite can be reacted with zircon to produce calcium zirconylsilicate, calcium zirconate and calcium silicate or mixtures of zirconium dioxide and calcium or magnesium silicate.[20,21] The end products depend upon the ratio of calcium oxide to zircon and the temperature e.g.

$$CaO + ZrO_2.SiO_2 \xrightarrow{1100^\circ C} CaZrSiO_5$$

$$2CaO + ZrO_2.SiO_2 \xrightarrow{1600^\circ C} ZrO_2 + Ca_2SiO_4$$

Zircon can also be decomposed by reaction with silicofluoride to form fluorozirconates and silica.[22]

$$K_2SiF_6 + ZrO_2.SiO_2 \xrightarrow{600^\circ C} K_2ZrF_6 + 2SiO_2$$

A number of methods for the chemical decomposition of baddeleyite are quoted in the literature but many refer to the Brazilian ore zirkite. South African baddeleyite is very unreactive and attacked only slowly by acids and acid salts and this is used to advantage as a means of upgrading baddeleyite by removal of impurities.[23] Nevertheless, dissolution of baddeleyite by sulphuric acid is used commercially, the product being zirconium sulphate tetrahydrate, $Zr(SO_4)_2.4H_2O$, which is crystallised from the extract. Baddeleyite can also be chlorinated, as for zircon, to produce zirconium tetrachloride or decomposed with alkali and alkaline earths to produce sodium or calcium zirconate. Russian workers claim that baddeleyite can also be attacked by ammonium sulphate at 500°C.[24]

## 4.3. Treatment of Zircon and Baddeleyite Decomposition Products.

The zirconium containing products obtained from the decomposition of zircon can either be treated by physical or chemical means, to remove impurities, and then sold directly or further treated to give eventually acidic solutions from which pure zirconium chemicals can be produced.

The primary aim of the plasma arc route for dissociating zircon is the production of zirconium oxide. The reaction product is, therefore, treated with boiling aqueous caustic soda to remove the silica as sodium metasilicate, leaving partly porous spheres of zirconia containing low levels of silica.[13] The resulting zirconia has an unusual physical form derived from its method of manufacture. The process is environmentally acceptable since sodium metasilicate is a useful by-product.

The 'zirconium cyanonitride' resulting from decomposition with carbon can be roasted in air to give a low purity zirconia for use in refractories. The advantage of this method of producing zirconia is that no wet processing is involved, although the environmental problems of containing and extracting airborne silica cannot be neglected. Alternatively the 'zirconium cyanonitride' can be chlorinated at 400°C to produce $ZrCl_4$.[25] This process was used for many years for the production of $ZrCl_4$, but has been increasingly replaced by the direct chlorination of zircon in spite of the higher temperatures required.

The direct chlorination of zircon produces zirconium tetrachloride and silicon tetrachloride, both useful raw materials. Zirconium tetrachloride is contaminated by $FeCl_3$, $TiCl_4$, $SiCl_4$ and $AlCl_3$ and is usually treated with water to produce zirconium oxychloride solution, from which pure zirconium chemicals can be produced (see later). Alternatively the zirconium tetrachloride can be resublimed and used to produce zirconium metal[25] or zirconium compounds, such as the alkoxides.

Sintering with alkalis is most commonly used where the primary aim is the production of zirconium chemicals. In these processes the silica is leached from the decomposition products, to a greater or lesser extent, by washing with water or insolubilised by treatment with acids. Where the decomposition products contain sodium silicate, this can be removed by treatment with water, which also hydrolyses any sodium zirconate to a crude hydrous zirconia. This crude hydrous zirconia is then dissolved in hydrochloric or sulphuric acid for further purification. Where the decomposition product is sodium zirconium silicate, treatment with hydrochloric or sulphuric acid is used to dissolve the zirconium content and insolubilise the silica. Various modifications of the acid treatment to remove most of the silica in a filterable form have been used.[26, 27] The main problem with the alkali sintering route is the disposal of the silicaceous effluent present as either an alkaline silicate solution or a voluminous silica cake.

The methods for treatment of zircon with alkaline earth oxides can be classified according to whether after decomposition the zirconia is combined as calcium zirconate/calcium zirconylsilicate or occurs as discrete zirconia crystals. The main process used industrially is that leading directly to zirconium oxide. The decomposition product is leached with hydrochloric acid to remove the calcium silicate and the remaining crystalline zirconia is dried for sale.[28] The quality of the zirconia depends upon the control of the decomposition reaction, with the main impurity being silica. For the production of zirconium chemicals, the routes leading to calcium zirconate or calcium zirconylsilicate can be used. Methods based on physical separation of calcium zirconate and calcium silicate, followed by dissolution of the calcium zirconate in acid,[21] or direct attack of the decomposition products with acid[29] have been described but have not reached commercial significance.

The use of fluorides to decompose zircon is mainly used for the production of fluorozirconates. Potassium fluorozirconate is leached from the silica with hot water and crystallised from solution on cooling. However, the use of fluorozirconate for the production of zirconium oxide has been described.[11]

Alkali and alkaline earth zirconates produced from baddeleyite have none of the attendant silica problems associated with zircon. However, the higher price and lower availability of baddeleyite have restricted its use as a raw material for production of zirconium chemicals and its main use is as a natural form of zirconium oxide. Sintering of baddeleyite with alkali or alkaline earth oxides produces sodium or calcium zirconate which can be treated with hydrochloric or sulphuric acid to dissolve the zirconium. Chlorination of baddeleyite yields $ZrCl_4$ which can be dissolved in water to produce zirconium oxychloride. The only significant industrial route to zirconium chemicals from baddeleyite is, however, by direct dissolution in sulphuric acid (see section 4.2.).

4.4. Isolation of Substantially Pure Zirconium Chemicals. Zirconium in its mineral forms is associated with variable quantities of impurities such as silica, titanium, iron and aluminium. Removal of these major impurities, and also minor but sometimes troublesome impurities such as phosphorus, is most readily achieved by chemically precipitating pure zirconium compounds from solutions in which the impurities are soluble. Precipitation processes also have the advantage, especially in the case of zirconia production, of yielding solid products whose physical form can be carefully controlled.

The most common outcome of the treatment of the decomposition products of zircon and baddeleyite is a sulphuric or hydrochloric acid solution of zirconium. These solutions make ideal intermediates from which to produce pure zirconium chemicals.

Zirconium oxychloride crystals can be produced from zirconium oxychloride solutions by addition of excess hydrochloric acid.[30] This effects separation from impurities such as titanium, iron and aluminium. However, it is only useful with solutions low in silica, such as those produced from zirconium tetrachloride, since silica tends to be carried down under highly acidic conditions.

Similarly, zirconium compounds such as the mandelate and phthalate can be precipitated in pure form from zirconium oxychloride solution, but the precipitate tends to be difficult to handle and the process more costly than other purification processes.

The most popular route for the production of pure zirconium chemicals is by the precipitation of zirconium basic sulphate from acid solution.[31] The process is usually operated by addition of sulphuric acid or a sulphate, e.g. ammonium sulphate, to zirconium oxychloride solution followed by heating. Metallic impurities remain in solution and are removed, but very little separation from titania is achieved.

The process can also be operated starting from sulphuric acid solutions by removal of part of the sulphate followed by hydrolysis. This is, however, less practicable because conditions for optimum precipitation are more difficult to control.

Finally the crystallisation of zirconium orthosulphate crystals from sulphuric acid solutions can also be used for production of pure zirconium chemicals. This process is the most popular route to production of zirconium chemicals from baddeleyite.

The processes used commercially for the production of pure zirconium chemicals have depended upon a detailed knowledge of the aqueous solution chemistry of zirconium in order to produce handleable and reactive zirconium chemicals. This knowledge has also been essential in understanding and thus expanding the industrial applications of zirconium chemicals.

## 5. Aspects of the Chemistry of Zirconium

A useful review of the chemistry of zirconium has been published by Clearfield;[32] however, it is worthwhile highlighting industrially important features of its aqueous solution chemistry. The preferred oxidation state in aqueous solution is 4 and no redox chemistry is known under these conditions. As to stereochemistry, zirconium displays no stereochemical preferences and, because of the high charge to radius ratio, it exhibits high co-ordination numbers (6, 7 and 8) as well as a great variety of co-ordination polyhedra.[33] A further consequence of the high charge to radius ratio is that the aqueous solution chemistry is characterised by hydrolysis and the presence of polymeric species. Studies of these species are complicated by sensitivity to their environment and the slowness with which equilibrium is attained. In practice this means that good chemical process controls are needed in order to manufacture zirconium chemicals with consistent properties. Although little work has been reported on the size distribution of zirconium species in aqueous solution, industrial experience has shown that this has an important bearing on the performance of zirconium chemicals in certain applications.

As mentioned earlier, the most common outcome in the extraction of zirconium from its ores is an aqueous solution of zirconium oxychloride. Fortunately this system has been extensively studied and it is now widely accepted that in concentrated solutions the species present is a tetramer $\left[Zr_4(OH)_{12}(H_2O)_{12}\right]^{4+}$. In this species Zr lies in a distorted square, linked by pairs of hydroxy bridges, and bound to three water molecules so that the geometry is dodecahedral. Hydrolytic polymerisation of these tetramers can be readily achieved by simply ageing, heating or by a reduction in acidity. The latter when carried out to an extreme brings about the precipitation of zirconium hydroxide at about pH 2. From a manufacturer's point of view the rate at which polymerisation is effected is important in determining whether gelatinous or readily handleable solid zirconium chemicals are produced. Generally speaking randomly structured gelatinous products are produced by the addition of base and more ordered easily handleable chemicals by heating. Where the $Cl^-$ anion is replaced by $NO_3^-$ or $ClO_4^-$ the solution chemistry is qualitatively the same; however, with $SO_4^{2-}$ considerable differences exist, namely the presence of neutral and anionic complexes. This is attributed to the ability of $SO_4^{2-}$ to complex strongly with zirconium. The tendency for inorganic ligands to form complexes can be placed in the series:

$$OH^- > F^- > PO_4^{3-} > CO_3^{2-} > SO_4^{2-} > NO_3^- > Cl^- > ClO_4^-$$

where it is clear that oxygen is an important ligand donor atom in the chemistry of zirconium.

In the production of pure zirconium chemicals, zirconium basic sulphate is of particular interest because of the considerable control over the form in which it is precipitated. Although there is considerable debate as to its structure, it is generally considered to consist of single strands of $\left[Zr(OH)_2^b\right]_n^{2n+}$ joined by bridging sulphates. A typical formula for basic zirconium sulphate is shown in Figure 1.

Figure 1. Basic Zirconium Sulphate.

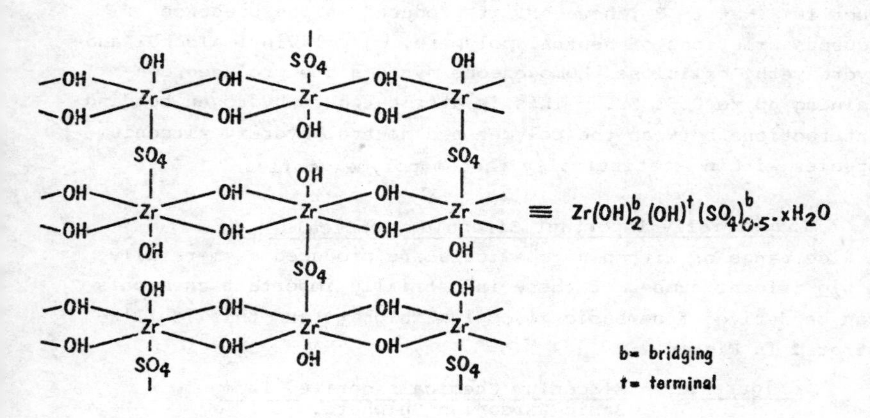


The formation of this material from zirconium oxychloride solution is considered to involve sulphate attack on a tetramer, followed by opening of the tetramer and polymerisation.

Zirconium basic sulphate is particularly useful since it can be used for the production of zirconium hydroxide and basic carbonate. These materials are used respectively for the production of zirconia and a variety of zirconium chemicals and are produced and sold as moist pastes. This is done to minimise the effects of ageing, which if allowed to occur to a significant extent can seriously affect chemical reactivity. The chemical reactions involved in ageing are generally described by the following equation.

$$2Zr\text{-}OH^t \longrightarrow Zr\text{-}O\text{-}Zr \quad + \quad H_2O$$

and in the case of basic zirconium carbonate,

$$Zr\text{-}CO_3\text{-}Zr \longrightarrow Zr\text{-}O\text{-}Zr \quad + \quad CO_2$$

Drying out of the moist pastes facilitates these reactions.

Clearly, there are two features of the aqueous solution chemistry of zirconium which are important in terms of industrial applications. The first is that the solution species are invariably polymeric, and the second that they may be cationic, anionic or neutral. One neutral system which is worthy of special mention is zirconium hydroxide. Zirconium hydroxide is usually regarded as being highly insoluble (solubility product $10^{-54}$ at $19^{o}C$); however, if produced in the presence of aqueous solutions of neutral polymers, eg polyvinyl alcohol and hydroxyethylcellulose, homogeneous systems are produced containing up to 0.5% Zr. This is attributed to hydrogen bonding interactions between the polymer and neutral hydroxy zirconium species within statistically random polymer coils.

## 6. Industrially Important Zirconium Chemicals

A wide range of zirconium chemicals are produced commercially. A significant number of these industrially important chemicals can be derived from basic zirconium sulphate and this is illustrated in Figure 2.

Figure 2. Zirconium Chemicals derived from Basic Zirconium Sulphate.

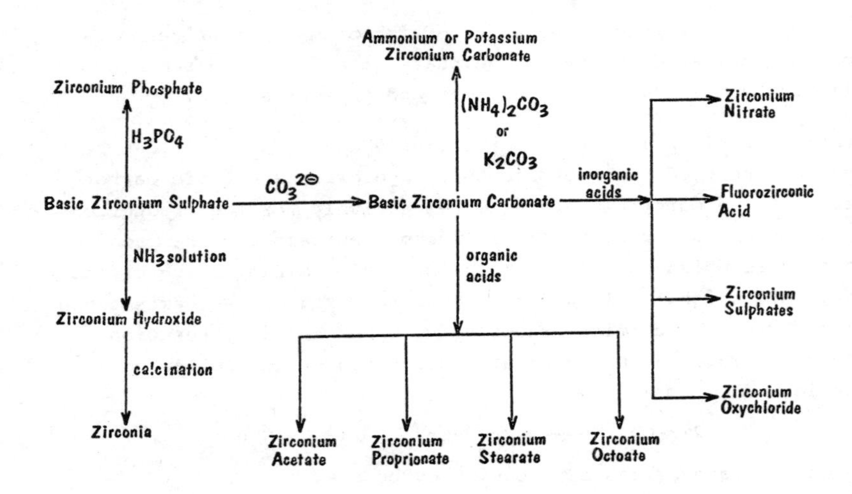

6.1. Zirconium Oxy- and Hydroxy-chlorides and Nitrates. These are usually produced as highly acidic solutions, but structurally well defined solids eg $ZrOCl_2.8H_2O$ are also available. The solutions are easily prepared from zirconium basic carbonate and hydrochloric or nitric acid but the solid $ZrOCl_2.8H_2O$ is normally prepared from $ZrCl_4$.

A useful structural representation of cationic zirconium species in aqueous solution, whether derived from zirconium oxychloride or nitrate is

OH OH OH OH OH Zr⊕ OH Zr⊕ OH Zr⊕

6.2. Zirconium Acetate. This is readily produced from zirconium basic carbonate as a mildly acidic solution having a mole ratio of acetate to zirconium of 2 to 1. It can also be produced as a water soluble solid having the composition $Zr(OH)_{2.5}(OAc)_{1.5}.H_2O$. Unfortunately the position with regard to the nature of zirconium acetate in aqueous solution is far from resolved.

Reaction of zirconium basic carbonate with other organic acids gives products such as zirconium 2-ethylhexanoate, zirconium stearate, zirconium propionate, which are soluble in organic solvents.

6.3. Ammonium and Potassium Zirconium Carbonates. These products are of industrial importance because they are soluble alkaline salts of zirconium. The ammonium salt is produced and sold as a solution but the potassium salt is more conveniently sold as a water soluble solid.[34] Anionic solution species of zirconium are present in these products and can be represented as

OH OH OH OH OH Zr OH Zr OH Zr $OCO_2^{\ominus}$ $OCO_2^{\ominus}$ $OCO_2^{\ominus}$

Where hydrolytic stability of these solutions is of particular importance this can be achieved by partial replacement of carbonate by chelating ligands such as tartrate ions.[35]

6.4. Zirconium Sulphates. Zirconium orthosulphate solution is highly acidic and can be made from zirconium basic sulphate or carbonate. However, the structurally well defined solid $Zr(SO_4)_2.4H_2O$ can be produced directly from baddeleyite (see section 4.2.). Alkali metal double sulphates can also be produced to give slightly higher pH (~2) solution.

Zirconium sulphates exist in solution as anionic species and these can be represented in a similar manner to the previously described soluble carbonates.

6.5. Fluorozirconates. Numerous fluorozirconate salts have been prepared in which 5, 6, 7 or 8 fluorine atoms are present per zirconium atom. Commercially the most important is potassium hexafluorozirconate $K_2ZrF_6$, which is a structurally well defined solid having limited solubility in cold water. The principal species in solution is the anion $ZrF_6^{2-}$. The fluorozirconates can be made by reacting zircon with a silicofluoride (see section 4.2.) or by the reaction of zirconium oxide, hydroxide or basic carbonate with hydrofluoric acid and an alkali fluoride.

6.6. Zirconium Phosphate. This is a very insoluble solid which has ion-exchange properties. As produced on an industrial scale it is a white, free flowing powder which is amorphous to X-rays and has the typical empirical formula

$$Zr(OH)_{0.5}(HPO_4)_{1.75}(H_3PO_4)_{0.07}$$

Structurally well defined phosphates of zirconium[36, 37] can be produced by addition of phosphoric acid to zirconium oxychloride solution and boiling; however, they have not yet achieved commercial importance.

6.7. Zirconium Oxide. Three polymorphs are well established, monoclinic (baddeleyite), tetragonal and cubic. The transition temperature for monoclinic $\longrightarrow$ tetragonal and tetragonal $\longrightarrow$ cubic are respectively 1100°C and 2370°C. These transition temperatures are, however, sensitive to impurities. The oxide is fairly unreactive, but is attacked by hot concentrated sulphuric and hydrofluoric acids. Its high melting point (2700°C), high refractive index (monoclinic 2.15, tetragonal >2.15), low coefficient of thermal expansion (monoclinic $7 \times 10^{-6}$ $°C^{-1}$, tetragonal $13 \times 10^{-6}$ $°C^{-1}$) and high electrical resistivity ($10^{-14}$ ohm cm) account for its use in a wide variety of applications. A sometimes troublesome feature of the monoclinic-tetragonal phase change is the associated volume change; however, the incorporation of a few per cent of calcium , magnesium or yttrium oxide stabilises the cubic form down to room temperature, and this has useful refractory and electrical properties.

A wide range of zirconium oxides are available with different purities and physical properties depending upon their method of preparation. Methods of production of zirconium oxide by the dissociation of zircon, decomposition of zircon in the presence of carbon and calcination, reaction of lime and zircon, and merely physical beneficiation of baddeleyite have been discussed in sections 4.1 and 4.2. These products are widely used in for example refractories, ceramics and abrasives. However, the control over the high temperature properties of zirconia that is possible by calcining zirconium hydroxide, derived from basic zirconium sulphate, is very important for many applications.

Apart from zircon, zirconium oxide is sold in the largest tonnage of any zirconium chemical. However, the tonnage of other reactive zirconium chemicals has increased recently because of a number of new applications. The price of zirconium chemicals, in the main, ranges from around £1000/tonne for some of the zirconias to £5000/tonne for special oxides and chemicals.

## 7. Uses of Zirconium Chemicals

As mentioned earlier zirconium chemicals find use in a wide range of applications and examples of these are shown in Table 3. It is however, convenient for the purpose of discussion to classify them on the following basis:

(a) Applications for zircon

(b) Applications for zirconia

(c) Applications for 'reactive' zirconium chemicals

### 7.1. Zircon Applications.

Zircon ($ZrSiO_4$) has been previously discussed as an important raw material for the preparation of zirconium chemicals; however, it does find considerable use in its own right. Space does not permit us to discuss the applications in any detail; however, a useful review is given by Baldwin.[38] Important industrial applications include: refractories, and here one disadvantage is its limited corrosion resistance under alkaline conditions; general metallurgical uses,

TABLE 3. Uses for Zirconium Chemicals

| Chemical | | Uses |
|---|---|---|
| Zircon | - | refractories (bricks, cements), foundry molds, glaze opacifiers, alkali resistant glass, abrasives. |
| Zirconia | - | refractories, abrasives, ceramic colours (tiles, sanitary ware), piezoelectrics (spark ignitors, sonar devices, ultrasonics), capacitors, pyroelectrics (fire alarms), ceramics(furnace linings, crucibles, nozzles, extrusion dies), ceramic heating elements, thermal barrier coatings, ceramic fibres, solid electrolytes (oxygen sensors, fuel cells), glass (optical glasses, photochromic lenses, glass polishing), gemstones, catalysts. |
| Zirconium oxy & hydroxychloride | - | oil industry (thixotropic cements, clay stabilisation) wool flameproofing, antiperspirants. |
| Zirconium acetate | - | wax emulsion water repellents (textiles, paper), photography (gelatin hardening), refractory binders. |
| Ammonium zirconium carbonate | - | paper (binder insolubiliser), textiles (fungicidal treatment), adhesives (water resistance), floor polishes, emulsion paints, non-wovens, metal conversion coatings. |
| Potassium zirconium carbonate | - | thixotropic paints. |
| Zirconium sulphates | - | leather (tanning), photography, pigment coating (titania). |
| Zirconium phosphate | - | kidney dialysis, ion exchange, catalysts. |
| Potassium fluorozirconates | - | wool flameproofing, metal conversion coatings, metallurgical (Mg and Al alloys). |
| Zirconium basic carbonate and Zirconium basic sulphate | - | zirconium 'soaps' (paint driers), preparation of other zirconium chemicals |
| Zirconium tetrachloride | - | production of zirconium metal, zirconium alkoxides. |

for example in mold compositions and washes, and tundish nozzles; glaze opacifiers; and alkali resistant glass for glass fibre reinforced cement.

7.2. Zirconia Applications. Next to zircon, zirconia ($ZrO_2$) commands the largest market of all zirconium chemicals. As with zircon, the literature on zirconia and its applications is immense and only a general indication of its usefulness can be given here. It should be emphasised that each application for zirconia makes its own specific demands on physical and chemical properties. On the one hand thermally cracked oxides or baddeleyite may be satisfactory, and on the other hand, ultra high purity material derived from zirconium alkoxides may be needed.

7.2.1. Ceramic Colours. The production of ceramic colours is a major industrial use for zirconia. These colours are used for ceramic tiles and sanitary ware, where it is essential to have resistance to high temperatures and to attack from molten glasses such as those used in glazes and enamels. Useful reviews of this subject have been written by Bell,[39] Eppler[40] and Wildblood.[41] In essence these colours are produced by heating at 850-1000°C, under oxidising conditions, an intimate mixture of $ZrO_2$ and $SiO_2$ with $V_2O_5$, $Pr_2O_3$ or $Fe_2O_3$ in the presence of alkali metal halides. Zircon is formed with the colouring metal ion trapped in the lattice, producing the vanadium blue, praseodymium yellow or iron pink stains. The alkali metal halide aids the transport of reactants throughout the mixture and the formation of zircon at low temperatures. The movement of reactants within the mixture at the time of colour formation has been studied by the "marker technique" and this has shown that the reaction site is $ZrO_2$ with vanadium, praseodymium, and silicon species being transported to it in the vapour state. It is strongly suspected that the volatile silicon species is $SiX_4$, generated as a result of the reaction between $SiO_2$ and alkali metal halides, and that zircon formation can be represented by the equation:

$$SiX_4 + ZrO_2 + O_2 \longrightarrow ZrSiO_4 + 2X_2$$

In the case of iron colours, vapour phase transport of iron does not occur; consequently good particle contact between $Fe_2O_3$ and $ZrO_2$ is important. In general the production of iron colours is more difficult than those of vanadium or praseodymium.

Ceramic colours based on zirconia rather than zircon are also known and find use. When a mixture of vanadium pentoxide and zirconia is heated to about 1500°C a yellow colour is produced. If an addition of yttrium oxide is made a deeper yellow shade is produced, whereas if indium oxide is used an orange colour results. These colours are stable, as are the zircon colours, to glazes up to 1350°C. As to the nature of zirconia pigments they are considered to be mordant colours in which vanadium species are attached to the surface of monoclinic zirconia. Only monoclinic zirconia produces these colours.

If colours of consistent quality are to be produced then it is essential that the manufacturer of zirconia carefully controls the chemical purity and physical properties.

7.2.2. Ceramic Materials. Despite its relatively high cost, zirconia finds use as induction furnace linings, as crucibles for molten metals and as a material for the manufacture of metering nozzles used in the continuous casting of steel. These uses depend upon the high temperature stability and chemical inertness of zirconia. However, monoclinic zirconia is not an easy material to use, due to its low thermal conductivity and the volume change, about a 9% contraction, on reaching 1000-1100°C. Further heating leads to the formation of a cubic phase. One way of eliminating the troublesome phase changes, so as to give more or less linear expansion characteristics, is to introduce a few percent of $MgO$, $CaO$, or $Y_2O_3$. This yields the thermally irreversible cubic form, which is known as stabilised cubic zirconia. The nature of stabilised zirconia has been extensively considered by Garvie.[42]

Stabilised zirconia still unfortunately has a high coefficient of linear expansion and this militates against high thermal shock resistance. In practice improved thermal shock resistance is best attained by adding insufficient $CaO$, $MgO$ or $Y_2O_3$ to completely stabilise the zirconia and allowing the material to cool slowly from its sintering temperature so that monoclinic zirconia is precipitated within the grains of stabilised zirconia. Such materials are referred to as partially stabilised zirconias; as well as having high thermal shock resistance, they have high strength. This has led to new uses, for example as dies for the hot extrusion of metals, as metal cutting tools, and as tappet facings in high power diesel engines.[43] Another area of application is as thermal barrier coatings to prolong the life of turbine blades and other engine components and to increase efficiency at high operating temperatures. Impurities can have a profound effect on the performance of stabilised cubic zirconia, especially $SiO_2$ and $TiO_2$ because they combine with MgO and CaO; consequently commercial zirconia needs to be of high purity, typically $ZrO_2$+$HfO_2$ 99.9%, $SiO_2$ < 0.1% and $Fe_2O_3$ < 0.02%.

A new and increasing use[44] of unstabilised zirconia, which takes advantage of the monoclinic-tetragonal phase change, is in improving the fracture toughness of other ceramics, for example those based on alumina ($Al_2O_3$). This is achieved by developing microcracks in the ceramic. These are formed as a result of the volume increase associated with the tetragonal-monoclinic phase change, which occurs as the ceramic body is cooled from the sintering temperature. As well as improving fracture toughness the thermal conductivity of the composite is significantly reduced.

7.2.3. Solid Electrolytes. The introduction of cations such as $Ca^{2+}$, $Mg^{2+}$, and $Y^{3+}$ into the zirconia lattice stabilises the cubic fluorite structure. This results in the formation of anion vacancies, which are necessary to maintain charge neutrality, and accounts for oxygen ion conduction, via the defect structure, at about 1000°C. Divalent cations give solid solutions of formulas $Zr_{1-x}M^{2+}_xO_{2-x}$, and trivalent cations formulas of $Zr_{1-2x}M^{3+}_{2x}O_{2-x}$, for additions of the appropriate oxide in the range 8-16 mole %. An excellent review has been written by Hagenmuller and Van Gool,[45] who describe both the theory and applications for these materials. Interesting applications are as (a) oxygen-sensors to control the composition of gases entering car exhaust catalytic convertors,[46] (b) probes for determining the oxygen content of liquid metals,[47] (c) solid electrolytes in high temperature fuel cells and high temperature water electrolysers and (d) high temperature heating elements in oxidising atmospheres.

7.2.4. Electrical Ceramics.[38] Zirconia is widely used to produce crystalline materials which show the piezoelectric effect. This effect results in the development of an electric charge when the material is stressed and is due to the displacement of atoms within the crystal lattice. The converse also occurs, namely when subjected to a voltage gradient the material undergoes a proportional strain. Important piezoelectric materials are the lead zirconate-titanates $Pb(Zr_xTi_{1-x})O_3$ which have a crystal structure based on perovskite. They are prepared by heating a mixture of PbO, $ZrO_2$ and $TiO_2$ and their formation is controlled by solid state diffusion across inter-particle contacts. It is essential that the reaction goes to completion and this depends on the particle size and calcination temperature, as well as upon the perfection of the initial mixing of reactants. Lead zirconate-titanate electrical ceramics find use as sonar devices, strain gauges, ultrasonic cleaners and spark ignitors.

The alkaline earth zirconates also find application in electrical ceramics, and are produced from the reaction of zirconia with alkaline earth oxides or carbonates. In the preparation of ceramic capacitor bodies the alkaline earth zirconates can comprise 5-20% of the dielectric composition, the remainder being barium titanate. Their function is to modify the dielectric properties of barium titanate. As yet no significant applications for the alkaline earth zirconates in their own right exist.

7.2.5. Catalysts. Although the use of zirconia is small in comparison to alumina, interest in zirconia is growing, for example where the catalyst system may have to operate at high temperatures. Under such circumstances alumina may be unsatisfactory because of reactions with the catalyst to form, for example, a spinel. The use of zirconia can avoid this problem.[48]

Where zirconia is stabilised it has recently been found that the behaviour of heterogeneous catalysts can be greatly modified. The decomposition of nitric oxide to nitrogen and oxygen is receiving considerable attention in the USA.[49] When platinum metal is employed to catalyse the dissociation of nitric oxide difficulties arise because of the inhibiting effect of oxygen. The use of scandia stabilised zirconia has been suggested as a means of 'pumping'oxygen away from the catalyst surface so as to enhance the rate of nitric oxide decomposition. Investigation of the system has shown a considerable enhancement in the rate of decomposition, and this is believed to involve the zirconia surface itself. It is suggested that F-centres on the zirconia surface, formed by the applications of a potential, are responsible for the enhanced catalysis.

7.2.6. Abrasives, Glass and Gemstones. Zirconia is used in the abrasives industry, usually with alumina, for high quality grinding wheels. The ability of zirconia to modify the toughness of alumina and reduce its thermal conductivity accounts for its use. Zirconia is also used for polishing glass as an alternative to ceria.

The glass industry uses zirconia in the manufacture of special optical glasses and this is an important area for high purity zirconia. High refractive index and low absorption in the visible part of the spectrum are the properties of zirconia which are essential for this application. Other transition elements such as iron and chromium need to be kept at very low levels, and hence chemically precipitated grades of zirconia are required.

A new application for zirconia, also requiring high purity, is in the production of synthetic gemstones. Cubic stabilised zirconia gemstones are being produced which are very similar in properties to diamonds. Consequently they are being used in costume jewellery as a substitute for diamonds.

7.3. 'Reactive' zirconium chemicals applications. Although zircon and zirconia make up the largest tonnage applications for zirconium chemicals the greatest diversity of applications is produced by the other, in the main termed 'reactive', zirconium chemicals (see Table 3). Space does not permit us to cover each of the application areas in depth; therefore, wherever possible every effort has been made to link the applications on the basis of a common chemistry.

7.3.1. Applications involving zirconium chemicals and organic polymers. Proteinaceous substances. In recent years the increasing emphasis on safety in the home and at the workplace has led to an interest in flame-resist treatments for wool - a fibre generally regarded as being of low flammability. A well established flame-resist treatment for wool is based on the use of potassium hexafluorozirconate ($K_2ZrF_6$) under aqueous acidic conditions. This treatment is particularly useful in that it is durable to washing and does not impart any colour to the wool, unlike titanium and chromium treatments. Whilst the precise way in which $K_2ZrF_6$ brings about the improvement in flame-resist properties is not known, it is generally considered to arise from activity in the solid state.[50] The initial reaction with $K_2ZrF_6$ is thought to involve the disruption of salt cross-linkages between acidic and basic side groups on polypeptide chains (see Figure 3).

Figure 3.

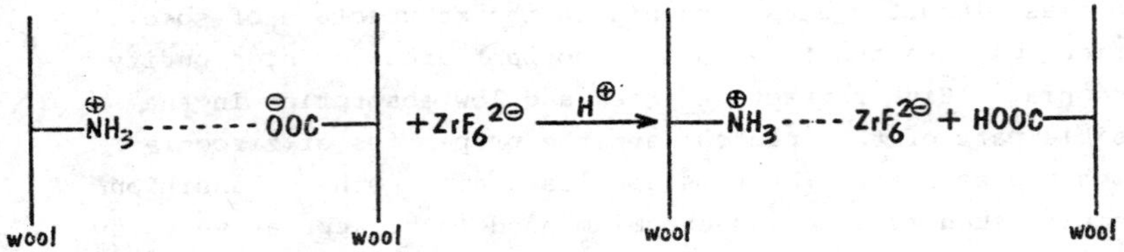

$$\text{wool}-\overset{\oplus}{N}H_3 \cdots\cdots \overset{\ominus}{O}OC-\text{wool} + ZrF_6^{2\ominus} \xrightarrow{H^{\oplus}} \text{wool}-\overset{\oplus}{N}H_3 \cdots ZrF_6^{2\ominus} + HOOC-\text{wool}$$

When the treated wool is subsequently washed, further chemical reactions take place which result in the reduction of the fluorine to zirconium mole ratio from about 5 to 0.2 after 20 washes. The loss of fluoride leads to the formation of a complex with wool carboxyl groups, and this accounts for maintained performance of the flame resist effect.[51]

The tanning of raw animal hides to produce leather is associated with the formation of cross-linkages between polypeptide chains making up collagen fibres, which in turn make up the bulk of the fibrous portion of animal hides. An important traditional method of tanning uses chromium chemicals under highly acidic conditions and it is thought that the properties associated with leather stem from the ability of Cr(III) to cross-link polypeptide chains through carboxyl groups on reactive side chains. Zirconium oxychloride, which produces polycationic species under acidic conditions, can theoretically replace chromium chemicals; however, in practice the hydrothermal stability of the treated hide, measured by determining the shrinkage temperature, falls short of that desired. Zirconium chemicals are however used to re-tan hides previously treated with Cr(III),[52, 53] to produce special leathers, and this requires that zirconium species react with the remaining reactive side chains containing basic groups. (see Figure 4).

Figure 4.

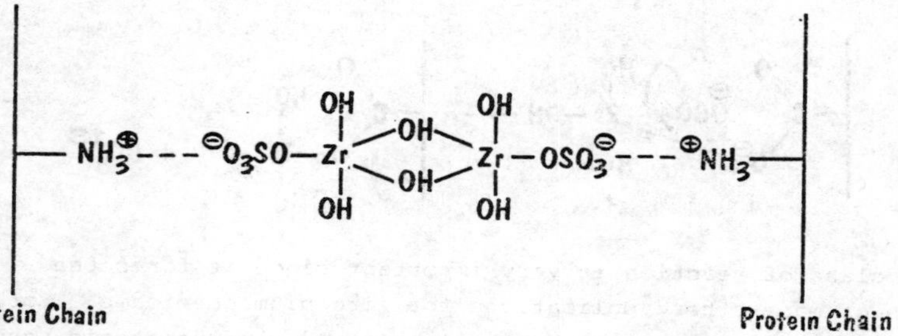
OH
OH
$NH_3^{\oplus}$ - - - $^{\ominus}O_3SO$—Zr
OH
OH
Zr—$OSO_3^{\ominus}$ - - - $^{\oplus}NH_3$
OH
OH
Protein Chain
Protein Chain

In order to do this anionic species are required, and it has been found that those derived from zirconium sulphate systems are suitable and produce leathers with improved tightness of grain and uniformity. Zirconium sulphate systems are traditionally used instead of chromium for tanning of white leathers and, with the concern over the effect of chromium-containing effluents on the environment, zirconium systems generally are receiving renewed interest.

An industrially useful product obtained by the selective hydrolysis of collagen is gelatin, which consists of a mixture of water soluble proteins of high average molecular weight. Aqueous zirconium species react with gelatin solutions by a mechanism similar to that described for leather, but in this instance the effect of the reaction is to produce coagulation of the gelatin or the hardening of films subsequently produced. These effects are important in the photographic industry[54] and the most useful zirconium chemicals are: zirconium oxychloride, zirconium sulphate, zirconium acetate, and the ammonium salt of basic zirconium carbonate. The ability of the last mentioned chemical to produce these effects needs further comment since the polyanionic species obtained in solution exist at a pH greater than 7 and they cannot cross-link protein molecules by the same mechanism as those derived from acidic zirconium sulphate solutions. Here the effect seen is almost certainly due to the ability of protein carboxyl groups to displace co-ordinated carbonate from the anionic zirconium species (see Figure 5).

Figure 5.

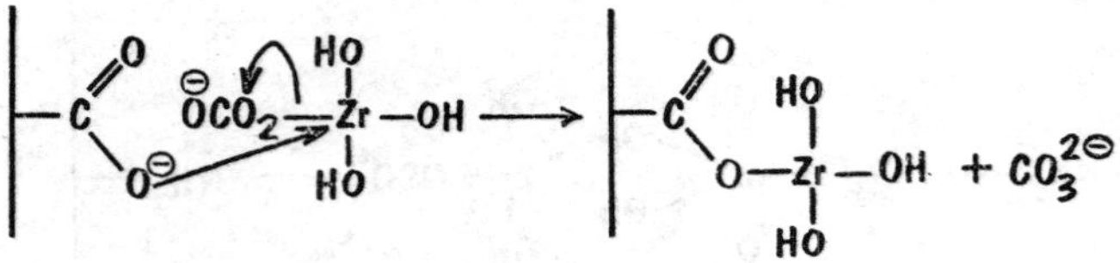

$$-C(=O)O^{\ominus} \;+\; {}^{\ominus}OCO_2-Zr(OH)_3 \longrightarrow -C(=O)O-Zr(OH)_3 + CO_3^{2\ominus}$$

This class of reaction is very important since it forms the basis of many other applications for zirconium chemicals. For example, the ammonium salt of basic zirconium carbonate is used to improve the water resistance of soya protein and casein (a phospho-protein) in paper coating.[55]

The action of zirconium chemicals, most notably the zirconium chlorides, on human skin to produce an antiperspirant effect is well established; however, the precise nature of the chemical reactions involved is unknown. Normally the high acidity of zirconium oxychloride would preclude its use in skin applications; however, this in practice[56] is overcome by using (a) a basic zirconium chloride, eg zirconium hydroxychloride which has a Cl to Zr mole ratio of 1 to 1; or (b) by forming a mixed metal complex with aluminium; or (c) complexing part of the zirconium with a simple amino acid, for example glycine.

Polysaccharides. For many years zirconium chemicals have been employed in conjunction with polysaccharides in the areas of paper coating[57] and in the waterproofing of cotton textiles.[58] More recent developments include bacteriostatic,[59] fungicidal[60] and flame-resist treatments for cotton,[61] and the modification of the rheology of systems containing hydroxyethylcellulose.[62,63]

Paper coating using starch binders is a well established use for the ammonium salt of basic zirconium carbonate. In this application starch is used to bind the pigment coating which is applied to improve the printing properties of the paper. However, starch is sensitive to moisture and requires insolubilisation. This is necessary to prevent removal of the coating when subjected to water in the offset litho printing process.

The ammonium salt of basic zirconium carbonate can insolubilise starch and this is ascribed to the ability of the ammonium ion to form a pseudo cationic starch, which is then ionically cross-linked by basic zirconium carbonate ions as loss of ammonia and carbon dioxide occurs during drying (see Figure 6).

Figure 6.

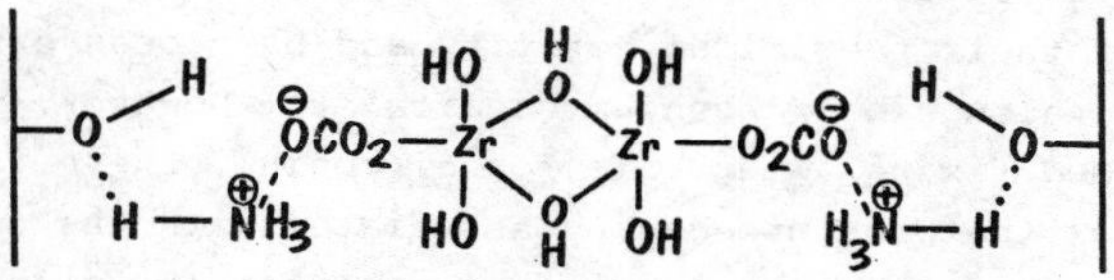

If the starch is chemically modified, for example so as to contain carboxyl groups, then a further improvement in the water resistance of the coating can be achieved by a mechanism similar to that shown in Figure 5. One major advantage of the zirconium system over its rivals, such as urea and melamine-formaldehyde resins, is the rapid speed of insolubilisation, especially under alkaline conditions. Normally the resins condense under acidic conditions in the presence of heat. In addition because the zirconium chemical reacts with the starch rather than self polymerising, the effect is more evenly distributed. Recently, concern over formaldehyde emissions from the resins systems has caused increased interest in zirconium, especially for food packaging applications.

Within the last few years zirconium-starch systems have also found use as binders for woven glass fibre wall coverings. The binder system strengthens the fabric making it easier to handle especially in the presence of water.

When paper or textiles are treated with a solution of zirconium acetate they become water-repellent on drying. This is associated with a hydrogen bonding interaction with cellulose fibres and results in the formation of a hydrophobic surface. Such surfaces are exploited in the waterproofing of heavy duty outdoor cotton textiles with carboxylated waxes. Clearly the ability of zirconium acetate to produce hydrophobic cotton fibres is important; however, this is not the sole reason for the use of zirconium acetate. It is almost certain that carboxyl groups associated with the wax react with zirconium. Evidence for this comes from the fact that the ammonium salt of basic zirconium carbonate can replace zirconium acetate. In this case carboxyl group displacement of carbonate (see figure 5) accounts for the fixation of the wax to the textile.

The role of zirconium chemicals, and in particular that of the ammonium salt of basic zirconium carbonate, in the fungicidal treatment of cotton cloth destined for outdoor use is often claimed to be one of chemically fixing the biologically active component to the cotton. Inorganic copper (II) chemicals are well known for their fungicidal activity and have been extensively studied alongside zirconium chemicals. However, despite the claimed fixing agent role for zirconium, there is good reason to think that the processing associated with the copper-zirconium carbonate treatment results in the precipitation of basic zirconium species within and on the surface of fibres which then act to 'screen' the cloth from photochemical degradation. In this way the loss of copper fungicidal activity, which is known to be due to reaction with the products of the photochemical degradation of cellulose, is reduced. The improved performance achieved with zirconium does not, however, render the treatment wash fast even though a substantial proportion of the zirconium is resistant to loss on washing. It is worth digressing at this point to mention that titanium dioxide pigments are sometimes coated with zirconium chemicals[64] so as to minimise their photochemical activity, thereby, for example, increasing significantly the life-time of paint films.

Zirconium acetate with hydrogen peroxide has recently been evaluated as a bacteriostatic finish for hospital linen. Regular laundering is important and the treatment has been found to be persistent over twenty washing cycles. This is attributed to the formation of peroxo complexes with zirconium which slowly hydrolyse, as a result of the presence of absorbed water in the linen, to release hydrogen peroxide. The washfastness of the treatment also suggests that zirconium acetate is capable of penetrating the infra-structure of cotton cellulose fibres and polymerising.

Aqueous solutions of hydroxyethylcellulose find wide use as a means of altering the rheology of systems into which they are introduced. Further special rheological modifications can sometimes be achieved by the introduction of zirconium chemicals. The most effective way of doing this is to generate zirconium hydroxide in the presence of the polymer (see Figure 7).

Figure 7.

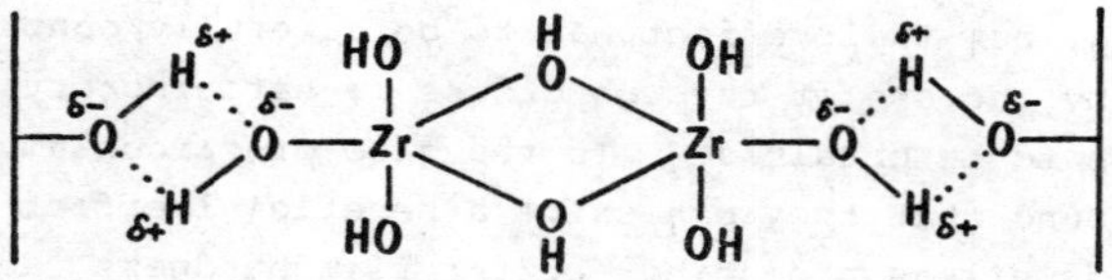

The manner in which this is done is important. For example, when rendering water borne polyvinyl acetate emulsion paints non-drip, it is essential that zirconium hydroxide is generated slowly so that the paint can be readily handled in the factory. In practice this is achieved by the use of the potassium salt of basic zirconium carbonate which slowly hydrolyses in the alkaline environment of the paint. Thixotropic cement, which finds use for sealing fissures encountered in oil drilling, provides another example. Using zirconium oxychloride, zirconium hydroxide is rapidly generated in the presence of the polymer at the time the cement mix is made up with water. Similar effects can be achieved with other hydroxylic polymers which are not necessarily polysaccharides, eg polyethylene glycol and polyvinyl alcohol. In the case of the latter, it is possible to produce clear solutions of polyvinyl alcohol containing significant levels of zirconium hydroxide, which, on drying yield films with improved tensile properties.

Other polymers. The functional groups which have been found to be important in applications involving proteins and polysaccharides will clearly be of relevance in other organic polymer systems.

The paint industry,[65] particularly in the USA, has recently been subjected to legislation affecting architectural paints. This limits the lead content in the paint films to less than 0.06%, and more recently aims to control the level of emission of volatile organic matter to the atmosphere. As a consequence of the lead legislation, formulators of air drying organic solvent alkyd resin paints (an alkyd is in essence a polyester modified by the presence of fatty acid) cannot use lead driers. Replacement of lead driers by zirconium driers has, however, adequately resolved this situation. The formation of an alkyd resin paint film containing an air drying fatty acid involves an oxygen initiated free-radical polymerisation, which is usually catalysed by redox driers based on cobalt or manganese. Redox driers promote rapid surface drying, but their effect has to be carefully controlled and balanced by the use of through-driers if satisfactory film properties are to be obtained. In the case of zirconium driers it has been found that they can exert a beneficial effect on the redox drier as well as producing better film hardness, adhesion and gloss at zirconium concentrations as low as one tenth of the lead addition. The driers in question are essentially metal derivatives of organic acids, eg naphthenic or octanoic acid, which are soluble in solvents such as white spirit. Zirconium driers are readily prepared by treating basic zirconium sulphate or carbonate with the appropriate organic acid in the presence of a suitable organic solvent, eg white spirit. As to the role of the zirconium drier there is little experimental evidence available; however, it seems most likely that it will involve reaction with carboxyl groups, either those already present in the resin or generated during the course of oxidative polymerisation.

The effect of the organic solvent emission legislation has also been significant, and it has been predicted that the market share for conventional solvent industrial coatings will drop from 50% in 1978 to 5% in 1990. Over the same period it is predicted that water borne coatings, which include water reducible alkyds and polyesters, vinyl acetate, vinyl acrylic, and acrylic latices, will increase from 15% to 25%. Whilst there are technical disadvantages associated with the use of water reducible (soluble) alkyds, particularly with regard to viscosity control, they are now used in significant quantities in the USA. Water reducible alkyds, which still require the use of driers, are simply obtained by increasing the carboxyl group content to such a level that they can be rendered water soluble by the use of organic amines. In the long term these systems may themselves be restricted due to concern over the toxicity of the amines employed. Consequently it is not surprising that a great deal of effort is being directed towards developing gloss emulsion paints. As yet no breakthrough has occurred, and this in itself may not be due to an insufficient level of gloss being obtained, but rather because of deficiencies in other properties associated with gloss, namely, body, washability and durability. Where the polymers forming these emulsions contain carboxyl groups, zirconium chemicals can contribute towards improved solvent and wet abrasion resistance. This requires the use of a zirconium chemical which will not react until the time of application. The ammonium salt of basic zirconium carbonate, where part of the carbonate has been replaced with tartrate ion to ensure stability prior to film formation, has been found most useful in this connection, more so if the film forming process is carried out at temperatures in excess of ambient. The important reaction, which results in cross-linking of the polymer, involves displacement of carbonate by carboxyl groups. Polyacrylic and polyvinyl acetate emulsions also find wide use in adhesives, printing ink and floor polish formulations, and as binders for non-woven textiles. Not surprisingly zirconium chemicals are being considered as a means of improving performance or polymer cost-effectiveness. In each case the choice of zirconium chemical is governed principally by the nature of the functional groups on the polymer, the stabilising surfactant and the final pH of the system.

The use of zirconium oxychloride with hydroxyethylcellulose in the oil drilling industry was described earlier. With the increasing depths to which oil wells are being drilled, there is a need to continually improve the temperature stability of the dispersants used in the drilling mud. Lignosulphonates are very effective dispersants and their performance at high temperature has been improved by preparing chromium and iron derivatives. However, environmental concern over the effects of chromium has led to an interest in zirconium, and this is now an area of active development.[66] Enhanced oil recovery has assumed topicality and importance as the likelihood of a world oil shortage looms nearer. Many processes for recovering the large proportion of the original oil-in-place are currently under consideration, one of which is the displacement of crude oil by polymer solutions. Polyacrylamide solutions have received attention and the finding that the reaction of zirconium oxychloride with low levels of residual carboxyl groups alters the rheology, namely to produce viscoelastic systems, is of interest.

7.3.2. Metallurgical applications. Almost all of the zirconium tetrachloride manufactured today is used to produce zirconium metal by the Kroll process. Western world capacity for zirconium metal is at present estimated to be ~ 7,500 tonnes. Zirconium metal has many interesting physical and mechanical properties not least of which are its resistance to hot water and liquid sodium and its transparency to thermal neutrons. This has led to its main application as a protective cladding for uranium fuel elements for use in nuclear reactors.[7] However, in this role it is essential to remove hafnium, which is invariably associated with zirconium in natural sources. This is because, unlike zirconium, hafnium has a very high neutron capture cross-section.

Other applications, not requiring hafnium free material, include its use as shredded foil in camera flash bulbs, as a component of artifical joints and limbs, and as a grain refiner to improve the strength of magnesium and aluminium metals. Magnesium-rare earth - zirconium and magnesium-zinc-zirconium alloys are particularly important in the aircraft and space industries. Instead of using zirconium metal in the production of these alloys, it is possible to generate zirconium in situ by use of alkali fluorozirconates, for example $KZrF_5$. This method of approach forms the basis of making magnesium-zirconium master alloys for use by magnesium alloy producers.[6]

7.3.3. Conversion coatings. Traditionally the appearance, corrosion resistance and paint adhesion properties of an aluminium metal surface has been improved by forming a chromate conversion coating. Growing concern in recent years about the effect on the environment of chromium containing effluents has directed attention towards finding alternatives. One such alternative is based on fluorozirconate species, and here the so-called hexafluorozirconic acid ($H_2ZrF_6$) has been found especially useful. The treatment,[67] which is colourless, is particularly useful for food contact applications, namely beer and beverage cans, where it has been given FDA approval. The polymer coatings used to exploit the properties of the zirconium conversion coatings are usually water soluble or dispersible polyacrylic acids or esters.[68] From what has been discussed earlier in this chapter it would seem reasonable to propose that reaction between polymer carboxyl groups and the aluminium bound zirconium species is important. Whilst little is known about the chemical reactions leading to the fixation of zirconium species on aluminium surfaces, it is suggested that it involves an aluminium-fluorine-zirconium bridge. Diagramatically the protective coating might be represented as:

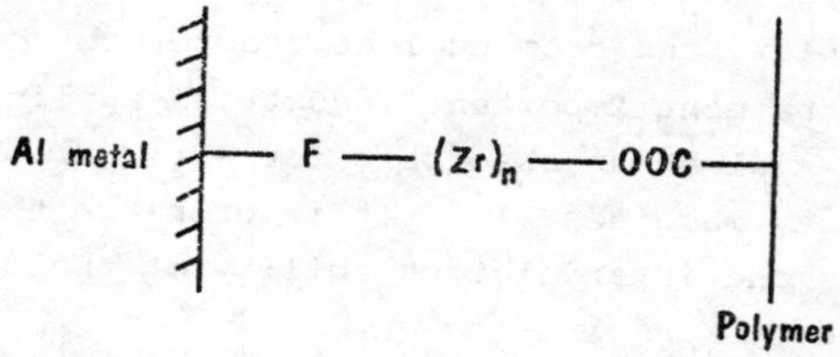

Al metal — F — $(Zr)_n$ — OOC — Polymer

Zirconium chemicals such as zirconium acetate and the ammonium salt of basic zirconium carbonate have also found use as rinse solutions for improving the corrosion resistance and receptivity of phosphated steel[69] towards paints. Knowing how good a ligand phosphate is for zirconium, one can readily envisage reaction between the phosphated steel surface and zirconium species, followed by reaction with, for example, polymer carboxyl groups.

7.3.4. Ion Exchange. Problems associated with the stability of organic ion exchange resins in aqueous solution at high temperatures and in the presence of ionizing radiation brought about an interest in inorganic ion exchange materials in the 1950's.[70] Zirconium phosphate is one such material whose cation exchange properties arise from the replacement of hydrogens on monohydrogen phosphate groups. Whilst the early preparations were amorphous gels, it was soon discovered that refluxing in strong phosphoric acid resulted in the formation of crystalline material, eg $Zr(HPO_4)_2.H_2O$. By altering the preparative procedures several zirconium phosphates with different degrees of crystallinity can be obtained, and this provides a means of controlling properties for specific applications.

The production of an amorphous gel can present manufacturing difficulties; however, this can be overcome by preparing material of controlled particle size by use of the sol-gel method[71] or by reacting basic zirconium sulphate with phosphoric acid.[72, 73] Unlike the amorphous material, cation separations using crystalline forms are not necessarily straightforward. This is because mixtures of ions behave differently to the ions when exchanged individually. For example the addition of sodium chloride to a solution of magnesium chloride and hydroxide allows $Mg^{2+}$ to be exchanged, whereas in the absence of $Na^+$ the exchange does not occur.[74] Whilst there is a considerable need for further study of multi-ion solutions, certain possible applications present themselves, such as the use of the sodium form as a water softener or the use of an amine intercalated form to remove mixtures of polyvalent or complex ions from nuclear reactor corrosion products. Today, the most important industrial application is for the amorphous material as an absorbent for $NH_4^+/NH_3$ in portable recirculatory kidney machines.[75] It is probable that the combined ion exchange and intercalation ability of the material accounts for its use.

Zirconium phosphates are also strong acid catalysts and readily dehydrate or dehydrogenate alcohols. In the transition metal ion form they also exhibit high catalytic activity for hydrogenation, dehydrogenation, dehydrohalogenation and oxidation reactions. Recently a commercial process for the one step preparation of methyl isobutyl ketone has been developed.[76]

Another zirconium chemical which finds use as an ion exchanger is the hydroxide.[70] Depending upon the pH of the system it possesses both cation (alkaline conditions) and anion (acidic conditions) exchange properties.

7.3.5. Soil treatments. During the course of producing oil from an oil well a dramatic reduction in the permeability of a geological formation can result when trying to stimulate further output by means of fracturing or water flooding. The loss of permeability, and hence output, under such circumstances is due to the dispersion of 'sensitive' clays, which then migrate and cause blocking. Where the nature of the geological formation suggests that stimulation procedures might cause a reduction in permeability, it has been found that the incorporation of zirconium oxychloride into the stimulation fluid has a significantly beneficial effect. As a consequence of this discovery by Dowell,[77] the use of zirconium oxychloride for clay stabilisation is now well established. The zirconium chemical is believed to produce the desired effect, due to the enhanced ability of polymeric cations to flocculate clay particles. The stability of polymeric cationic zirconium species under highly acidic conditions is also utilised if it is decided to stimulate the oil well by means of acid. Acidising is frequently employed when sandstone formations are encountered.

## 8. The Future

In many respects zirconium chemicals are in their infancy in that many of the special properties of zirconium are only now becoming of industrial importance. Zirconia in particular can be seen to have the high temperature stability, resistance to corrosion and electrical properties required for tomorrow's ceramics and electronics industries. Other zirconium chemicals are likely to find increased outlets in polymeric coatings, textiles, paper and oil recovery, aided by the strong affinity of zirconium for oxygen containing species, low toxicity and lack of colour. These properties, together with the abundance of zirconium ores and the ready availability of zirconium chemicals make zirconium very much an element with a future.

## REFERENCES


1. R.L. Parker, in 'Data of Geochemistry' 6th Ed., (M. Fleischer, Tech. Ed.), Geol. Survey Professional Paper 440-D, U.S. Government Printing Office, Washington, D.C., 1967, pp D1-D19.
2. M.H. Klaproth, Ann. chim. phys., 1789, 8.
3. J.J. Berzelius, Ann. chim. phys., 1824, 26, 43.
4. A.E. Van Arkel, Z. anorg. chem., 1925, 148, 345.
5. G. von Hevesy, Chem. Rev., 1925, 2, 1.
6. Magnesium Elektron Ltd., Brit. P. 652,227 (1951).
7. H. Etherington, R.C. Dalzell & D.W. Lillie in 'The Metallurgy of Zirconium', (B. Lustman and F. Kerze, Eds.), McGraw-Hill, New York, 1955, p1.
8. C.A.Seabright (to Harshaw Chemical Co.), Brit.P. 625,448(1949)
9. B.I.O.S. Final Reports Nos. 420, 421. Brit P. 517,638 (to I.G. Farben).
10. P.D.H. Gadsden, Mining Annual Review, 1979, p97.
11. J.D. Hancock, Minerals Sci. Engng., 1977, 9, 25.
12. Norton Grinding Wheel Company, Brit.P. 709,882 (1954).
13. P.H. Wilks, P. Ravinder, C.L. Grrat, P.A. Pelton, R.J. Downer & M.L. Talbot, Chem.Eng. World, 1974, 9, 59.
14. A.M. Evans and J.P.H. Williamson, Trans. J. Br. Ceram. Soc., 1979, 78, 68.
15. C.J. Kinzie and D.S. Hake, U.S.P. 2,072,889, (1937) 2,168,603 (1939).
16. G.L.Miller 'Metallurgy of the Rarer Metals - 2.Zirconium', Butterworths, London, 1957, pp 34-36.
17. J.W. Marden & M.N. Rich, U.S.Bur. Min. Bull., 186, 1921.
18. G.H. Beyer, D.R. Spink, J.B. West & M.A. Wilhelm, U.S. Atomic Energy Commission ISC-437 (Rev), 1954.

19. V.G. Chukhlantsev, Yu.M. Polezhaev and K.V. Alyamovskaya, Izv. Akad. Nauk SSSR Neorg. Mater., 1968, 4, 745.
20. R.A.Schoenlaub, U.S.P. 2,578,748 (1951); 2,721,115(1955).
21. R.A.Schoenlaub, U.S.P. 2,721,117 (1955); 3.832,441(1974).
22. H.C.Kawecki, U.S.P. 2,418,174 (1947).
23. E.Greinacher and Wilhelm Brugger, (to Th. A-G.Goldschmidt) Brit. P. 1,147,738 (1969).
24. V.N. Shumenko, M.A. Kolenkova, N.T. Rybachuk, and A. Kh. Nazirov, Izv. Vyssh. Ucheb. Zaved., Tsvet. Met., 1972, 15, 68
25. A.J. Kaufmann & E.D. Dilling in 'The Metallurgy of Zirconium' (B. Lustman and F. Kerze, Eds.), McGraw-Hill, New York, 1955, p66.
26. C.de Rohden, M.Kastner, & M.Paquet, U.S.P. 2,564,522(1951).
27. Deutsche Gold-und Silber-Scheideanstalt, Brit. P. 971,594 (1964).
28. R.F. Tatnall, Ceramic Age, 1967, 83, 32.
29. M.A. Kolenkova, A.I. Lainer, V.A. Sazhina and A.I. Popov, Tsvetn. Metal., 1968, 41, 109.
30. F.H. McBerty, 'Anhydrous Chlorides Manufacture' F.I.A.T. Rev., Germ. Sci., Final Rep. No. 774, 1946.
31. R.H. Nielsen and R.I. Govro, US Bureau of Mines RI 5214,1956.
32. A.Clearfield, Rev. Pure and Appl. Chem., 1964, 14, 91.
33. T.E. MacDermott, Coord. Chem. Rev., 1973, 11, 1.
34. Magnesium Elektron Ltd., Brit. P. 1,373,634 (1974).
35. D.T. Stewart & I. McAlpine Brit. P. 1,337,983 (1973).
36. A. Clearfield & J.A. Stynes, J.Inorg. Nucl. Chem. 1964 26, 117.
37. A. Clearfield & J.M. Troup, Inorg. Chem., 1977, 16, 3311.
38. W.J. Baldwin in 'High Temperature Oxides Part II', (A.M. Alper, Ed.), Academic Press, London, 1970, pp 167-184.
39. B.T. Bell, Ceramic Industries Journal, 1980 (April), 29.
40. R.A. Eppler, Ceramic Bulletin, 1977, 56, 213.
41. N.C. Wildblood in 'High Temperature Chemistry of Inorganic and Ceramic Materials' (E.P. Glasser and P.E. Potter, Eds.), Chemical Society Special Publication No. 30, 1977, p12.
42. R.C. Garvie in ref 38, pp 117-166.
43. R.C. Garvie, R.H.J. Hannink & N.A. McKinnon, European P. 0,313,599 A1.
44. D. Greve, N.E. Claussen, D.P.H. Hasselman, G.E. Youngblood, Ceramic Bulletin, 1977, 56, 514.

45. P. Hagenmuller and W. Van Gool, 'Solid Electrolytes, General Principles, Characterisation, Materials, Applications', Academic Press, London, 1978.
46. W.J. Fleming, J. Electrochem. Soc., 1977, 124, 21.
47. C.B. Alcock in ref. 41, p 116.
48. J.P. Van Hook & J.C. Yarze, U.S.P. 4,026,823 (1977).
49. T.M. Gür, 'Electrochemical Reduction of Oxygen and Nitric Oxide on Stabilised Zirconia', PhD Thesis, Stanford University, 1976.
50. P.E. Ingham 'Thermal Studies of Wool and the Action of Flame Retardants', PhD Thesis, Bradford University, 1977.
51. L. Benisek and P.E. Ingham, J. Text. Inst., 1977, 68, 176.
52. D.E.A. Williams-Wynn, 'Theoretical Aspects of the Reaction of Zirconium Compounds and Vegetable Tannins with the Chromium-Collagen Complex', PhD Thesis, Rhodes University 1968.
53. A.L. Hock, J. Soc. Leather Tech. & Chem., 1975, 59, 181.
54. Magnesium Elektron Ltd., Brochure No. 109 'Zirconium Chemicals in Photography' 1980.
55. Oxford Paper Co.,Brit. P. 956,748 (1964).
56. G.W. Owens, 'Cosmetics, Toiletries and Health Care Products, Recent Developments', Noyes Data Corporation, 1978.
57. A.E. Staley Manufacturing Co., Brit.P. 911,500 (1962).
58. W.B. Blumenthal, Ind. Eng. Chem., 1950, 42, 640.
59. G.F. Danna, T.L. Vigo, & C.M. Welch, Text. Chem and Colorist, 1977, 9, 77.
60. C.J. Conner, A.S. Cooper, W.A. Reeves, B.A. Trask, Text. Res. J., 1964, 34, 347.
61. T.A. Calamari, J.R. Harper and B.J. Trask, J. Fire Retardant Chem. 1978, 5, 123.
62. Halliburton Co. U.S.P. 3,804,174 (1974); 3,959,003 (1976).
63. Magnesium Elektron Ltd., Brit. P. 1,343,032 (1974).
64. British Titan Ltd., Brit. P. 1,365,412 (1974).
65. L.A. O'Neill, Polymers Paint and Colour J., 1979, 169, 746.
66. Dresser Industries, U.S.P. 4,220,585, (1980).
67. Amchem Products Inc., Brit. P. 1,570,041 (1980).
68. Amchem Products Inc., U.S.P. 3,912,548 (1975); 4,191,596 (1980).
69. Amchem Products Inc., Brit. P. 1,326,146 (1973).
70. C.B. Amphlett, 'Inorganic Ion Exchangers', Elsevier Publishing Co., Amsterdam, 1964.

71. V. Baran, R. Caletka, M. Tympl & V. Urbanek, J. Radioanalytical Chem., 1975, 24, 353.
72. Magnesium Elektron Ltd., Brit. P., 1,499,805 (1978).
73. CCI Life Systems, U.S.P. 3,850,835 (1974).
74. G. Alberti, Accts. Chem. Res. 1978, 11, 164.
75. CCI Life Systems, U.S.P. 3,669,880 (1972).
76. Y. Onoue, Y. Mizutani, S. Akiyma, Y. Izumi and Y. Watanabe, Chemtech, 1977 (Jan), 36.
77. C.D. Veley, J. Petroleum Tech., 1969 (Sept.), 1111.

# The Preparation, Properties, and Industrial Uses of Inorganic Tin Chemicals

By P. A. Cusack and P. J. Smith
INTERNATIONAL TIN RESEARCH INSTITUTE, FRASER ROAD, PERIVALE, GREENFORD, MIDDLESEX UB6 7AQ, U.K.

## A. Introduction

The world's tin requirements, currently running at ca. 200,000 tons, are derived mainly from the mineral Cassiterite ($SnO_2$), and, to a much lesser extent, the sulphide ore Stannite ($Cu_2S.FeS.SnS_2$). Rarer tin-bearing minerals, such as Malayaite ($CaSnSiO_5$), are also found[1] in Perak, Malaysia and in Devon, England. Elemental tin has the largest number of naturally occurring isotopes of any element: $^{112}Sn$ (0.96%), $^{114}Sn$ (0.66%), $^{115}Sn$ (0.35%), $^{116}Sn$ (14.30%), $^{117}Sn$ (7.61%), $^{118}Sn$ (24.03%), $^{119}Sn$ (8.58%), $^{120}Sn$ (32.85%), $^{122}Sn$ (4.72%) and $^{124}Sn$ (5.94%). The bulk of tin is used in metal products, e.g. tinplate (ca. 40%) and solder (ca.28%), although, at the present time, at least 8,000 tons of tin are used in the manufacture of inorganic tin chemicals, this figure representing some 16,000 tons of products (TABLE 1).

TABLE 1 APPROXIMATE ANNUAL WORLD CONSUMPTION OF INORGANIC TIN CHEMICALS (TONS)

| Chemical | Tons |
|---|---|
| $SnO_2$ and metastannic acid | 4,000 |
| Stannous octoate | 2,500 |
| $SnCl_4$ and $SnCl_4.5H_2O$ | 3,300* |
| $SnCl_2$ and $SnCl_2.2H_2O$ | 1,500** |
| $Na_2Sn(OH)_6$ | 2,100 |
| $K_2Sn(OH)_6$ | 1,600 |
| Other metal stannates | 50 |
| $SnSO_4$ | 600 |
| Miscellaneous tin(II) salts ($SnF_2$, $Sn(BF_4)_2$, $Sn_2P_2O_7$, SnO and various stannous carboxylates)*** | 400 |
| | 16,050 |

*Excludes tonnage for organotin production
**Excludes tonnage for organotin and stannous octoate production
***Excluding stannous octoate

The production of inorganic tin chemicals is expected to rise still further during the present decade, owing to their wide range of industrial applications, described in Section E, and their generally accepted non-toxicity. In addition, both anhydrous stannous and stannic chlorides are used as starting materials for the production of organotin chemicals, the consumption of which is currently running at about 30-35,000 tons[2] and is also increasing rapidly.

The use of inorganic tin chemicals in electrolyte solutions for tin and tin alloy plating ($Na_2Sn(OH)_6$, $K_2Sn(OH)_6$, $SnSO_4$, $SnCl_2$ and $Sn(BF_4)_2$) and as intermediates in the manufacture of organotin chemicals ($SnCl_4$ and $SnCl_2$), where the tin in the finished product is in the metallic or organometallic, as distinct from the inorganic, form, represent "indirect" uses of these compounds and will not be discussed further in this article. Before describing the "direct" industrial outlets for inorganic tin chemicals, the principal methods of manufacture and the chemical and physical properties of the most important derivatives are discussed.

## B. Structure and Bonding in Inorganic Tin Compounds

The outer electron configuration of the tin atom is $5s^2 5p_x^1 p_y^1$ and the element can form compounds in both the +4 (stannic) and +2 (stannous) oxidation states.

Ionic inorganic tin(IV) compounds result from the loss of all four valence electrons to form the stannic ion, $Sn^{4+}$, which has an outer electron configuration, $5s^0 5p^0$, e.g. $SnO_2$. However, if one of the 5s-electrons is promoted to the vacant $5p_z$ orbital, the four atomic orbitals may be $sp^3$-hybridised to form tetrahedral covalent compounds, e.g. $SnCl_4$. Inorganic tin(IV) complexes with higher coordination numbers of 5(trigonal bipyramidal or square pyramidal), 6(octahedral), 7(pentagonal bipyramidal) and 8 (dodecahedral or square antiprismatic) are formed[3] when the 5d orbitals are incorporated in the hybridisation scheme, e.g. $SnCl_4 \cdot 2$ pyridine. In tin(II) compounds, the outer electron configuration of the element must contain a completely filled 5s orbital and the environment of the tin atom in most of its tin(II) derivatives has a very low symmetry because of the distorting effects of the non-bonding lone pair of electrons[4]. The two most common geometries found for inorganic tin(II) compounds are pyramidal (3-coordinate), e.g. $SnCl_3^-$, or square pyramidal (4-coordinate), e.g. SnO, with a stereochemically active lone pair occupying a position directed away from the strongly bonded coordination sites. Coordination numbers of five, six and seven for the nearest-neighbour atoms are also found[3] in tin(II) compounds, but these are relatively uncommon.

## C. Chemical and Physical Properties

### 1. Inorganic Tin(IV) Compounds

#### 1.1 TIN(IV) HALIDES

$SnF_4$, a white powder, is prepared either by fluorination of $SnCl_4$ or by direct interaction of its elemental constituents. It has a polymeric structure[5], with distorted $SnF_6$ octahedra in infinite layers, sharing four equatorial fluorine atoms. $SnF_4$ is very soluble in cold water and, in the presence of excess fluoride, forms $SnF_6^{2-}$ anions. It is insoluble in most organic solvents, but will dissolve to a certain extent in hot tetrahydrofuran.

$SnCl_4$ is a colourless liquid, which fumes in moist air due to hydrolysis to basic chlorides (with the evolution of HCl), and reacts violently with water to form the white crystalline pentahydrate, $SnCl_4.5H_2O$, which is best regarded[6] as $SnCl_4(OH_2)_2.3H_2O$. Aqueous solutions of $SnCl_4$ are slowly hydrolysed, although excess HCl stabilises the $SnCl_6^{2-}$ anion. Both anhydrous and hydrated $SnCl_4$ are very soluble in methanol and the former is miscible with most organic solvents. $SnCl_4$ dissolves exothermically in donor organic solvents, due to the formation of neutral adducts, e.g. $SnCl_4.2(C_2H_5)_2O$ and $SnCl_4.2CH_3OH$.

$SnBr_4$ exists as colourless, deliquescent rhombic pyramids, which also fume in air. Chemically, the bromide is similar to $SnCl_4$; $SnI_4$, a red cubic material, is much less reactive than either $SnCl_4$ or $SnBr_4$ and it does not fume in air.

The tin tetrahalides show Lewis Acid behaviour ($SnCl_4 \gg SnBr_4 > SnI_4$) and form a large number of adducts with organic donor molecules, e.g. $SnCl_4$.2pyridine. However, with the possible exception[7] of $SnCl_4.2Ph_3P{:}O$, these compounds are of little importance industrially. The tin(IV) oxyhalides, $SnOX_2$ (X = F, Cl, Br or I), also have no commercial outlets at the present time.

Some physical properties of the tin(IV) halides are listed in TABLE 2.

TABLE 2 PHYSICAL PROPERTIES OF TIN(IV) HALIDES

| Compound | m.p. (°C) | b.p. (°C) | Specific gravity at 20°C |
|---|---|---|---|
| $SnF_4$ | 705(subl.) | — | 4.78 |
| $SnCl_4$ | -33 | 114 | 2.23 |
| $SnCl_4.5H_2O$ | 56(d.) | — | 2.04 |
| $SnBr_4$ | 31 | 202 | 3.34 (35°) |
| $SnI_4$ | 144.5 | 364.5 | 4.47 |

## 1.2 TIN(IV) OXIDE

$SnO_2$ occurs naturally as the mineral Cassiterite, which has a rutile-type lattice[8]. The oxide is amphoteric but is unattacked by mineral acids, except concentrated $H_2SO_4$ (in which it dissolves). $SnO_2$ dissolves in caustic alkalis to give stannates:

$$SnO_2 + 2OH^- + 2H_2O \longrightarrow Sn(OH)_6^{2-}$$

The stannic acids, which can be regarded as hydrated forms of tin(IV) oxide, are of indefinite composition[9], approximating to $SnO_2.1.8H_2O$. There are two modifications: "$\alpha$-stannic acid" or "ortho-stannic acid", which is prepared by the action of ammonia on $SnCl_4$ or by the action of acids on $Na_2Sn(OH)_6$, and "$\beta$-stannic acid" or "metastannic acid", which is obtained by the reaction of tin metal with concentrated nitric acid (see Section D). The latter form is much more inert to chemical attack. $SnO_2$ has a specific gravity of 6.95 at 20°C, and melts at 1637°C.

### 1.3 THE STANNATES

The most important of these are the sodium and potassium salts, $Na_2Sn(OH)_6$ and $K_2Sn(OH)_6$, both of which are white powdered materials, giving alkaline solutions in water, due to free hydroxyl ions. On heating, three molecules of water are lost, to form the metastannates, $M_2SnO_3$, e.g.

$$Na_2Sn(OH)_6 \xrightarrow{140^\circ} Na_2SnO_3 + 3H_2O$$

Insoluble metal stannates may be prepared by a double decomposition reaction in aqueous solution:

$$M^{2+} + Na_2Sn(OH)_6 \longrightarrow MSn(OH)_6\downarrow + 2Na^+$$

(where M = Mg, Ca, Sr, Ba). The divalent transition metal hydroxystannates, i.e. M = Mn, Fe, Co, Ni, Cu, Zn and Cd, are precipitated by double decomposition from ammoniacal solution[10]. These hydroxystannates all contain the octahedral $Sn(OH)_6^{2-}$ ion[3]. The thermal decomposition of these compounds has been studied[11] and it appears that all dehydrate in a similar manner (at temperatures $< 460°C$):

$$MSn(OH)_6 \xrightarrow{\Delta} MSnO_3 + 3H_2O$$

Several undergo a further phase change at higher temperatures, to form[11] the orthostannate and $SnO_2$:

$$2MSnO_3 \xrightarrow{\Delta} M_2SnO_4 + SnO_2$$

The specific gravities at 20°C of $Na_2Sn(OH)_6$ and $K_2Sn(OH)_6$ are 3.03 and 3.30 respectively.

### 1.4 TIN(IV) SULPHIDE

$SnS_2$ is prepared by the direct interaction of tin and sulphur. The compound, which exists as golden-yellow hexagonal scales, is insoluble in water but dissolves in yellow ammonium sulphide:

$$SnS_2 + (NH_4)_2S \longrightarrow (NH_4)_2SnS_3$$

$SnS_2$ decomposes without melting at 865°C and has a specific gravity of 4.50 at 20°C.

### 1.5 TIN(IV) SALTS OF OXY-ACIDS

These compounds are relatively unimportant industrially. The sulphate crystallises in white, hexagonal prisms as a dihydrate, $Sn(SO_4)_2.2H_2O$, and is extensively hydrolysed in water. The nitrate, prepared by the action of $N_2O_5$ on $SnCl_4$, exists as white needles and the tin atom has dodecahedral coordination to four aniso-bidentate nitrate groups[12]. $Sn(NO_3)_4$ decomposes at 50°C and has a specific gravity of 2.65 at 20°C. There are no simple forms of tin(IV) borate, although discrete materials of composition, $MSn(BO_3)_2$ (where M = Mg, Ca, Sr, Ba, Mn, Fe, Co, Ni and Cd), and with dolomite-type lattices[13,14], are known.

### 1.6 TIN(IV) HYDRIDE

Tin forms a very unstable gaseous hydride, stannane, $SnH_4$ (m.p. -150°; b.p. -52°C), which is prepared by the reaction of $SnCl_4$ and $LiAlH_4$ in ether at -30°C. It decomposes rapidly at 0°, giving $\beta$-Sn, and is very readily oxidised in air.

## 2. Inorganic Tin(II) Compounds

### 2.1 TIN(II) HALIDES

$SnF_2$ crystallises in both rhombic and monoclinic modifications and is readily soluble in water, but insoluble in organic solvents. In aqueous solution, $SnF_3^-$ is the major species but $SnF^+$ and $Sn_2F_5^-$ ions can also be detected.

$SnCl_2$ is by far the most important tin(II) compound used industrially. The anhydrous salt is a white, rhombic crystalline material, soluble in water and in several organic solvents (alcohol, acetone and very slightly soluble in ether). The chloride crystallises from aqueous solution as the dihydrate, $SnCl_2.2H_2O$, which exists as colourless, monoclinic crystals. On standing, aqueous solutions of $SnCl_2$ hydrolyse to give an insoluble basic salt; this can be redissolved by addition of HCl to give a colourless solution containing the $SnCl_3^-$ ion. Addition of alkali to $Sn^{2+}$ solutions initially precipitates the hydrous tin(II) oxide as a white solid, which, with excess alkali, redissolves to give the stannite ion, $Sn(OH)_3^-$.

$SnCl_2$ is a good reducing agent and is used for inorganic ($Ag^+ + e^- \longrightarrow Ag°$) and organic reductions (Stephen Reagent) alike. In common with all other tin(II) compounds, $SnCl_2$ is easily oxidised to tin(IV) and, as a result, most tin(II) compounds, unless stringently protected from air, normally contain some tin(IV) impurity.

$SnBr_2$ and $SnI_2$ exist as yellow rhombic and orange-yellow monoclinic crystals respectively. The bromide is quite soluble in water but the solution is very cloudy due to hydrolysis. $SnI_2$ is only slightly soluble in water, but both $SnBr_2$ and $SnI_2$ are fairly soluble in methanol, giving clear solutions. Chemically, the bromide and iodide much resemble anhydrous $SnCl_2$. Some important physical properties of the tin(II) halides are listed in TABLE 3.

TABLE 3 PHYSICAL PROPERTIES OF TIN(II) HALIDES

| Compound | m.p. (°C) | b.p. (°C) | Specific gravity at 20°C |
|---|---|---|---|
| $SnF_2$ | 219.5 | 853 | 5.33*, 4.87** |
| $SnCl_2$ | 246 | 652 | 3.95 |
| $SnCl_2.2H_2O$ | 38(d.) | — | 2.71 |
| $SnBr_2$ | 215.5 | 620 | 5.12 |
| $SnI_2$ | 320 | 717 | 5.29 |

*Rhombic form **Monoclinic form

## 2.2 TIN(II) OXIDE

When aqueous tin(II) solutions, e.g. freshly prepared $SnSO_4$, are treated with ammonia, a white precipitate of hydrous tin(II) oxide forms. On heating in ammoniacal solution to 60-70°C, this precipitate turns black due to dehydration[15]. The final product, SnO, consists of blue-black cubic crystals which decompose without melting at 1080°C and have a specific gravity of 6.45 at 20°C. The oxide is amphoteric, dissolving in acids to give tin(II) salts and in alkalis to give stannites.

## 2.3 TIN(II) SALTS OF OXY-ACIDS

These salts are prepared by dissolving precipitated hydrous tin(II) oxide in the appropriate acids. Most important of these compounds are the white amorphous materials: tin(II) sulphate, $SnSO_4$ (m.p. 360°(d.), specific gravity 4.18 at 20°), tin(II) pyrophosphate, $Sn_2P_2O_7$ (m.p. 400°(d.), specific gravity 4.01) and tin(II) arsenate, $SnHAsO_4.0.5H_2O$ (specific gravity 4.29). Of these, only the sulphate is water-soluble. Several basic salts of oxy-acids can be obtained[4] from aqueous solutions at low pH, e.g. $Sn_3(OH)_4(NO_3)_2$ and $Sn_3(OH)_2OSO_4$.

## 2.4 TIN(II) CARBOXYLATES

Several tin(II) carboxylates have been prepared and some are used commercially. Various complex carboxylate anions, e.g. $Sn(O.CO.CH_3)_3^-$, have also been studied[16]. The physical properties of some tin(II) carboxylates are given in TABLE 4.

TABLE 4 PHYSICAL PROPERTIES OF TIN(II) CARBOXYLATES

| Compound | m.p.(°C) | b.p.(°C) | Specific gravity at 20°C |
|---|---|---|---|
| $Sn(O.CO.H)_2$ — FORMATE | 198(d.) | — | 2.96 |
| $Sn(O.CO.CH_3)_2$ — ACETATE | 182 | 240 | 2.11 |
| $\overline{Sn.O.CO.CO.O}$ — OXALATE | 280(d.) | — | 3.56 |
| $\overline{Sn.O.CO.CH(OH).CH(OH).CO.O}$ — TARTRATE | 280(d.) | — | 2.60 |
| $Sn(O.CO.CH_2.CHEt.C_4H_9)_2$ — OCTOATE | — | * | 1.26 |
| $Sn(O.CO.(CH_2)_{16}CH_3)_2$ — STEARATE | 90 | — | 1.05 |

*Decomposes on distillation at atmospheric pressure

All of the compounds listed in TABLE 4 are white solids, with the exception of tin(II) octoate (tin(II) 2-ethylhexoate), which is a fairly viscous, pale yellow liquid. Tin(II) formate is slightly soluble, the acetate decomposes, and the others are almost completely insoluble, in water. Tin(II) octoate is soluble in most organic solvents, tin(II) acetate is slightly soluble in alcohol or acetone, but tin(II) formate shows negligible solubility in these solvents.

### 2.5 OTHER TIN(II) COMPOUNDS

A few other simple tin(II) compounds are known, including tin(II) fluoroborate, $Sn(BF_4)_2$, which only exists in solution, and tin(II) isothiocyanate, $Sn(NCS)_2$[17]. A large number of adducts of the type, $SnX_2.nL$, (n= 1,2), are known but,[7] apart from $SnCl_2.2Ph_3P:O$, these are of little industrial importance. Trichlorostannane dietherate, $H^+SnCl_3^-.\ 2(C_2H_5)_2O$, a dense, pale-green oily liquid, which may be prepared[18,19] by passing dry HCl gas through a suspension of anhydrous $SnCl_2$ in diethyl ether at room temperature:

$$SnCl_2 + HCl \xrightarrow{ether} H^+SnCl_3^-.\ 2(C_2H_5)_2O$$

is an important intermediate[19] in the industrial manufacture of 2-alkoxy-carbonyl-ethyltin compounds from organic acrylates:

$$HSnCl_3 + RO.CO.CH{:}CH_2 \xrightarrow{ether} RO.CO.CH_2.CH_2SnCl_3$$

## D. Industrial Manufacture

### 1. Inorganic Tin(IV) Chemicals

### 1.1 STANNIC CHLORIDE, $SnCl_4$ and $SnCl_4{\cdot}5H_2O$

Anhydrous stannic chloride is prepared by the direct reaction of dry chlorine gas with metallic tin at 110-115°C, the reaction being highly exothermic. The chlorine is passed through shallow trays of granulated tin in a vertical mild steel reaction vessel and, at the end of the reaction, the anhydrous liquid product is recovered at the base of the column. During the chlorination, some anhydrous stannous chloride may also be formed, but the tetrachloride is easily separated by an enclosed distillation at atmospheric pressure. Since anhydrous stannic chloride reacts violently with water, it is very important that all moisture should be excluded in the manufacturing process.

The pentahydrate, $SnCl_4.5H_2O$, is prepared by the addition of anhydrous stannic chloride to water under carefully controlled conditions. The product forms as a solid white mass, which is then broken up into suitably sized lumps for packing. Both anhydrous and hydrated stannic chloride are shipped in mild steel containers.

## 1.2 STANNIC OXIDE, $SnO_2$, AND HYDRATED STANNIC OXIDES

For the large-scale economical production of anhydrous stannic oxide, two methods, by thermal oxidation, are commonly used:-

a) Introduction of molten tin metal into a suitable furnace at elevated temperature. Physical characteristics are controlled by the metal viscosity, rate of feed, oxygen ratios, etc.

b) Conversion of tin metal powder, which is obtained by atomising a stream of molten tin metal with a jet of air or water under carefully controlled conditions of temperature and pressure. Once in fine form, the powder is thermally oxidised by blowing it through a furnace at high temperature (ca. 700$^0$C) in a stream of air.

Both methods produce stannic oxides which are suitable for most ceramic and polishing applications (see Section E).

The industrial manufacture of hydrated stannic oxide ("metastannic" or "$\beta$-stannic acid") involves the reaction of metallic tin (as a powder or fairly small particles) with concentrated nitric acid. The exothermic reaction can be maintained at $\geq$100°C by the controlled addition of the tin to the acid. The white precipitate is then separated, washed with water to remove excess acid, filtered and dried at 120-150°C. The product is a stannic oxide of high surface activity, but, because of its higher cost of production, it is only used in a minority of applications where the reactivity is of importance. These include the preparation of certain ceramic pigments (e.g. the tin/antimony blue-greys) or $SnO_2$-based heterogeneous catalysts.

"α-" or "ortho-stannic acid" gels, which are not significant commercially, result from the cold hydrolysis of redistilled $SnCl_4$ with aqueous ammonia or the acidification of cold solutions of purified sodium hydroxystannate (see below) with sulphuric or nitric acids[9].

### 1.3 SODIUM AND POTASSIUM HYDROXYSTANNATE, $Na_2Sn(OH)_6$ AND $K_2Sn(OH)_6$

Sodium and potassium hydroxystannates are usually manufactured by the dissolution of tin in KOH or NaOH using a graphite anode, and, as the concentration of stannate increases in the solution, the hydrated material will precipitate out and may be removed by decantation or filtration. The stannates require the presence of ca. 1.5-2% of free alkali to ensure water solubility. An alternative source is from alkaline de-tinning processes[20], in those countries where chemical solution or electrolytic de-tinning of tinplate scrap in sodium hydroxide is carried out on a large scale.

$Na_2Sn(OH)_6$ may be used in the manufacture of insoluble metal hydroxystannates (e.g. those of Ca and Sr) by an aqueous double decomposition reaction with a soluble salt of the appropriate metal.

## 2. Inorganic Tin(II) Chemicals

### 2.1 STANNOUS CHLORIDE, $SnCl_2$ AND $SnCl_2.2H_2O$

Anhydrous stannous chloride, which is more stable and has better storage characteristics than the dihydrate, $SnCl_2.2H_2O$, is prepared industrially by the reaction of hydrogen chloride gas or, more commonly, stannic chloride vapour, with molten tin. If the reaction is carried out below the boiling point of $SnCl_2$ (i.e. 600-650°C), the molten salt has to be continuously swept away from the surface of the molten tin, whereas, at temperatures above this, the product may be volatilised from the tin surface and obtained as a condensate or sublimate.

The dihydrate is normally obtained industrially by the action of hydrochloric acid on tin metal. It should be noted here that, whereas dry chlorides or $SnCl_4$ or chlorine gas do not corrode mild steel reaction vessels, the aqueous $SnCl_2/HCl$ system is particularly aggressive towards the latter, since the tin(II) tends to strip off the oxide layer. Therefore, more resistant (and more expensive) metal substrates must be used in this process.

### 2.2 STANNOUS SULPHATE, $SnSO_4$

Stannous sulphate may be prepared by the reaction of granulated tin metal with an excess of sulphuric acid at elevated temperatures (ca. 100°C) for several days, followed by vacuum evaporation and removal of the excess acid with alcohol. Alternative methods of synthesis include the reaction of stannous oxide with sulphuric acid, the replacement of copper in an aqueous solution of copper sulphate by metallic tin and the anodic dissolution of tin metal in a sulphuric acid electrolyte.

### 2.3 STANNOUS OXIDE, SnO

A convenient method of synthesis of blue-black tin(II) oxide is from a solution of stannous sulphate (prepared by boiling an aqueous solution of copper sulphate, acidified with sulphuric acid, with excess granulated tin metal). The resulting solution is neutralised with caustic soda and then boiled to crystallise out the stannous oxide product[15].

### 2.4 STANNOUS FLUORIDE, $SnF_2$

This compound is usually prepared by the reaction of stannous oxide with aqueous hydrofluoric acid or by a direct reaction of metallic tin with anhydrous hydrogen fluoride.

## 2.5 STANNOUS OCTOATE, $Sn(O.CO.CH_2CHEt(CH_2)_3Me)_2$

The industrial manufacture of stannous octoate is carried out by two routes, namely those using either stannous oxide or stannous chloride as a starting material.

Where stannous oxide is employed, this is reacted directly with free 2-ethyl-hexanoic acid (usually in excess) by heating the mixture in an oxygen-free atmosphere. After completion of the reaction, the excess free acid is removed by distillation at reduced pressure. A more rapid reaction would result from using a precipitated "stannous hydroxide" rather than anhydrous (i.e. blue-black) stannous oxide.

Where stannous chloride is used as a starting material, the reaction is usually carried out in a two-phase (aqueous-organic) solvent system, involving a double-decomposition reaction in which the stannous chloride and sodium 2-ethyl-hexanoate are reacted to form stannous octoate (soluble in the organic phase) and sodium chloride (soluble in the aqueous phase).

The stannous octoate resulting from the above preparations is stabilised with suitable additives (e.g. hydroquinones) to prevent subsequent oxidation of the liquid tin(II) compound and the formation of an undesirable precipitate in the product.

Similar methods may be used to prepare other stannous carboxylates, e.g. stannous formate[21], acetate[15], oxalate[22] and stearate[23].

## 2.6 STANNOUS PYROPHOSPHATE, $Sn_2P_2O_7$

This compound is prepared industrially by precipitation from solutions of stannous chloride dihydrate and a soluble pyrophosphate, e.g. $Na_4P_2O_7$ or $H_4P_2O_7$, following neutralisation with sodium carbonate or ammonia.

## 2.7 STANNOUS FLUOROBORATE, $Sn(BF_4)_2$

Stannous fluoroborate is usually prepared by the reaction of stannous oxide with aqueous fluoroboric acid and is available commercially as a 50% aqueous solution.

## E. Uses.

### 1. Ceramic Glazes and Pigments.

#### 1.1. OPACIFIED GLAZES

Anhydrous $SnO_2$ is used as an opacifier in high-quality artware glazes in which it gives a highly attractive appearance at levels[24] of 4-8%. Stannic oxide is employed when the highest reflectance and purest colours, as well as the greatest strength and abrasion resistance, are required. The opacifier is suspended in the glaze as fine solid particles and the light falling on them is scattered and reflected from the surface (due to the different optical properties of these and the surrounding glass).

#### 1.2. COLOURED PIGMENTS

The rutile-type structure of stannic oxide is able to accommodate transition metal colourant ions in the lattice and these form the basis of the commercial pigments for ceramic tiles and pottery. The two established pigments of this type[25] - the Sn/V yellows and the Sn/Sb blue-greys - which contain 2-5% V and 3-8% Sb, are prepared by thermal reaction between $SnO_2$ and either $NH_4VO_3$ or $Sb_2O_3$ at $1200^0$-$1300^0$C. A third commercial tin-based pigment, which results from firing a mixture of $SnO_2$, $K_2Cr_2O_7$, $CaCO_3$ and $SiO_2$ at ca. $1150^0$C, is pink in colour at chromium levels of 0.5-2%, and probably contains Malayaite ($CaSnSiO_5$) as the host lattice[1,26,27]. $CaSnSiO_5$ has also been used as an encapsulating lattice for unstable coloured pigments, e.g. Cd(S,Se) reds[28].

A less common tin-containing pigment - "Purple of Cassius" - is prepared by reduction of aqueous gold(III) chloride by stannous chloride[29]. Neutralisation of the solution affords a pink-purple precipitate, which consists of a colloidal dispersion of metallic gold on hydrated $SnO_2$. The precipitate is fired at $800^0$-$835^0$C and the resulting pigment, which is, of course, expensive (typical gold content ca. 20%), is used mainly as an on-glaze decoration for high grade tableware and related products[30,31].

## 2. Glass Applications.

### 2.1. USE IN GLASS MANUFACTURE

Another important application for stannic oxide is in the manufacture of lead glass, where, in the form of sintered electrodes, it is used in the electromelting process[32]. At temperatures in excess of about 800°C, glass melts become electrically conductive and further heating is achieved by passage of a high current through the melt *via* these electrodes. This method is preferred to external heating, which leads to extensive loss of lead by volatilisation. A typical electrode contains about 98% $SnO_2$ with small additions of other oxides (such as those of antimony and copper) and can weigh between 5-50 kg[25].

A small outlet for tin chemicals in the glass industry is the use of stannous oxide as an additive in the production of gold-tin and copper-tin ruby glass.

Finally, it should be mentioned that tin metal itself is used in the molten state, in the manufacture of high quality, flat "float glass" for shop windows and mirrors, *etc.*, which has replaced the traditional mechanically polished "plate" glass.

### 2.2. SURFACE COATINGS

In the production of silver mirrors, the flat glass surface may be sensitised with a solution of $SnCl_2$, prior to treatment with a soluble silver salt. $SnCl_2$ is also a very useful sensitising agent for the metallisation of plastics[33].

A second, more extensive use of tin salts for glass coating, involves the formation of thin surface films of $SnO_2$[34]. These are usually deposited by bringing the hot glass surface in contact with a mixture of dry air and anhydrous stannic chloride, either as the vapour or as a spray in alcohol solution. At the elevated temperature (500°-600°C), pyrolysis of the tin compound occurs, resulting in the formation of $SnO_2$ films, which can be varied in thickness according to the application.

At the low end of the thickness scale (<1000Å), transparent $SnO_2$ films are used for strengthening glasses, bottles and jars, whereas, if the depth of the surface layer of $SnO_2$ is of the same order as the wavelength of visible light (*i.e.* 1000-10,000Å), thin-film interference occurs, to give the article a decorative iridescent lustre. Relatively thick films of $SnO_2$ (>10,000Å) are employed where electrical conductivity combined with optical transparency is required, *e.g.* for de-icing aircraft windscreens. In the latter case, indium oxide is often added, to give improved electroconductivity.

## 3. Catalysts.

### 3.1. HETEROGENEOUS OXIDATION CATALYSTS.

Stannic oxide is one of the components of a number of binary oxide systems which find extensive use in industry as heterogeneous oxidation catalysts[25]. The following reactions are representative:

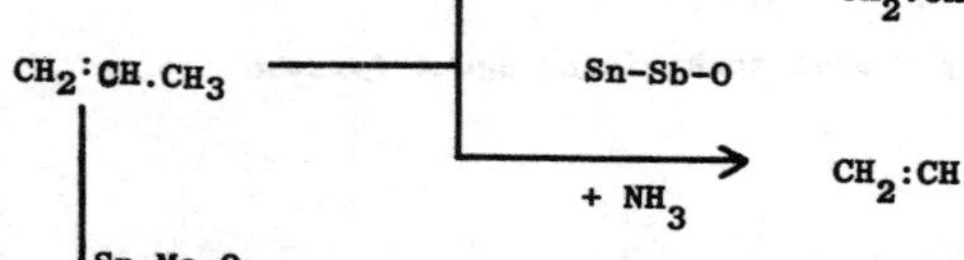

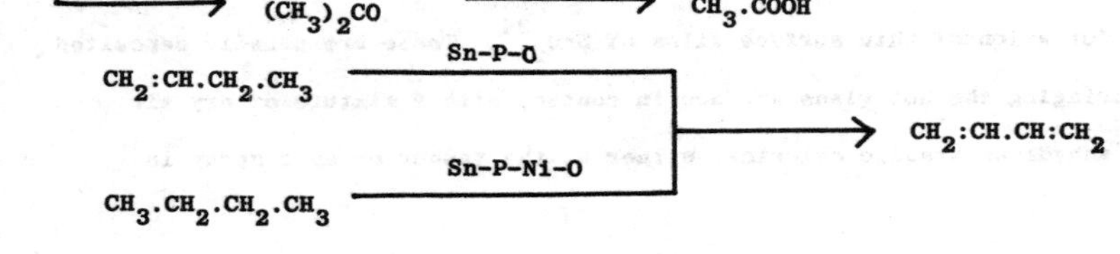

Benzene —(Sn-V-O)→ maleic anhydride

$CH_2{:}CH.CH_3$ —(Sn-Sb-O)→ $CH_2{:}CH.CHO$ → $CH_2{:}CH.COOH$

$CH_2{:}CH.CH_3$ —(Sn-Sb-O, + $NH_3$)→ $CH_2{:}CH.CN$

$CH_2{:}CH.CH_3$ —(Sn-Mo-O)→ $(CH_3)_2CO$ → $CH_3.COOH$

$CH_2{:}CH.CH_2.CH_3$ —(Sn-P-O)→ $CH_2{:}CH.CH{:}CH_2$

$CH_3.CH_2.CH_2.CH_3$ —(Sn-P-Ni-O)→ $CH_2{:}CH.CH{:}CH_2$

Certain heavy transition metals (Pt, Pd and Re), in combination with $SnO_2$, and usually supported on alumina, appear to have an increased activity when used as catalysts for dehydrogenation and related reactions of hydrocarbons in the petroleum industry. A disadvantage of this type of catalyst when used for promotion of the low temperature oxidation of carbon monoxide by nitrogen oxides and oxygen (e.g. the removal of toxic gases from motor vehicle exhaust emissions, tobacco smoke and air-purification systems), is that the activity is reduced in the presence of water vapour. Recent work[35] has shown that, if $SnO_2$ itself is used as a support for palladium, the activity of the catalyst is enhanced in the presence of water vapour.

## 3.2. HOMOGENEOUS CATALYSTS.

Stannic chloride is a strong Lewis Acid and may be used as a Friedel-Crafts catalyst for alkylation and acylation reactions, e.g.

$$C_6H_6 + CH_3Cl \xrightarrow{SnCl_4} C_6H_5CH_3 + HCl$$

It also finds use as a polymerisation catalyst[36].

The most important industrial homogeneous tin catalyst, however, is stannous octoate (stannous 2-ethylhexoate) which is used in the production of rigid and flexible polyurethane foams[37]. Reaction between a long-chain diol and an aromatic di-isocyanate forms a prepolymer with urethane linkages. Treatment with water leads to further polymerisation accompanied by the evolution of large amounts of carbon dioxide gas which causes the polyurethane to foam[38]. The use of the tin catalyst improves efficiency and enables the process to be carried out in one step. Stannous octoate is also employed as a curing agent in room-temperature-vulcanising (RTV) silicones[39], the products being flexible, elastomeric solids with a host of different uses, such as dental-impression moulds.

Another tin(II) carboxylate - stannous oxalate - is used[40] as a catalyst in esterification, transesterification and polyesterification reactions, e.g. the esterification of phthalic anhydride with iso-octanol:

$$C_6H_4(CO)_2O + 2\,{}^{i}C_8H_{17}OH \longrightarrow C_6H_4(C(=O){-}O{}^{i}C_8H_{17})_2 + H_2O$$

Its advantages include the limitation of (a) destructive side reactions, which are responsible for degradation of esters at preparation temperatures, and (b) acidic corrosion of the reaction vessel (the tin compound being neutral in character). Other carboxylates, such as tin(II) undecanoate,are also used.

## 4. PVC Stabilisers.

A further application of tin(II) carboxylates is as heat stabilisers for rigid polyvinyl chloride. These function by inhibiting the loss of HCl from the PVC at the processing temperature ($180^0$-$200^0$C), and thereby prevent discolouration and embrittlement of the plastic.

Stannous stearate[23] is approved in the U.S.A. as a non-toxic additive for PVC food containers[41] and, along with stannous oleate, appears in the list of tin compounds recommended by the Japan Hygenic PVC Association. It has been found[42] that zinc stearate gives a synergistic improvement in stabilising effectiveness when used in conjunction with the tin compound and this blend is therefore employed in some proprietary formulations.

## 5. Fabric Applications.

For many years, hydrated stannic chloride has found extensive use as an aqueous treatment (which also contains[43] $Na_2HPO_4.12H_2O$ and $Na_2SiO_3.9H_2O$)

for the weighting of natural fibres, particularly silk. Recent studies using $^{119m}$Sn Mössbauer spectroscopy have shown[43] that the tin species present in the fibres is hydrated $SnO_2$. Stannic chloride has also been employed as a mordant in the dyeing of silk. In this application, the tin compound is reacted with a dyestuff to precipitate an insoluble, coloured compound - a "lake" - resulting in permanent colouring of the impregnated fabric.

However, the most important outlet for inorganic tin chemicals in the field of fabrics is as flame-retardants. The earliest reported work on tin-based treatments was by Versmann and Oppenheim[44] who, in 1859, developed a process based on the precipitation of $SnO_2$ (by neutralisation of stannic chloride with sodium hydroxide) in cotton fabric. This concept was modified and extended by Perkin[45] at the turn of the century, as a double-bath technique in which deposition of $SnO_2$ in the cotton was effected by immersion of the fabric in sodium hydroxystannate solution, followed by an aqueous ammonium sulphate dip. This impregnation was claimed to be the first durable flame-retardant treatment, although little interest was shown in the process. In the 1930's and during World War II, more research was carried out on similar methods with some success, particularly with the stannic chloride - sodium tungstate system[46], in which coprecipitation of $SnO_2$ and $WO_3$ confers both flameproofing and glowproofing properties to the fabric. $^{119m}$Sn Mössbauer studies on treated cotton[47] have shown that, as in weighted silk[43], the tin species present is indeed hydrated $SnO_2$.

Although these processes have long fallen into disuse, there has been much interest in flame-resist treatments for wool recently and two tin-containing formulations are used commercially. The Wool Research Organisation of New Zealand (W.R.O.N.Z.) Tin Spray Treatment for woolly sheepskins[48,49] is an aqueous formulation containing hydrated stannic chloride, ammonium bifluoride, isopropanol and a silicone polishing agent. The active tin species appears to be a hydroxyfluorostannate dianion (possibly $SnF_5OH^{2-}$)[50], and repeated washing of $SnCl_4/NH_4HF_2$-treated fabric leads to complete hydrolysis to $SnO_2$, thereby rendering the treatment inactive. The treatment is, however, unaffected by dry-cleaning and is used for treating approximately 4,000 sheepskins per year for certain export markets[51]. The second formulation, based on stannous chloride and potassium hexafluorozirconate[52], is currently used in Greece[53] for the treatment of all-wool flokati rugs.

During the seventies, inorganic tin compounds have been prepared for use as flame-retardant additives in synthetic fibres such as nylon[54], and in various plastics[55]. It has also been found[56] that certain anhydrous stannates (particularly $ZnSnO_3$), when incorporated into glass-reinforced-polyester test panels, show significant smoke suppressant properties.

## 6. Pharmaceuticals.

Tin and its salts have a number of applications in the pharmaceuticals industry. In dentistry, stannous fluoride[57] and possibly stannous pyrophosphate[58] are widely used in certain toothpastes and dentifrices[59]; the fluoride has a marked anticaries effect[60]. Stannous octoate, as mentioned previously, is used as a catalyst for the cross-linking of silicones for dental impression moulding and tin metal is an important constituent of conventional dental amalgams based on $Ag_3Sn$. Some newer amalgams contain both tin and copper[61].

A pharmaceutical preparation known as Stannoxyl*, which contains a mixture of tin powder and stannous oxide, is marketed in the U.K. in tablet form for

*Trade name.

the treatment of various skin complaints, such as acne, boils and carbuncles. Stannous arsenate-based pellets have veterinary applications for the control of parasitic worms in sheep.

Stannous fluoride and "stannous hydroxide" are important constituents of radiopharmaceuticals based on technetium -99m, which are used in diagnostic nuclear medicine for lung, bone and liver scintigraphy. The stannous salt reduces the technetium in aqueous solution (see following Section) to form a $^{99m}$Tc-labelled tin colloid suitable for injection. X-ray studies have indicated that an Sn-O-Tc linkage is involved in the product[62].

## 7. Reducing Agents.

### 7.1. INORGANIC REDUCTIONS.

Stannous salts are used extensively as reducing agents in inorganic chemistry, as illustrated by the following equations:

$$Tc^{7+} \text{ (as } TcO_4^-) + Sn^{2+} \longrightarrow Tc^{5+} + Sn^{4+}$$

$$2Ag^+ \text{ (as AgX)} + Sn^{2+} \longrightarrow 2Ag^0 + Sn^{4+}$$

$$2Au^{3+} \text{ (as } AuCl_4^-) + 3Sn^{2+} \longrightarrow 2Au^0 + 3Sn^{4+}$$

The Tc reduction is utilised in the production of radiopharmaceuticals; the Ag reduction in the manufacture of silver mirrors (Section E.2.2.) and in photography[63]; and the Au reduction in the preparation of the "Purple of Cassius" pigment (Section E.1.2.).

### 7.2. REDUCTION OF ORGANIC COMPOUNDS.

The reduction of aromatic nitro compounds to primary amines and of nitriles to aldehydes (Stephen reaction)[64] may be achieved using Sn/concentrated aqueous HCl and $SnCl_2$/anhydrous HCl respectively:

$$RNO_2 \xrightarrow[H_2O]{Sn/conc.\ HCl} RNH_2$$

$$RC{\equiv}N \xrightarrow[(C_2H_5)_2O]{SnCl_2/anhyd.\ HCl} (RCH{=}NH_2)_2SnCl_6 \xrightarrow{H_2O} RCHO$$

The Stephen reaction probably involves the initial formation[18,19] of $HSnCl_3{\cdot}2(C_2H_5)_2O$. Selective reduction of aromatic nitro compounds containing other reducible functional groups, e.g. $-C{\equiv}N$, is readily effected[65] using ethanolic $SnCl_2/NaBH_4$:

$$N{\equiv}C{-}C_6H_4{-}NO_2 \xrightarrow[C_2H_5OH]{SnCl_2/NaBH_4} N{\equiv}C{-}C_6H_4{-}NH_2$$

Stannous chloride is known[66] to react with sodium borohydride at elevated temperatures (about 200$^0$C) to form metallic tin and diborane. Finally, the reducing properties of the $SnCl_4/NaBH_4$ system have recently been reported[67].

## 8. Miscellaneous.

There are a number of miscellaneous applications of inorganic tin chemicals, which, although not accounting for a particularly large tonnage, are of specialised interest. $SnCl_4$ and $SnCl_2$ are used to stabilise perfumes in toilet soaps.

Stannic oxide (anhydrous) is a useful material for stone polishing[68]; in the hydrated ($\alpha$-) form, it is used as a stationary phase in column chromatography for the separation of radionuclides, e.g. $^{97}$Ru from irradiated molybdenum[69]. It is also of interest to note that three radioactive tin isotopes are used commercially - $^{119m}$Sn ($t_{\frac{1}{2}}$ = 245 d ) and $^{121}$Sn ($t_{\frac{1}{2}}$ = 76 y ) in sealed sources for $^{119}$Sn and $^{121}$Sb Mössbauer spectroscopy respectively; $^{113}$Sn ($t_{\frac{1}{2}}$ = 115 d ) is used in radiopharmaceutical generators for the production of the short-lived $^{113m}$In radioisotope ($t_{\frac{1}{2}}$ = 1.7 h

**Other miscellaneous uses include ceramic capacitor bodies and, possibly, corrosion inhibitors[70] (certain metal stannates), activators in certain U.V. fluorescent phosphors[71](tin(II) salts) and as a bronzing agent ($SnS_2$) for wood and gypsum - "tin bronze" [25].**

## F. Acknowledgements.

**The International Tin Research Council, London, is gratefully acknowledged for permission to publish this review. The authors also wish to express their appreciation to Mr. H. Morriss (Keeling and Walker Ltd., Stoke-on-Trent), Dr. J. Thorpe (William Blythe Ltd., Accrington) and Mr. R. Haddick (Durham Chemicals Ltd., Birtley) for helpful comments on the manuscript, and B. Patel (I.T.R.I.), for density measurements of $Na_2Sn(OH)_6$ and $K_2Sn(OH)_6$.**

Tin oxide based pigments are used in the tiling of the London Underground's Victoria Line

## References


1. S. Takenouchi, Mineral. Deposita (Berl.), 1971, 6, 335.
2. A.G. Davies and P.J. Smith, Adv. Inorg. Chem. Radiochem., 1980, 23, 1.
3. P.A. Cusack, P.J. Smith, J.D. Donaldson and S.M. Grimes, I.T.R.I. Publicn. No.588, 1981.
4. J.D. Donaldson, Prog. Inorg. Chem., 1967, 8. 287.
5. R. Hoppe and W. Dähne, Naturwissenschaften, 1962, 49, 254.
6. J.C. Barnes, H.A. Sampson and T.J.R. Weakley, J. Chem. Soc., Dalton Trans., 1980, 949.
7. M & T Chemicals Inc., Chem. Eng. News, March 1975, 21.
8. W.H. Baur, Acta Crystallogr., 1956, 9, 515; 1970, B27, 2133.
9. M.J. Fuller, M.E. Warwick and A. Walton, J. Appl. Chem. Biotechnol., 1978, 28, 396.
10. T. Dupuis, C. Duval and J. Lecomte, C.R. Hebd. Seances Acad. Sci., 1963, 257, 3080.
11. P. Ramamurthy and E.A. Secco, Can. J. Chem., 1971, 49, 2813.
12. C.D. Garner, P. Sutton and S.C. Wallwork, J. Chem. Soc. (A), 1967, 1949.
13. D. Schultze, K.T. Wilke and C. Waligora, Z. Anorg. Allg. Chem., 1971, 380, 37
14. J. Vicat and S. Aléonard, Mater. Res. Bull., 1968, 3, 611.
15. J.D. Donaldson, W. Moser and W.B. Simpson, J. Chem. Soc. (Suppl.), 1964, 5942.
16. J.D. Donaldson and J.F. Knifton, J. Chem. Soc.(A), 1966, 332.
17. B.R. Chamberlain and W. Moser, J. Chem. Soc.(A), 1969, 354.
18. E.J. Bulten and J.W.G. Van den Hurk, J. Organomet. Chem., 1978, 162, 161.
19. R.E. Hutton, J.W. Burley and V. Oakes, J. Organomet. Chem., 1978, 156, 369.
20. 'Kirk-Othmer Encycl. Chem. Technol.', 2nd edn., John Wiley, New York, 1969, vol. 20, p.294.
21. J.D. Donaldson and J.F. Knifton, J. Chem. Soc.(A), 1964, 4801.
22. Y.A. Ugai, J. Gen. Chem. USSR, 1954, 24, 1297.
23. G.J. Chertoff, U.S. Patent, 2,629,700 (1953).

24. K. Shaw, 'Ceramic Glazes', Applied Science, London, 1971.

25. M.J. Fuller, Tin Its Uses, 1975, 103, 3.

26. J.B. Higgins and F.K. Ross, Cryst.Struct. Commun., 1977, 6, 179.

27. F. Hund, U.S. Patent, 3,753,754 (1973).

28. A. Broll, H. Beyer, H. Mann and E. Meyer-Simon, U.S. Patent, 3,847,639 (1974).

29. K. Shaw, 'Ceramic Colours and Pottery Decoration', Applied Science, London, 1968.

30. R.R. Dean and C.J. Evans, Tin Its Uses, 1977, 113, 12.

31. R.R. Dean and C.J. Evans, Tin Its Uses, 1977, 114, 9.

32. W.B. Hampshire and C.J. Evans, Tin Its Uses, 1978, 118, 3.

33. C.J. Evans, Tin Its Uses, 1973, 98, 7.

34. C.J. Evans, Glass, 1974, 51, 303.

35. G.Croft and M.J. Fuller, Nature (London), 1977, 269, 585.

36. B.M. Badran, I.M. El-Anwar, M.S. Ibrahim and W.M. Khalifa, J. Oil Colour Chem. Assoc., 1979, 62, 199.

37. S. Karpel, Tin Its Uses, 1980, 125,1.

38. D.J. Walsh, in J.M. Buist, 'Developments in Polyurethane-1', Applied Science, London, 1978, p.9.

39. W. Noll, 'Chemistry and Technology of Silicones', Academic Press, New York, 1968.

40. M & T Chemicals Inc., Tech. Data Sheet No.227, 1975.

41. Anon, Fed. Reg., April 1977, 21, 615, part 181.29.

42. K.C. Bergman and T.C. Jennings, U.S. Patent, 3,639,319 (1972).

43. Zw. Busowa and D. Christov, Melliand Textilber., 1979, 9, 773.

44. F. Versmann and A. Oppenheim, Eng. Patent, 2077 (1859).

45. W.H. Perkin, Eng. Patent, 9620 (1902).

46. S. Coppick and W.P. Hall, in R.W. Little, 'Flameproofing Textile Fabrics', A.C.S. Monogr. Ser. No.104, Reinhold, New York, 1947, p.221.

47. P.A. Cusack, P.J. Smith and J.S. Brooks, Unpublished work, 1979.

48. P.E. Ingham, Wool Res. Org. New Zealand Commun., No.30., 1974.

49. P.E. Ingham, Tin Its Uses, 1975, 105, 5.

50. P.A. Cusack, P.J. Smith, J.S. Brooks and R. Smith, J. Text. Inst., 1979, 70, 308.

51. P.E. Ingham, Unpublished report, 1976.

52. L. Benisek, Brit. Patent, 1,385,399 (1975).

53. L. Benisek, Personal communication, International Wool Secretariat, Ilkley, Yorks., 1978.

54. Toray Industries Inc., Brit. Patent, 1,382,659 (1975).

55. I. Touval, J. Fire Flammability, 1972, 3, 130.

56. P.A. Cusack, P.J. Smith and L.T. Arthur, J. Fire Retardant Chem., 1980, 7,

57. Procter and Gamble Ltd., Brit. Patent, 939,230 (1963).

58. B. Svatun and A. Attramadal, Acta Odontol Scand., 1978, 36, 211.

59. 'Kirk-Othmer Encycl. Chem. Technol.', 3rd. edn., John Wiley, New York, 1980, vol.10, p.819.

60. D.C. Picton, Dental Update, 1977, 4, 9.

61. M.L. Malhotra and K. Asgar, J. Am. Dent. Assoc., 1978, 96, 444.

62. E. Deutsch, R.C. Elder, B.A. Lange, M.J. Vaal and D.G. Lay, Proc. Natl. Acad. Sci. USA, 1976, 73, 4287.

63. S.S. Collier, Photog. Sci. Eng., 1979, 23, 113.

64. H. Stephen, J. Chem. Soc., 1925, 127, 1874.

65. T. Satoh, Ventron Alembic, ed. R.C. Wade, Issue 14, Ventron Corp., Beverly, Ma USA.

66. W. Jeffers, Chem. Ind. (London), 1961, 431.

67. S. Kano, Y. Yuasa and S. Shibuya, J. Chem. Soc. Chem. Commun., 1979, 796.

68. E.S. Hedges, Tin Its Uses, 1972, 91, 17.

69. P.J. Pao, J.L. Zhou, D.J. Silvester and S.L. Waters, Radiochem. Radioanal. Lett., In press.

70. M.E. Warwick and J.H. Hancox, Personal communication, I.T.R.I., 1980.

71. K.H. Butler, 'Fluorescent Lamp Phosphors', Pennsylvania State University Press, 1980, pp.49-51, 185-210, 314-5.

# Trends and Technology in Industrial Chromium Chemicals

By W. H. Hartford
BELMONT ABBEY COLLEGE, BELMONT, NORTH CAROLINA 28012, U.S.A.

## Introduction

An accident of timing, in that chromium appears early in the alphabetical order of the elements, dictates the areas of emphasis of this presentation. In 1979, Volume 6 of the Third Edition of Kirk-Othmer Encyclopedia of Chemical Technology appeared. This contained[1] a considerably longer discussion of chromium compounds than is possible in this chapter. In particular, it contained a fairly complete summary of the chemistry of the element, mentioning a substantial number of compounds which have been sold, but as fine chemicals in small volume. In accordance with the definition established in the opening chapter of this book, discussion of these minor commercial compounds will be omitted, and the reader is referred elsewhere.[1,2,3] In these earlier references, a comprehensive discussion of analytical methods is given, and here, too, other references should be consulted.[1,4,5].

In the 3½ years that have intervened since the writing of the Kirk-Othmer article[1], two strong international trends have emerged with sufficient clarity that they will serve as the theme for this chapter. The first is the growing realization, not only that chromium is an essential element in the manufacture of steel, refractories, and industrial chemicals, but that none of the nations of the Atlantic community have self-sufficient deposits of chrome ore or chromite within their own borders. Conservation of the element is therefore essential, despite the fact that it is a fairly common element in the earth's crust. The second trend is the growing concern over the environmental hazards associated with the manufacture and use of chromium compounds. It is no longer sufficient to employ manufacturing process or foster uses where the benefits derived do not substantially exceed any environmental debits.

With this in mind, this chapter will emphasize the trends which appear to be developing in the chromium chemical industry in response to these two overriding influences. The geochemistry, distribution and composition of ores, particularly as they bear on the major uses, need to be examined in more detail than in previous descriptions of the subject. The manufacturing process must be studied to determine those processes which will be both cost and environmentally effective. Finally, we will be seeing a shift of use patterns into fields where chromium performs a unique and utilitarian function.

## Historical

In 1763, Klaproth described the brilliant orange mineral crocoite (chemically, $PbCrO_4$) from the vicinity of Ekaterinburg (now Sverdlovsk) on the Asian slope of the Ural Mountains in Siberia. The mineral was ground for use as an artist's pigment, so it was not surprising that Vauquelin in 1797 named the element chromium, from χρομά, the Greek word for color. Crocoite came from the same mineralization which today forms the basis of the Soviet Union's important position in chromite, the only commercial chromium mineral. At first, the chromite was rafted to the Ob and the Arctic, so that it took two years for the ore to reach Andreas Kurtz of Manchester, who manufactured lead chromate, or chrome yellow, for pigment use. In 1811, deposits of the ore were found near Baltimore, Maryland, and until about 1860, Maryland and Eastern Pennsylvania supplied chromite to the world.[6] Over the years, deposits in Turkey, New Caledonia, the Philippines, Cuba, Finland, the Balkans, Zimbabwe, the Union of South Africa, and the Soviet Union, have all been of major importance, with the latter two in the most commanding positions today.

The manufacture of chromium chemicals has centered on the manufacture of potassium, and later sodium, dichromate as the basic compound from which all other industrial and fine chromium chemicals can be made. Andreas Kurtz's factory was soon joined by that of John and James White of Glasgow, and the United Kingdom has remained an important producer. In 1845, the Tyson family of

Baltimore started manufacture in Baltimore; the present plant of Allied Chemical Corporation is on the site. It was the world's largest facility from 1950 until 1971, when the plant of Diamond-Shamrock Chemical Corp. at Castle Hayne, N.C. achieved that distinction. Germany, under the aegis of the former I. G. Farbenindustrie, was an important producer at the time of World War II. After the war, their expertise became a matter of public record[7] and aided the building of new plants in the U.S. and U.K. The older technology has been summarized in detail in the German literature.[8]

Environmental concerns in the U.S. have figured largely in the building of chromium chemical plants since 1950. These considerations will play a larger part in the future.

## Chromium Ores and Chromium Supplies

According to the U.S. Bureau of Mines,[9] the world production of chromium ore in 1977 was 10,804,000 tons. Countries producing more than 500,000 tons were:

| Country | Production (thousand tons) |
|---|---|
| Union of South Africa | 3656 |
| U. S. S. R. | 2400 |
| Albania | 970 |
| Turkey | 700 |
| Zimbabwe (Rhodesia) | 660 |
| Finland | 655 |
| Philippines | 592 |
| | 9633 |

These seven countries account for almost 90% of world production. However, no summary of chrome ore production would be accurate without consideration of the $Cr_2O_3$ content of the ore and the uses to which it is put. This is best illustrated by the U.S. statistics for the same year, when the U.S. had a net consumption of 1,149,000 tons, made up as follows:

| Type of ore | % | Average $Cr_2O_3$ content, % |
|---|---|---|
| Metallurgical | 57.8 | 41.3 |
| Refractory | 20.8 | 36.0 |
| Chemical | 21.4 | 44.7 |
| Overall | 100.0 | 40.9 |

Metallurgical Chrome Ore. The situation in 1977 was confused by the U.N. sanctions against Rhodesia, which are now removed with the formation of the Zimbabwe government. Normally, Zimbabwe would be a much larger producer of metallurgical grade ore, which is traditionally over 48% $Cr_2O_3$ in tenor, and has a Cr:Fe weight ratio of 3:1 or more. Because this ore is used for the production of ferrochromium in the electric furnace (an expensive way to produce iron) the available premium ore from the U.S.S.R. and Turkey is diluted with refractory grade ore from other sources, and chemical ore from the Union of South Africa to maintain the 3:1 Cr:Fe ratio, even at the expense of ore tenor.

The nature of this problem is further illustrated by Table 1, which lists the analyses of some typical chrome ores, and Figure 1, which shows the relationship of ore grade to molecular composition of the pure chromite mineral.

In 1977, the U.S. price of Russian metallurgical ore was $136/ton, and of Turkish metallurgical ore $120 - $130/ton.

Refractory Chrome Ore. In this use, dimensional stability on heating, chemical neutrality, and high melting point, rather than $Cr_2O_3$ tenor and iron content, are of prime importance. Refractory use does not employ the ore as a source of chromium. In refractory ores, the $Al_2O_3$ is desirably high, over 25%; $Cr_2O_3$ is usually under 40%; MgO is high; and iron should be low.

Chemical Chrome Ore. Ore for the manufacture of chromium chemicals is subject to none of the above chemical restrictions, and hence the cost per pound of chromium is the main determinant for use. Chemical chrome ore today is almost entirely Grade B ore from the Transvaal area in the Union of South Africa, of a composition similar to D in Table 1, although concentrates of lower $SiO_2$ content have limited use. In 1977, about 246,000 tons of Transvaal ore were used in U.S. chromium chemical manufacture;

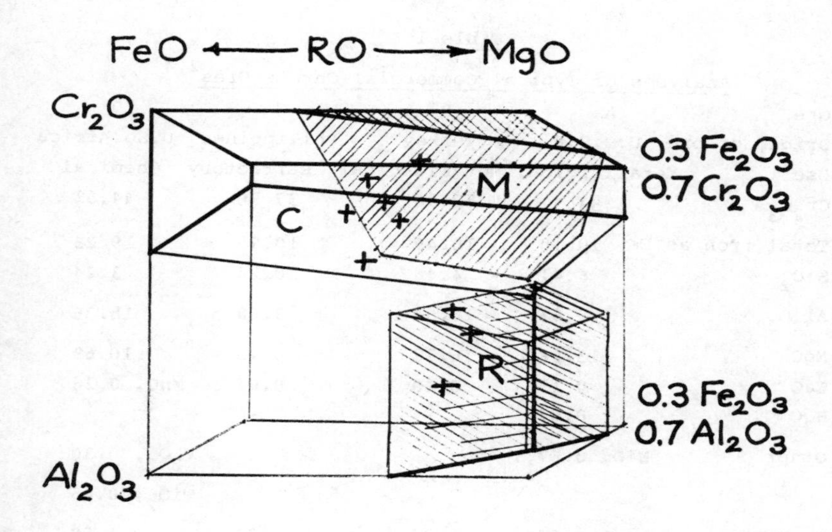


Fig. 1. Composition of chromite in ores. C = chemical, M = metallurgical, R = refractory. After Thayer.[10]

the price was $56-61/ton. Chemical ore of lower grade, not now economic for use, is found in the U.S. and elsewhere, but the South African deposits are the largest deposits now known.

Chrome ore is one of the world's mineral resources which is in short supply. The supply of good metallurgical ore is especially short, as evidenced by the adjustments made in the U.S. during the U.N. sanctions against Rhodesia. The South African deposits, which are used mainly for chemicals manufacture, will probably eventually have to be adapted for metallurgy, and even lower-grade ores, such as those in Montana, brought into use as supplies diminish. From purely theoretical considerations, the poorer ores may be the most abundant, and the chance of locating such low-grade sources for the future seems favorable, as the necessity for their general use develops.

Processes are also known for the preferential removal of iron from non-metallurgical ores. These may become important in the future, as supplies diminish.

Table 1

Analyses of Typical Commercial Chrome Ores[2,4]

| Ore | A | B | C | D |
|---|---|---|---|---|
| Origin of ore | U.S.S.R | Turkey | Philippines | U.So.Africa |
| Use | Metallurgical | Metallurgical | Refractory | Chemical |
| $Cr_2O_3$ | 53.57 | 50.96 | 37.98 | 44.51 |
| Total iron as Fe | 10.10 | 10.49 | 10.95 | 19.28 |
| $SiO_2$ | 6.41 | 4.46 | 0.24 | 3.24 |
| $Al_2O_3$ | 9.67 | 10.94 | 8.68 | 15.36 |
| MgO | 13.08 | 16.78 | 16.45 | 10.68 |
| CaO | 0.15 | 0.40 | 0.04 | MnO, 0.26 |
| $H_2O^+$ | 0.67 | - | - | - |
| Other | NiO, 0.67 | | | $V_2O_3$, 0.30 |
| | | | | $TiO_2$, 0.45 |
| Cr:Fe | 3.63 | 3.32 | 2.37 | 1.58 |

## Geochemistry and Industrial Chemistry of Chromium Ores

The chromium-containing mineral in chromium ores is chromite, ideally $FeCr_2O_4$, a member of the isomorphous spinel family. The unit spinel lattice consists of 32 cubic close-packed oxide ions, in which structure there are 8 tetrahedral "holes" and 16 octahedral "holes " of ligand field stabilization energies for the various metallic cations determine the structure. Chromium is invariably found in the octahedral holes. The pure compound $FeCr_2O_4$ is unknown except in meteorites. Chromite, as found in nature, has replacement of $Fe^{2+}$ by $Mg^{2+}$, and of $Cr^{3+}$ by $Al^{3+}$ and $Fe^{3+}$. The pure species of the replacing spinels are known as:

$MgCr_2O_4$, magnesiochromite

$FeAl_2O_4$, hercynite

$MgAl_2O_4$, spinel

$FeFe_2O_4$, magnetite

$MgFe_2O_4$, magnesioferrite

As may be seen in the analyses in Table 1, there is also trace replacement of $Fe^{2+}$ by Ni and Mn, and of $Cr^{3+}$ by V and Ti. Very careful analyses of the pure crystal show occasional excess of trivalent over divalent oxide. This excess $M_2O_3$ is in the $\gamma$-form, which also crystallizes in a spinel-like lattice in which some of the holes are vacant.

Replacement of chromium by ferric iron does not take place to a very large extent. Hence the triangular prism which could be used to show all possible mixtures of the six species given is not needed, and a truncated prism can be used, as in Figure 1. It is helpful to locate the various domains defining the use to which chromium ores are put, and the composition of several chemical, metallurgical, and refractory ores is shown.

Thayer[10] has given a more complete description of the mineralogy and geology of chromium.

However, the ore as supplied is not a pure chromite spinel, but invariably contains gangue from 5 to 15%. This gangue is a magnesium silicate containing very little iron and chromium. The nature of the gangue is related closely to the formation geochemistry.

Chromite is formed by magmatic segregation from both pyroxenite, in which the gangue mineral is enstatite, $MgSiO_3$ (60% $SiO_2$); and peridotite, in which the gangue material was originally olivine, $Mg_2SiO_4$ (42% $SiO_2$). The pyroxenites are commoner rocks than the peridotites, and because of their higher $SiO_2$ content are less basic. High-iron, or noritic, chromites, such as the Transvaal ore used in chemical manufacture, are found in pyroxenite. The more valuable high-magnesium ores are found in peridotite, which alters to serpentine.

The chemical ore given as example D in Table 1 may be calculated to be chromite 94.6%, enstatite 5.4%. Such ores, because of their high content of $Fe^{2+}$, are more difficult to convert to chromates in the manufacturing process than the high-Mg chromites used prior to World War II, but now needed for metallurgical use.

The Manufacture of Sodium Dichromate

The key chemical from which all major industrial chromium compounds are made is sodium dichromate, usually sold and carried in trade statistics as the dihydrate, $Na_2Cr_2O_7 \cdot 2H_2O$. The basic principles of the operation have been unchanged for many years. They are:

1. Roasting of the ore under alkaline oxidizing conditions and leaching of the calcine to yield a solution of alkaline chromate.
2. Conversion of the chromate to dichromate with acid.
3. Purification of the sodium dichromate. The operation is shown diagrammatically in Figure 2.

The operation is carried out today in plants having a capacity of 30,000-100,000 tons $Na_2Cr_2O_7 \cdot 2H_2O$/year. The newest facility in the U.S. is that of the Diamond-Shamrock Chemical Corporation at Castle Hayne, N.C.

The most important secondary product is chromic acid. This and other chemicals such as potassium dichromate, prepared chromium tanning formulations, sodium chromate and ammonium dichromate are all made from sodium dichromate and will be discussed in later subsections.

Roasting and Leaching of Chrome Ore. Although the process of mixing pulverized chrome ore with sodium or potassium carbonate and a diluent such as lime has been practised since early in the 19th century, there are still a number of variations practised in modern times. Environmental considerations, relative costs of materials, overall yields, and ease of processing are all important. Also, as all plants have used Transvaal ore exclusively for the past four decades, this has affected procedures somewhat. Today, the technology depends on whether lime or recycled refuse is used as a diluent, and on the type of mechanical furnace used.

The chemistry of the roasting operation without lime is somewhat simpler, and the subsequent leaching presents fewer problems. When Transvaal ore is roasted with sodium carbonate in a rotary kiln at $1100^{\circ}$-$1200^{\circ}$ C, a suitable kiln mix is:

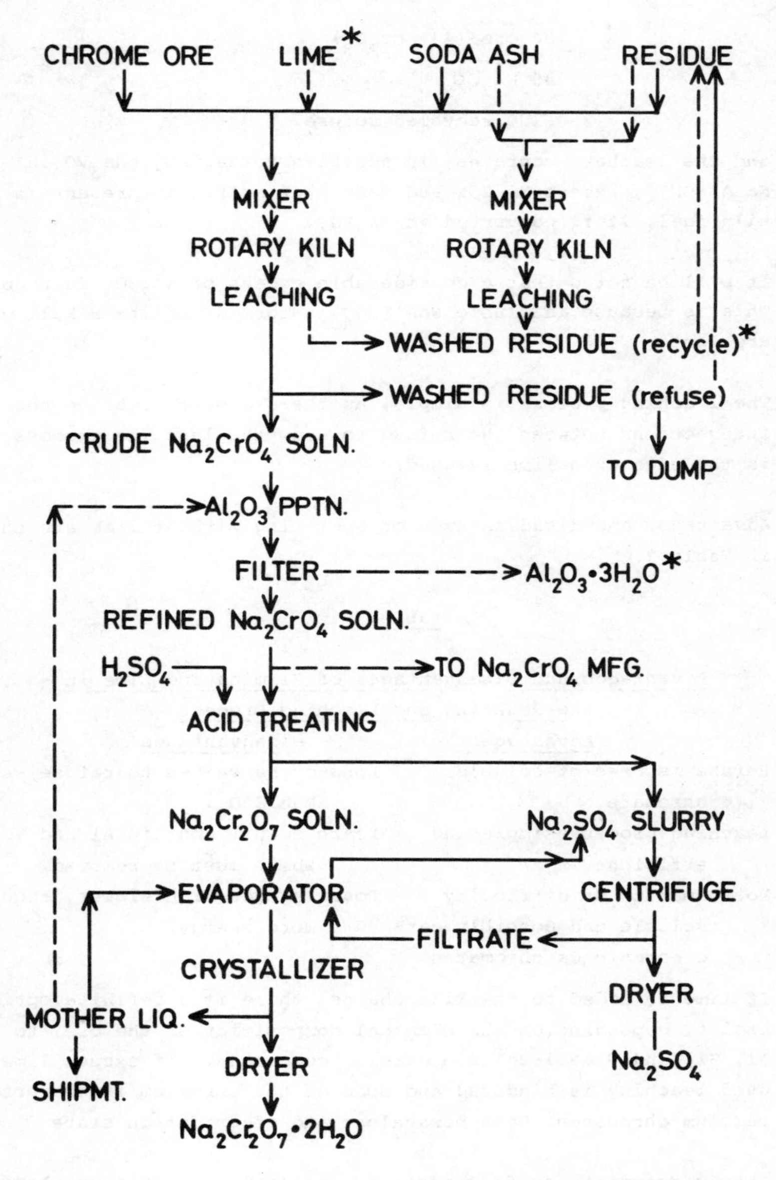


Fig. 2. Flow sheet for the production of sodium chromate and dichromate. Asterisks and dashed lines indicate optional steps.

100 ore (44% $Cr_2O_3$)

80 $Na_2CO_3$

250-300 recycled refuse.

and the leachate contains, in addition to $Na_2CrO_4$, $Na_3VO_4$, $Na_3Al(OH)_6$, excess $Na_2CO_3$ and NaOH. If sulfur is present in the kiln fuel, it is converted to $Na_2SO_4$.

It will be noted that a considerable excess of $Na_2CO_3$ is required. This is because insoluble $NaAlSiO_4$ is formed and its alkali values are lost.

The leaching process is simple, as there appear to be no chemical interactions between the refuse to slow the leaching process, as is the case when lime is used.

Advantages and disadvantages of operating without lime are shown in Table 2.

Table 2

Advantages and Disadvantages of Eliminating Lime in the Roasting and Leaching Process

| Advantages | Disadvantages |
|---|---|
| Refuse is free of soluble chromate. | Loss of Na values to refuse as $NaAlSiO_4$. |
| Leaching process simple and efficient. | Leach solns. contain Al and V which must be removed. |
| No formation of difficulty soluble and possibly carcinogenic Ca chromates. | Roasting reaction slower, requires more energy. |

If lime is added to the kiln charge, there is a definite optimum amount, depending on the chemical composition of the ore, to fix Al, Si, and V as insoluble calcium compounds. If excess lime is used leaching is hindered and some of the chromium is converted to calcium chromates, both hexavalent and of oxidation state 5 and 4.

The intermediate oxidation states available to chromium when lime is present stepped up the roasting process, but leaching is a time-consuming process and is customarily carried out in "filters," large leaching boxes in which weak leachate may be recirculated in a multi-stage operation.

Instead of rotary kilns, rotating-hearth furnaces have also been used. This is a much more capital-intensive installation, but the amount of recycling residue required is substantially less than with rotary kiln operation.

Still another variant of the roasting process which has been used is a two-step operation, in which ore, a leached residue from the first stage, lime (if used), and soda ash are used in a first roast, while residue from the first stage and ash are the feed to the final stage. A typical feed to the first stage of a two-stage roast is: ore 24, $Na_2CO_3$, 15, lime 12, residue 49 parts.

Additional variants of the roasting process involve the use of small amounts of lime, so that alumina is solubilized for recovery; or the use of dolomitic lime, bauxite, ferric oxide, and other diluents.[1,2,8].

Conversion of sodium chromate to dichromate with acid. This process is far simpler if a conventional lime addition is used in roasting and leaching, for the leach liquors then contain only $Na_2CrO_4$, excess alkali, and $Na_2SO_4$ if sulfur was present in the fuel used. Sulfuric acid is the customary conversion agent. The reaction is:

$$2Na_2CrO_4 + H_2SO_4 \rightarrow Na_2Cr_2O_7 + Na_2SO_4 + H_2O$$

This reaction is carried out either continuously or batchwise, using pH control. Much of the sodium sulfate formed is precipitated; it is combined with sodium sulfate slurry streams from the purification process described below. This is centrifuged, washed, and the $Na_2SO_4$ dried. It contains small quantities of dichromate and is used in the Kraft paper industry.

The use of carbon dioxide for conversion is an interesting variant if sodium dichromate is not converted in quantity into chromic acid, especially if lime is used. The reactions are:

$$CaCO_3 \rightarrow CaO + CO_2 \text{ (lime to roasting)}$$

$$2CO_2 + 2Na_2CrO_4 + H_2O \rightarrow Na_2Cr_2O_7 + 2NaHCO_3 \text{ (lime to roasting)}$$

$$2NaHCO_3 \rightarrow Na_2CO_3 + H_2O + CO_2 \text{ (}CO_2\text{ to conversion, }Na_2Co_3\text{ to roasting)}$$

Special kilns have been used to obtain high-purity $CO_2$, which is a necessity for the precipitation of $NaHCO_3$. The conversion step proceeds about 75% at moderate $CO_2$ pressures.

If low-lime or no-lime roasting is used, the leach solutions contain alumina and vanadium. Sulfuric acid, sodium dichromate, and carbon dioxide have all been used to precipitate alumina. Conditions must be very carefully controlled to precipitate alumina as the crystalline hydrate, gibbsite, $Al_2O_3 \cdot 3H_2O$. If $CO_2$ is used and the Na concentration is high, dawsonite, $NaAl(OH)_2CO_3$, may precipitate, with loss of alkali and $CO_2$.

Vanadium can be removed by treatment with lime, or by acidification of the mother liquors from sodium dichromate crystallization. Vanadium is an undesirable impurity in tanning and in pigment manufacture.

Purification of Sodium Dichromate. The feed for the purification process comes from two sources: (1) the direct conversion of chromate to dichromate by $H_2SO_4$ and/or $CO_2$ as described above; (2) the conversion of chromate to dichromate by the sodium bisulfate by-product from the manufacture of chromic acid. This process will be described in detail in the following section.

The feed to the evaporators consists of a mixture of sodium dichromate and sodium sulfate, containing about 25-40% sodium dichromate and saturated with sodium sulfate. The solution should be free of $Cr^{3+}$ and pH must be carefully controlled. The processing of this mixture is carried out in multiple-effect evaporators.

The evaporator output is saturated with sodium dichromate at about 65°C and contains 80-85% $Na_2Cr_2O_7 \cdot 2H_2O$. It is then sent to crystallizers to produce sodium dichromate crystals for commercial use. The mother liquor containing about 77% $Na_2Cr_2O_7 \cdot 2H_2O$, is filtered and used (1) for the production of secondary products such as potassium dichromate and prepared tanning compounds; (2) for shipment, after adjustment to 69% $Na_2Cr_2O_7 \cdot 2H_2O$ (the eutectic composition, f.p. -48°C).

The sale and use of mother liquor permits some purging of impurities, such as chloride from the use of Solvay process soda ash (the problem does not exist with the natural sodium carbonate from trona now in U.S. use), and vanadium.

## The Manufacture of Chromic Acid, $CrO_3$

Increasing markets in wood preservation and hard chromium plating have greatly increased the demand for chromic acid, which required 42.3% of sodium dichromate production in 1979, up from 27.0% in 1973.[11]

The fundamental equation for the manufacture of chromic acid is:

$$Na_2Cr_2O_7 + 2H_2SO_4 \rightarrow 2CrO_3 + 2NaHSO_4 + H_2O$$

Several variations of the process are in current commercial use, as shown in Table 3.

Table 3

Variations in the Chromic Acid Manufacturing Process

| Variant | Source of Cr | $H_2SO_4$ | Process |
|---|---|---|---|
| 1 | $Na_2Cr_2O_7 \cdot 2H_2O$ | 66°Be | Batch. Heat to fusion. $CrO_3$ and $NaHSO_4$ form two immiscible layers. Tap off and flake $CrO_3$. |
| 2 | $Na_2Cr_2O_7 \cdot 2H_2O$ | 66°Be | On mixing, a slurry of $CrO_3$ plus solution is separated and heated to fusion as in (1). The mother liquor is returned to $NaHSO_4$ conversion. |
| 3 | $Na_2Cr_2O_7$ | Oleum | An exothermic reaction produces $CrO_3$ and $NaHSO_4$ directly in the molten state.[13,14] |

Processes 2 and 3 may be considered in the light of Process 1, the original process. The time required to bring the mixture to complete fusion, and the temperatures necessary, cause substantial spontaneous decomposition of $CrO_3$ which is unstable at its melting point:

$$2CrO_3 \rightarrow Cr_2O_5 + \frac{1}{2}O_2 \text{ (approx)}$$

In the presence of $NaHSO_4$ the $Cr_2O_5$ reacts to form both soluble and insoluble $Cr^{3+}$ sulfates.

Under today's environmental pressures neither the acid nor the chromium values can be discarded. If the chromium is substantially all in the hexavalent state, it can be returned to the conversion operation directly. This is the purpose of the more sophisticated variations (2) and (3) in Table 3. If appreciable $Cr^{3+}$ is present in a bisulfate stream, as in variation (1) or the fusion portion of variation (2) then an alkaline precipitation and filtration step is required. The insoluble chromic oxide portion is returned to the roasting operation and the soluble hexavalent fraction returned to the appropriate point in the conversion process.

By this process, chromic acid is obtained as a flake produced by passing the molten $CrO_3$ over chilled steel rolls. The product contains less than 0.1% $SO_4$, as required by the plating industry, the major consumer.

## Other Chromium Chemicals

Potassium dichromate, $K_2Cr_2O_7$, is prepared by metathesis between sodium dichromate and potassium chloride:

$$2KCl + Na_2Cr_2O_7 \rightarrow K_2Cr_2O_7 + 2NaCl$$

Solubility relationships in the reciprocal salt system are such that NaCl separates on evaporation, while $K_2Cr_2O_7$ crystallizes in very high purity on cooling. Potassium dichromate is stable over a wide humidity range and is used in dry mixtures, brake linings, safety matches and similar dry formulations.

Sodium chromate, $Na_2CrO_4$, is prepared from the sodium chromate leach liquors after removal of alumina. The solution is brought to the neutral point with sodium dichromate, and then evaporated to dryness. Any sulfate present by virtue of sulfur present in the fuel used in roasting will be present in the product. The tetrahydrate, $Na_2CrO_4 \cdot 4H_2O$, is also available; it is prepared by crystallization of the neutralized and purified leach solution. It will still contain some sulfate by virtue of isomorphous replacement. Sodium chromate is used largely as a corrosion inhibitor for recirculating water systems and drilling muds.

Ammonium dichromate, $(NH_4)_2Cr_2O_7$, is prepared by a process similar to that for potassium dichromate, using ammonium sulfate and sodium dichromate as starting materials. It is used in photography, photo-engraving, and as a combustion moderator in solid rocket propellant mixtures.

Chromium tanning formulations have basic chromic sulfate, $Cr(OH)_x(SO_4)_{(3-x)/2}$, as their active ingredient. They are sold on the basis of basicity, which is defined as 100x/3 in the above formula. Two principal methods of production exist. In each case, sodium dichromate mother liquor is treated with a reducing agent. If the reducing agent is $SO_2$, the basicity of the solution is automatically 33%:

$$Na_2Cr_2O_7 + 3SO_2 + H_2O \rightarrow Na_2SO_4 + 2Cr(OH)SO_4$$

The other reducing agent customarily used is a carbohydrate. Molasses, glucose, corn syrup, and even sawdust have been used. The reaction is theoretically:

$$C_6H_{12}O_6 + 4\ Na_2Cr_2O_7 + (16-4x)H_2SO_4$$
$$4\ Na_2SO_4 + 8\ Cr(OH)_x(SO_4)_{(3-x)/2} + 6\ CO_2 + (22-8x)H_2O$$

but a certain amount of the sugar ends up as organic acid, so an empirical amount of $H_2SO_4$ is used to prepare commercial products with a basicity of from 33 to 52%. Additions, such as aluminum sulfate, syntans, and sodium formate, may now be made to the solution before it is spray-dried to produce the commercial dry product, which is sold under proprietary names.

## Health, Safety, and Environmental Considerations

As has been indicated in the Introduction, environmental considerations have played an increasing role in the technology of chromium compounds in the U.S. in recent years. Such recent legislation as the Resource Conservation and Recovery Act (RCRA), the earlier stipulations of the Occupational Safety and Health Administration (OSHA) and the Environmental Protection Agency (EPA), as well as the conservative trend shown in the 1980 elections, and the need for energy and material conservation which has now become evident, will serve to dictate the future uses of chromium compounds, as well as the production technology.

Perhaps the best fairly recent reference to the toxicology and industrial hygiene of chromium compounds is by Baetjer.[15] A fairly complete critical bibliography is given in [1].

Acute and Chronic Toxic Effects. From the practical standpoint, the only significant chromium compounds are those in oxidation states +3 and +6. No appreciable acute or chronic toxicity has been attributed to the trivalent compounds; hence we need be concerned only with the hexavalent oxidation state. As has been stated, these compounds are the ones of major commercial importance.

Acute systemic poisoning is rare. Fatalities have resulted from the accidental use of potassium chromate in an ointment. Non-fatal ingestion has caused kidney damage, yet continued use of up to 25 ppm $CrO_3$ in drinking water appears to be without harmful effect.[16] U.S. drinking water standards are set at 0.05 ppm, but it has been suggested that higher concentrations may be harmless.[17] The $LD_{50}$ is probably in the order of 3-5 g $CrO_3$ equivalent for adult humans.

Acute effects of chromates on the skin and mucuous membranes have been observed for 150 years. The breathing of chromate-containing dusts and mists leads to ulceration and eventual perforation of the cartilaginous portion of the nasal septum. As a result of studies of this phenomenon, exposure standards for air-borne particulates are set at 0.1 mg $CrO_3/m^3$.[18] Chromates may also produce ulcers when a cut or abrasion is contaminated.

Dermatitis and allergic reactions may result from prolonged exposure of the skin to chromates. In the majority of workers, dermatitis results only from prolonged exposure or conditions such as exposure to solvents. A few individuals exhibit an allergic response on exposure to very small amounts of chromates.

Lung cancer as a result of long-term exposure to high concentrations of chromate dusts was observed in German chromate plants as early as 1912. Several later studies on American experience demonstrate that work exposures in the plants in use prior to World War II produced a significant incidence of lung cancer.[19]

Under the conditions that existed prior to 1940, the average incubation period for chromate-induced cancers was 18 years. During World War II, both working hours and dust concentrations increased in the obsolete American plants. This was reflected in an increase in cancer cases during the 1950's and 1960's. There is definite evidence that the building of modern plants with adequate dust controls has eliminated the problem for the most part.[20]

The chromium compounds responsible for carcinogenesis are not clearly defined. Animal experiments have implicated only calcium, lead, zinc, and chromic chromates. These are all difficultly soluble chromates. Also, as encountered in the manufacturing operation, both calcium and chromic chromates can contain Cr(V) species such as $CrO_{2.4}$ and $Ca_3(CrO_4)_2$. It is possible that Cr(V), a powerful oxidant whose stereochemistry mimics P(V), and which could be generated continuously in very small amounts from a reservoir of difficultly soluble chromate, is the actual carcinogen.

Except for pigment plants where heavy concentrations of zinc or lead chromate dust may exist, the cancer hazard from chromates is confined entirely to those manufacturing plants which have not instituted modern pollution controls.

<u>Chromium in Nutrition</u>. Although chromium is widely distributed in soils, and hence in plant and animal tissues, it has only recently been recognized as an essential element in plant and animal nutrition.

Certain evidence for chromium as an essential element in animal and human nutrition has been reported.[21] Chromium plays a part in sugar metabolism, and intake levels of 50 μg/day improve glucose tolerance in the elderly. The U.S. daily average of 80 μg/day is probably marginal.

The case for plant nutrition is not well documented. Workers at Southwest Research Institute found that sugar-producing crops, such as sugar cane and pineapple, had a high $Cr_2O_3$ content and that $Cr_2O_3$ removed from the soil needed to be returned.[22] Circumstantial evidence is strong, for Hawaii and Cuba, both producers of sugar-rich crops, have basaltic or lateritic soils that are high in chromium.

Chromium in the Environment. Chromium is widely distributed in the earth's crust, but concentrates, as shown in the section on Chromium Ores and Chromium Supplies, in basic, ultramafic rocks. At an overall crustal concentration of 125 ppm Cr, it is the twentieth most abundant element. Chromium, especially in the 3+ state, is very insoluble and leaches very little into natural waters. In 15 North American rivers, the range was 0.7-84 ppb. Public water supplies had 0 to 36 ppb, with the median being 0.43 ppb. Food contains chromium in a range from 20 to 590 ppb, resulting in a daily intake for humans of 10-400 μg; the average is about 80 μg. About 20 U.S. cities, many of them industrial, show measurable amounts of chromium in air, generally about 0.01-0.02 $\mu g/m^3$. Baltimore, with a chromium chemicals and stainless steel plant, had 0.80 μg $Cr/m^3$ in 1960, and 0.10 $\mu g/m^3$ in 1969, showing a downward trend as more effective dust control equipment is installed. Rural air contains no detectable chromium.

Environmental Control and Public Health and Safety. In the light of the above facts regarding the biological activity of chromium and its compounds, and the wealth of knowledge regarding the chemistry of chromium, it would seem that the achievement of reasonable safety in the work place and in the environment would not be an extremely onerous task. Plants built with the best available technology of the 1950's have proved to be safe places to work, and the newer plants in the industry are exemplary.

If, as will be shown in the section on Uses of Chromium Compounds, the compounds of chromium have an important part to play in the conservation program which is equally important to environmental quality, then legislation should be so written as to be based on humanistic cost-benefit assessments. This is not the tenor of present American regulation, where there is a tendency to bar or very stringently regulate materials which may be carcinogenic, ontogenic, mutagenic, fetotoxic, or teratogenic, as demonstrated by animal experimentation with massive doses. Such action is purely legalistic and does not reflect sound scientific thought.

There is a growing body of evidence, supported in the case of chromium compounds with epidemiological evidence,[20] that exposures within the easy capability of good operating procedures [1] are giving good protection to workers and the general public. This reinforces the conclusion that the response of a population to a carcinogen is a form of the biological growth curve, concave upward, and so showing very little response at low concentrations.[23]

Finally, the user of chromium chemicals should know that $Cr^{6+}$, and $CrO_3$ particularly, is a very powerful oxidizing agent. In formulating proprietary mixtures, care should be taken that potential fire and explosion hazards do not exist. Ammonium dichromate also decomposes in a self-sustaining reaction when heated, the so-called "chemical volcano." The American Chemical Manufacturers' Association, Washington, D.C., has published chemical safety data sheets describing methods for the safe handling of chromates, dichromates, and chromic acid.[24]

## Uses of Chromium Compounds

It is axiomatic that the uses of any chemical depend on its properties. In 1980, the environmental restrictions being imposed on chemicals generally require that we look at the uniqueness and essentiality of the chemical for each use.

An obvious property of chromium compounds, which antedates in application the discovery of the element, is color. Pigments have long been an important use of chromium compounds. There are, however, other materials which can perform the same function. Pigments are still an important single use of chromium, but we can expect declining consumption.

Chromium compounds were frequently used as an oxidizing agent, especially in organic synthesis. This is now an extinct application. A few minor uses, as in oxidative dyeing, photography, lithography, and rocket propellants, still remain where the oxidative properties are called into play.

In dyeing, color is important, and occasionally the use of chromium is necessary to produce the desired shade. More importantly, chromium, as a mordant, and an essential part of some premetallized dyes, confers the property of durability or fastness on the color. In a general way, this concept of durability has permeated almost all the newer uses of chromium compounds. Obviously, the cause of conservation is served equally by using less, or by making existing material last longer. Other uses which manifest this concept of durability are hard chromium plating, chrome tanning of leather, metal finishes, corrosion prevention, drilling muds, and wood preservation. It is significant that these uses are increasing, while a largely decorative use, such as decorative chromium plating, can be expected to decline in the future.

A summary of current uses of chromium chemicals is given in Table 4. Each use will be discussed in the order given.

Pigments. Recent figures on the U.S. consumption of chromium chemicals in the production of pigments are shown in Table 5. The general decrease in share of the market is apparent. We now discuss the individual pigments.

The older chromium pigments are mainly compounds of lead, and lead chromate, $PbCrO_4$, both pure and in isomorphous mixtures with lead molybdate, $PbMoO_4$, and lead sulfate, are all important decorative pigments. Additives are used to improve physical properties such as working properties, hue, light fastness, and crystal size, shape, and symmetry. Manufacturing procedures are proprietary.

Medium chrome yellow, an orange-yellow, is essentially pure monoclinic $PbCrO_4$. This is the stable form, which occurs naturally as the mineral crocoite.

Table 4

Major Uses of Chromium Compounds, 1979[11,24,25,26,27]

| Use | D.E.* Consumed metric tons | % of Market | Property Involved | Trend |
|---|---|---|---|---|
| Pigments | 33150 | 21.6 | Color | - |
| Hard Chromium Plating | 27444 | 17.9 | Durability, Conservation | + |
| Tanning & Textiles | 22780 | 14.9 | Durability Color | ± |
| Misc. Metal Finishing | 20996 | 13.7 | Decorative, Corrosion Prevention | - |
| Decorative Chrome | 18296 | 11.9 | Appearance, Protection | - |
| Wood Preservation | 16783 | 11.0 | Durability, Conservation | + |
| Miscellaneous | 9177 | 6.0 | -- | ? |
| Drilling Muds | 4584 | 3.0 | Corrosion Protection Cost Reduction | ± |
| Total | 153210 | 100.0 | | |

*D.E. = Dichromate equivalent.

Table 5

Production of Chromium Pigments and Corresponding Consumption of Sodium Dichromate, 1974-1978, Metric Tons[11]

| Year | 1974 | 1975 | 1976 | 1977 | 1978 |
|---|---|---|---|---|---|
| Chrome yellow and orange | 37,942 | 26,091 | 35,335 | 35,207 | 37,028 |
| Molybdate orange | 14,586 | 9,559 | 16,883 | 13,974 | 14,403 |
| Chrome oxide green | 7,676 | 5,608 | 6,140 | 8,796 | 12,321 |
| Zinc yellow[a] | 4,100 | 3,300 | 2,500 | 1,700 | 1,500 |
| Chrome green and other | n.a. | n.a. | n.a. | n.a. | n.a. |
| Sodium dichromate equivalent | 42141 | 29374 | 36809 | 35078 | 36553 |
| % of production | 26.0 | 26.0 | 25.8 | 24.7 | 23.3 |

[a]Estimated from 1974-75 data n.a. = not available.

Lemon and primrose chrome yellows contain up to 40% $PbSO_4$, and some or all of the $PbCrO_4$ in the metastable orthorhombic form, which is stabilized by $PbSO_4$, which occurs naturally as the orthorhombic mineral anglesite. The higher the orthorhombic content, the greener and paler the shade.

Molybdate oranges are tetragonal solid solutions of $PbCrO_4$, $PbSO_4$, and $PbMoO_4$. An aging step is required in manufacture to permit development of the orange tetragonal form, which is isomorphous with natural $PbMoO_4$, the mineral wulfenite.

Chrome oranges are basic lead chromate, $PbCrO_4 \cdot PbO$.

Lead silicochromate, essentially medium chrome yellow precipitated on silica, has been developed for use in traffic paints where silica gives better abrasion resistance.

Chrome green. Lead chromate, precipitated or mixed with an iron blue such as Prussian blue, is known as chrome green. It has a strong tendency to change hue on exposure, generally becoming bluer. It has been largely displaced in recent years by phthalocyanine greens and by chromic oxide green.

The other important color pigments containing chromium are the chromium oxide greens, comprising anhydrous $Cr_2O_3$ and the hydrated Guignet's green.

Chromium oxide green, $Cr_2O_3$, is made by three manufacturing processes:

1. An alkali dichromate is reduced in a self-sustaining dry reaction by a solid reducing agent such as sulfur, carbon, starch, or wood flour. For pigment use, the reducing agent is sulfur. When a low-sulfur grade is needed as raw material for the manufacture of aluminothermic chromium, a carbonaceous reducing agent is employed:

$$Na_2Cr_2O_7 + S \rightarrow Na_2SO_4 + Cr_2O_3$$

$$Na_2Cr_2O_7 + 2C \rightarrow Na_2CO_3 + Cr_2O_3 + CO$$

After ignition with an excess of reducing agent in a reverberatory furnace or small kiln, the calcine is leached, washed, filtered, dried and pulverized. The product is 99+% $Cr_2O_3$; metallurgical grades contain less than 0.005% S.

2. Chromate-dichromate solutions are reduced by sulfur in a boiling alkaline suspension:[29]

$$2\ Na_2CrO_4 + Na_2Cr_2O_7 + 6S + 2xH_2O \rightarrow 2\ Cr_2O_3 \cdot x\ H_2O + 3\ Na_2S_2O_3$$

Excess NaOH is added to start the reaction, and not over 35% of the chromium can be added as dichromate. After filtration to recover thiosulfate, the hydrous oxide slurry is acidified to pH 3-4 and washed free of sodium salts. A fluffy oxide is obtained on calcining at 1200-1300°C; it may be densified and strengthened by grinding. The shade can be varied by additives and changing the chromate-dichromate ratio.

3. A dichromate or chromate solution is reduced under pressure to form a hydrous oxide, which is filtered, washed, and calcined at 1000°C. The calcined oxide is washed and dried. Suitable reducing agents are sulfur, glucose, sulfite, and reducing gases; temperatures of up to 210°C and pressures of 40-50 atm. are used.

Chromium oxide green is the most stable green pigment, and is non-toxic.

Guignet's green, $Cr_2O_3 \cdot XH_2O$, where x = about 2, is obtained from the heating of dichromates in a boric acid melt. It is permanent, but cannot be used in high-temperature ceramics. It has poor strength, but is a clear, transparent bluish-green, It is used mainly in automotive finishes.

Corrosion inhibiting pigments. These are all slightly soluble chromates. The major pigment of this group is zinc chromate; others includes zinc tetroxychromate, basic lead silicochromate, barium potassium chromate, and strontium chromate.

Zinc chromate (zinc yellow) became a major corrosion-inhibiting pigment for aircraft during World War II, and the production volume of that time has never been attained since. It is now used for protection of automobile bodies and light metals, and in combination with red lead and ferric oxide for structural steel.

Zinc yellow has the empirical formula $K_2O \cdot 4ZnO \cdot 4CrO_3 \cdot 3H_2O$. It is made by a variety of processes, all based on the reaction of zinc compounds, chromates, and potassium compounds in an aqueous medium. In one process to give a premium sulfate-free product, zinc oxide is "swollen" with potassium hydroxide and the chromium is added as potassium tetrachromate:[30]

$$4ZnO + K_2Cr_4O_{13} + 3H_2O \rightarrow K_2O \cdot 4Zno \cdot 4CrO_3 \cdot 3H_2O$$

Zinc tetroxychromate has a somewhat lower chromate solubility and has been used in wash primers.

Strontium chromate, $SrCrO_4$, works well on light metals, and is compatible with some latex emulsions which are coagulated by zinc chromate. It is also an ingredient of proprietary formulations for chromium plating (see below).

Basic lead silicochromate is a composite in which basic lead chromate (chrome orange) is precipitated on a lead silicate-silica base.

Recent price quotations for chromium pigments are shown in Table 6.

Good references on pigments are few. Further information is given in [1, 33, 34].

Table 6

Prices of Chromium Pigments, 1978-1980[31,32]

| Pigment | Price, U.S. dollars/lb. Sept. 1978 | Aug. 1980 |
|---|---|---|
| Chrome yellow | 0.83 | 1.09 |
| Chrome orange | 0.83 | 0.83 |
| Molybdate orange | 1.09 | n.q. |
| Chrome green, pure | 1.36 | 1.71 |
| Chromium oxide green | 0.78 | 1.35 |
| Guignet's green | 2.10 | n.q. |
| Zinc chromate | 0.86 | 1.12 |

n.q. = not quoted

Hard Chromium Plating is a successful process for reclaiming many industrial parts and tools, and for reducing wear and friction on others to prolong their life. In hard chromium plating, the plate is of appreciable thickness, from 1 to 300 μm. Unlike decorative plate, it is applied directly on a hard substrate such as steel.

Hard chromium plate is applied by electrolysis in a solution of chromic acid containing a small and carefully regulated amount of a catalyst ion such as sulfate or fluosilicate. It differs from decorative plating in operation at slightly higher temperatures, about 50$^{o}$C, and current densities, 10,000 A/m$^2$. Proprietary formulations are available, in which the sulfate and fluosilicate concentration are regulated by such salts as $K_2SiF_6$ and $SrSO_4$ in suspension.

The most important applications of hard chromium plating are cylinder liners and piston rings for internal combustion engines, and the rebuilding of worn cutting tools for machine tool equipment. There are excellent references[35,36,37,38] on chromium plating. During the period 1976-79, the percentage of chromium used in hard chromium plating has increased from 8.3 to 17.9.

Tanning and Textiles. Although compounds of hexavalent chromium are most important from the manufacturing standpoint, it is the ability of chromium in the trivalent state to form stable coordination complexes with proteins, cellulosic materials, dyestuffs, and polymers that is the common denominator underlying this use. Many of these reactions are still imperfectly understood.

The chrome tanning of leather is one step in a complicated series of operations leading from the raw hide to the finished product. Chrome tanning is the most important tannage for all hides except heavy cattle hides, which are usually vegetable tanned. Large tanneries usually make their own chrome tanning formulations (see above) from sodium dichromate. The tanning process is described empirically in a group of bulletins;[39] the chemistry has also been described.[40]

In the textile industry, sodium chromate and other chromium compounds find a variety of applications.[41] The former is used as oxidant and source of chromium; for example, to dye wool and synthetics with mordant acid dyes, oxidize vat dyes and indigosol dyes on wool, after-treat direct dyes and sulfur dyes on cotton to improve wash fastness, and oxidize dyed wool. Chromium also finds use in the manufacture of premetallized dyes, which are generally hydroxyazo or azomethine dyes in which a coordinating metal is directly incorporated. Still another use of chromium compounds is in the production of water- and oil-resistant coatings on textiles, plastics, and fiber-glass. Registered trade names are Quilon, Volan, and Scotchguard.

Metal Finishing and Corrosion Inhibition. In metal finishing, chromium compounds are used in the production of chemical conversion coatings. These are produced on non-ferrous metals, mainly magnesium, aluminum, zinc, and cadmium, by immersion in a hexavalent chromium solution containing an activating ion at a controlled pH. For color modification, other ingredients are added. The formulations are largely proprietary, and the largest use today is for colorless coatings on zinc.[42]

Oxide films on aluminum are produced by anodizing in a chromic acid solution. They impart exceptional corrosion resistance and paint adherence and are widely used on military aircraft assemblies. The films may be dyed. The usual procedure is to anodize at 35°C for 30 min. in a solution having 100 g total $CrO_3$/l and a pH of 0.8-0.9.[43]

Chromates are used to inhibit metal corrosion in recirculating water systems. Steel immersed in dilute chromate solutions does not rust, and corrosion of other metals is similarly prevented. The exact mechanism is unknown. Anode polarization plays a part, and in the inhibition on iron a film of $\gamma$-$Fe_2O_3$ containing Cr appears to form.

The concentration of chromate required to inhibit corrosion may range from 50 to 20000 ppm, depending on conditions. A pH of 8.9 is usually optimum.

Restrictions on the discharge of dilute chromate solutions into sewage have greatly limited this very effective procedure for preventing corrosion, conserving metal values, and improving heat transfer. It is not generally known that conventional sewage plants are not adversely affected by moderate dosages of chromate,[44] or that chromate, being an oxidant, is detoxified by the organic materials present in sewage.

Dichromates or chromic acid are used as sealers and after-dips to improve the corrosion resistance of various coatings on metals, such as phosphate coatings on zinc or steel, sulfuric acid anodic coatings on aluminum, and TFS plate on steel.

Dichromates are used for the etching and bright-dipping of copper and its alloys. A typical composition for the removal of scale after heat-treating contains 30 g $Na_2Cr_2O_7 \cdot 2H_2O$ and 240 ml $H_2SO_4$ per liter. It is used at 50-60$^{o}$C.

Decorative Chromium Plating. In decorative plate, the thickness of the chromium is 0.2-0.5 μm. It is widely used for automobile body parts, appliances, plumbing fixtures, and other products. When applied over steel, the plate is underlaid by electro-deposited nickel, frequently of more than one type. The plating conditions are also regulated to control the nature of the chromium. Thus, microcracked, microporous, crack-free or conventional chromium may be plated over simple, duplex, and triplex nickel undercoats.[35,36,37,38]

Wood Preservation. Wood preservation shares with hard chromium plating the distinction of being the fastest-growing use of chromium chemicals. Between 1976 and 1979, the percentage use in wood preservation increased from 5.4 to 11.0. This has been due to the excellent results achieved by chromated copper arsenate (CCA), available in three modifications under a variety of trade names. The treated wood is free from leaching of toxic materials, paintable, and of an attractive olive-green color. It is widely used, especially for utility poles, building lumber, and wood foundations.

Chromium-containing fire-retardants for wood are also used. Here the function of the chromium is to prevent leaching of the fire retardant from the wood and corrosion of the treating equipment.

The function of chromium in wood preservation is to prevent leaching of the water soluble preservatives, to reduce corrosion, and to reinforce the preservative value of the other ingredients.[45] In common with other preservatives, chromium-containing formulations, now representing 20-25% of all wood treated, share in saving about $7 billion and 660 square miles of forest land, as well as substantial quantities of energy, for the American economy each year.[46,47]

Drilling Muds. Since 1941, chromates have been used in the drilling of oil and gas wells to prevent fatigue corrosion cracking of drill strings; about 1 ton is required for an average West Texas well. More recently, however, proprietary drilling-mud formulations, specially designed to suit the ground water and rock strata in which the well is located, have been developed. In addition to both soluble and insoluble chromates, the formulations frequently contain chromium lignosulfonate, made like a tanning compound, but using lignosulfonate waste from sulfite pulp mills as the reducing agent.

Oil well drilling in the U.S. is now at a new high, so the present outlook for drilling mud consumption is favorable.

Miscellaneous Uses. The most important of these are:

Catalysts, consuming about 1500 tons of bichromate equivalent annually in the U.S. Chromia-alumina, zinc chromite, and copper chromite, used for polymerization, dehydrogenation, and oxidation are among the most important classes of chromium catalysts.

Photography and photoengraving uses depend on the ability of hexavalent chromium compounds to react with natural and synthetic organic polymers when exposed to light to form insoluble coatings. When a plate is coated, exposed, and rinsed, the bare metal can then be etched to produce a printing or lithographic plate.

Batteries. The shelf-life of dry batteries is increased 50-80% by the use of a few grams of chromate near the zinc anode. Also, since World War II, the U.S. military and space programs have used small quantities of barium and calcium chromate as depolarizers and activators in fused-salt batteries.

Magnetic Tapes. Chromium dioxide, $CrO_2$, made from chromic acid, is used as a ferromagnetic material in high-fidelity magnetic tapes. It has numerous advantages over the conventional magnetic iron oxides.

Fine and reagent chemicals. About 60 other chromium compounds have occasionally been used commercially in small volume. For listing and further details, as well as a more detailed exposition of uses, see references.[1,2]

## Market Statistics and Trends

On the international scale, the U.S. and Germany no longer dominate the world markets as they did before World War II. The U.K. has continued to manufacture chromium chemicals, and facilities have been added in Poland, the Union of South Africa, Mexico, Japan, and India, to mention just a few. Accordingly the export market in the U.S. has declined, and is now listed with the miscellaneous uses in the data to be presented.

As international statistics are not available, the remainder of this section will be concerned with U.S. facilities, use distribution, and future trends. Data have been compiled from a variety of sources and privately correlated. Unpublished information has been supplied,[25,26] and the assistance rendered is gratefully acknowledged. However, the supply and environmental problems which beset the U.S. are common in a general way to the industrial free world, so the analyses in this section should be helpful.

U.S. Production Facilities are concentrated in three plants: Diamond Shamrock at Castle Hayne, N.C., 94,000 tons $Na_2Cr_2O_7 \cdot 2H_2O$ annually; Allied Chemical at Baltimore, Md., 65,000 tons; and PPG Industries, Corpus Christi, Tex., 33,000 tons. The first two of these each have chromic acid facilities amounting to 20,000 tons $CrO_3$ (29,800 tons $Na_2Cr_2O_7 \cdot 2H_2O$) annually.[27,28] Increases in production at Castle Hayne to 109,500 tons are scheduled, and capacity demand for chromic acid has prompted the scheduling of expansions in these facilities. The profile articles cited stress supply and environmental considerations as important factors.

Market Trends. Statistics on the use and total production of chromium chemicals are tabulated in Table 7, while the total production is shown graphically in Figure 3. The solid lines of Figure 3 are plotted from a logarithmic regression of the data in Table 7, corrected for abnormally low production in 1975, when overproduction in 1974, a recession, and a strike all curbed output. The regression curves show the increase in sodium dichromate production to be 3.1% per year, while the increase in chromic acid is 8.0% per year. The increase in chromic acid is caused largely by the increase in hard chromium plating, for which no other chromium compound can substitute, and wood preservation, in which chromic acid is used for concentrated liquid preservation formulations.

Prices of sodium dichromate over the past few years are quoted as follows in U.S. dollars per pound:

| 1975 | 1976 | 1977 | 1978 | 1979 | 1980 |
|---|---|---|---|---|---|
| 0.23 | 0.24 | 0.28 | 0.30 | 0.37 | 0.37 |

In August, 1980, quotations for other chromium compounds were:

| Compound | Price $ per pound |
|---|---|
| Chromic acid | 1.03 |
| Sodium chromate, anhydrous | 0.39 |
| Potassium dichromate | 0.48 |
| Ammonium dichromate | 0.78 |

General Conclusions. Data on the geochemistry of chromium ores, production methods, environmental considerations, and uses have been developed, largely on the basis of current American information. The importance of the supply situation, the conviction that environmental and hygiene problems are thoroughly under control with present technology, and the direction of use patterns toward essential conservation practices are definitely emerging as forces in today's production and use of chromium compounds. Chromic acid will continue to occupy a larger share of the total market because of the need for this product in wood preservation and hard chromium plating.

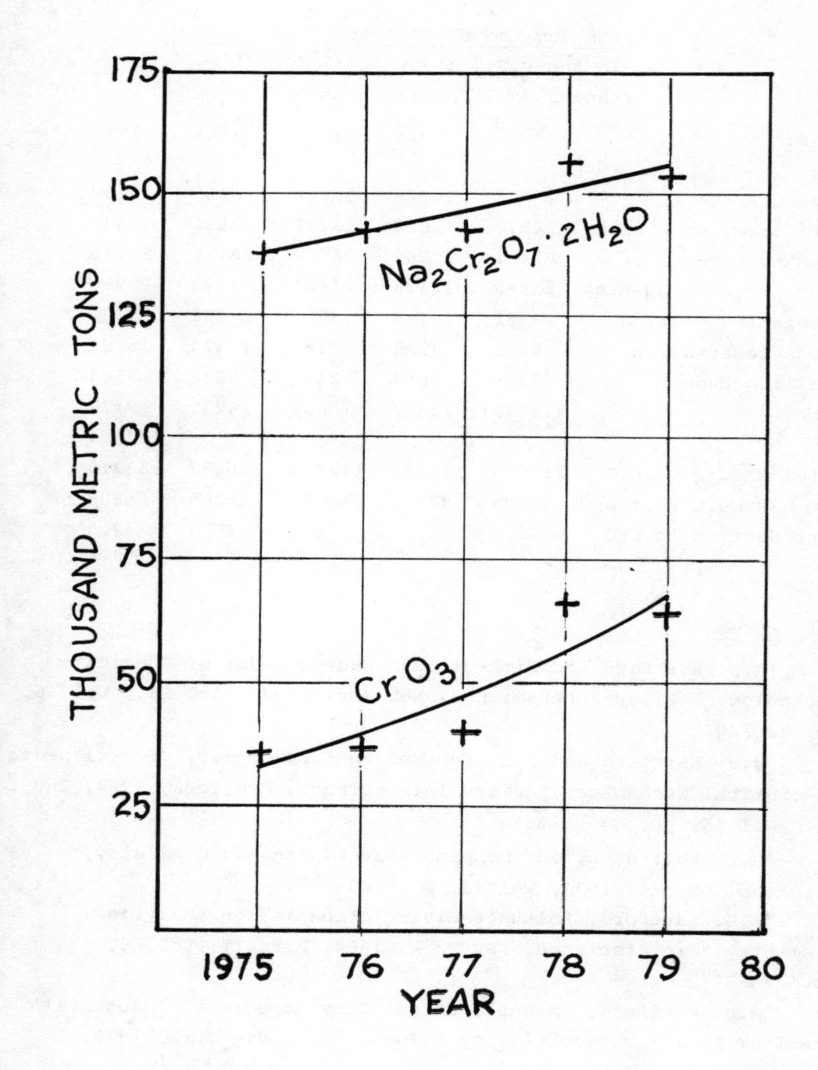


Fig. 3. U. S. production of sodium dichromate and chromic acid (as dichromate equivalent) 1975-79. 1975 figures have been adjusted to allow for abnormal conditions.

Table 7

Production and Use of Chromium Compounds in the U.S., 1975-79[11,25,26,48]

(thousands of metric tons)

| Year | 1975 | 1976 | 1977 | 1978 | 1979 |
|---|---|---|---|---|---|
| Use | | | | | |
| Pigments | 29374 | 36809 | 35078 | 36553 | 33150 |
| Hard chromium plating | 6970 | 11833 | 13298 | 25986 | 27444 |
| Tanning & Textiles | 22482 | 28269 | 26072 | 26110 | 22780 |
| Misc. metal finishing | 21480 | 22493 | 25426 | 17797 | 20996 |
| Decorative plating | 3351 | 18124 | 1393 | 14180 | 18296 |
| Wood Preservation | 6430 | 7732 | 12364 | 14574 | 16783 |
| Drilling Muds | 3389 | 5694 | 5219 | 5222 | 4584 |
| Other | 9498 | 11387 | 10859 | 16235 | 9177 |
| Total $Na_2Cr_2O_7 \cdot 2H_2O$ | 112974 | 142341 | 142209 | 156657 | 153210 |
| $CrO_3$ as $Na_2Cr_2O_7 \cdot 2H_2O$ | 25080 | 37151 | 40102 | 66266 | 64808 |
| % production to $CrO_3$ | 22.2 | 26.1 | 28.2 | 42.3 | 42.3 |

References


[1]W.H. Hartford, in Kirk-Othmer "Encyclopedia of Chemical Technology," Wiley-Interscience, New York, 1979, 3rd Ed., Vol. 6, pp. 82-120.

[2]W.H. Hartford and R.L. Copson, in Kirk-Othmer, "Encyclopedia of Chemical Technology," Wiley-Interscience, New York, 1964, 2nd Ed., Vol. 5, pp. 473-516.

[3]C.L. Rollinson, in "Comprehensive Inorganic Chemistry," Pergamon, Oxford, 1973, Vol. 3, pp. 623-700.

[4]W.H. Hartford, Kolthoff-Elving "Treatise on Analytical Chemistry," Interscience, New York, 1964, Part II, Vol. 8, pp. 273-377.

[5]W.H. Hartford, in Snell-Ettre "Encyclopedia of Industrial Chemical Analysis," John Wiley & Sons, Inc., New York, 1970, Vol. 9, pp. 680-709.

[6]W.H. Hartford, Rocks and Minerals, 55{2}, 52-59 (1980).

[7]F. McBerty and B.H. Wilcoxon, FIAT Rev. Ger. Sci., PB 22627, Final Report No. 796, Washington, 1946. A corresponding BIOS report exists.

[8]F. Ullmann, "Enzyklopädie der technishen Chemie," Urban and Schwarzenberg, Berlin, 1929, Vol. 3, pp. 400-433.

[9]U.S. Bureau of Mines, "1977 Minerals Yearbook," Washington, D.C., 1980, Vol. 1, pp. 257-267.

[10]T.P. Thayer, in Udy, "Chromium," Reinhold, New York, 1956. Vol. 1, pp. 14-52.

[11]U.S. Bureau of the Census, "Inorganic Chemicals," Reports M28A(78)-13, M28A(79)1-12, Washington, D.C. 1979-1980.

[12]P.R. Hines (Harshaw Chemical Co.) U.S. Patent 1,873,589, Aug. 23, 1932.

[13]T.S. Perrin, R.E. Banner (Diamond Alkali Co.) U.S. Patent 3,065,055, Nov. 20, 1962.

[14]T.S. Perrin, R.E. Banner, and J.O. Brandstaetter (Diamond-Shamrock Corp.) U.S. Patent 3,607,026, Sept. 21, 1971.

[15]A.M. Baetjer, Panel Chairman, Division of Medical Sciences, National Research Council. "Chromium" National Academy of Sciences, Washington, D.C., 1974.

[16]H.W. Davids and M. Lieber, Water Sew. Works 98, 528(1951).

[17]"Drinking Water and Health," U.S. Federal Register, 42(132) 35770 (1977).

[18]American National Standards Institute, Standard USAS Z37.7-1943(rev. 1971), New York, 1971.

[19]Federal Security Agency, "Health of Workers in the Chromate-Producing Industry," U.S. Public Health Service Publication No. 192. Washington, D.C., 1953.

[20]R.B. Hayes, A.M. Lilienfeld, and L.M. Snell, Int. J. Epidemiol. 8,365-74.

[21]W. Mertz, Physiol. Rev. 49, 163 (1969).

[22]Southwest Research Institute, private communication, 1956.

[23]W.H. Hartford, "Dosage-Response Curves and Chemical Carcinogens," presented to the SE-SW Regional American Chemical Society meeting, New Orleans, La., December 12, 1980.

[24]Manufacturing Chemists Association, "Chemical Safety Data Sheets," SD-44, 45, 46. Washington, D.C., 1952. (Now being revised).

[25]E.F. Foley, Jr., Diamond Shamrock Corp. Painesville, O., private communication, 1980.

[26]R. Mulholland, Allied Chemical Corp., Mornstown, N.J., private communication, 1980.

[27]Chemical Profile, "Chromic Acid," Chemical Marketing Reporter, 213(9), 9, March 6, 1978.

[28]Chemical Profile, "Sodium Bichromate," Chemical Marketing Reporter 216(2), 9, July 9, 1979.

[29]O.F. Tarr and L.G. Tubbs (Mutual Chem. Co. of America) U.S. Patent 2,245,907, July 30, 1940.

[30]O.F. Tarr and M. Darrin (Mutual Chem. Co. of America) U.S. Patent 2,415,394, Feb. 4, 1947.

[31]Chemical Marketing Reporter, 214{9}, (1978)

[32]Ibid., 218{5}, (1980)

[33]T.C. Patton, "Pigment Handbook," Wiley-Interscience, New York, 1973, Vol. 1.

[34]C.H. Love, "Important Inorganic Pigments," Hobart, Washington, D.C., 1947, pp. 351-389, 843-861.

[35]W.H. Hartford, "Chromium," in A.G. Bard, "Applied Electrochemistry of the Elements," Marcel Dekker, New York, (in press)

[36]G. Dupbernell, in Lowenheim, "Modern Electroplating," Wiley-Interscience, New York, 1974, 3d ed., pp. 87-151.

[37]P. Morriset, "Chromium Plating," Robert Draper, Teddington, Middlesex, 1954.

[38]Allied Chemical Corporation, Industrial Chemicals Division, "Practical Guide to Chromium Plating," Technical Service Report 17.60R, Syracuse, New York, 1971.

[39]Allied Chemical Corporation, Industrial Chemicals Division," Leather Group Technical Bulletins," nos. 77-I, 77-IV, 77-IX, Syracuse, New York, 1977.

[40]F. O'Flaherty, W.T. Roddy, and R.M. Lollar, "The Chemistry and Technology of Leather," Reinhold, New York, 1958-1962; vol.2, pp. 221-323, vol. 3, pp. 184-460.

[41]H.A. Lubs, "The Chemistry of Synthetic Dyes and Pigments," Robert E. Kriger Publishing Co., Huntington, New York, 1972, pp. 153, 160-161, 247, 258, 261, 284, 426.

[42]F.W. Eppensteiner and M.R. Jenkins in "Metal Finishing Guidebook and Directory," Metals and Plastics Publications, Inc., Hackensack, N.J., 1977, pp. 540-557.

[43]Allied Chemical Corporation, Industrial Chemicals Division, "Chromic Acid Anodizing of Aluminum." Technical Service Applications Bulletin 103, Syracuse, New York.

[44]W.A. Moore et al, Purdue Univ. Eng. Bull. Ext. Ser. No. 106, 158-182(1960).

[45]W.H. Hartford, in Nicholas "Wood Deterioration and its Prevention by Preservative Treatments," Syracuse Univ. Press, Syracuse, New York, 1973, Vol. 2, pp. 1-120.

[46]W.H. Hartford, Crossties, 58{4} 18-33(1977).

[47]W.H. Hartford, Proc. Am. Wood Preservers Assoc., 74, 88-91 (1978).

[48]Maloney and Pagliai, Proc. Am. Wood Preservers Assoc., 75, 342(1979).

# The Chemical Uses of Molybdenum and Its Compounds

By E. R. Braithwaite
CLIMAX MOLYBDENUM COMPANY LTD., 1/3 GROSVENOR PLACE, LONDON SWIX 7DL, U.K.

The current free-world production of molybdenum between 1974 and 1976 was estimated[1] to be approximately 75,000 tonnes p.a. increasing about 6 per cent annually; of this only about 10 per cent is used directly as feedstock for the chemical industry. Molybdenum is mined either as a primary ore, molybdenite, or as a secondary product (0.01-0.05% Mo) from some copper porphyry, of which Chile has the largest free-world reserve. Molybdenite (molybdenum disulphide $MoS_2$) contains about 60 per cent molybdenum and 40 per cent sulphur, and occurs as fine crystals which are generally so intimately intergrown with, embedded in, and enclosed by quartz and other minerals that they do not soil the hands when rubbed. The two most important molybdenum porphyry ore mines are the Amax mines at Climax and Henderson in Colorado. According to Vanderwilt[2] the Climax deposit was formed by intrusion of a quartz monzanite magma into granites, schists and porphyry; these porphyry deposits are the most important sources of molybdenum ore (0.3-0.4% Mo).

Some molybdenum is recovered during leaching treatment of uranium and other ores, but these are not major commercial sources. An excellent summary of molybdenum resources is given in Sutulov's International Molybdenum Encyclopedia.

$MoS_2$ is recovered from both the copper and molybdenum porphyry ores by the flotation process. Where the ore is primarily a copper sulphide ore, the $MoS_2$ and copper sulphides are floated together and separated later using techniques for depressing the copper sulphide mineral. Cleaning of the $MoS_2$ flotation concentrate is normally considered complete when the copper content has been reduced to less than about 1 per cent. To obtain the lowest possible copper content, it is necessary to resort to chemical leaching. In general the $MoS_2$ is more resistant to chemical decomposition, and the other sulphide minerals can be selectively dissolved away from the molybdenum.

Amax Mine at Climax Colorado

Crushing plant at Amax Henderson Mine where molybdenite ore is reduced in size before flotation

$MoS_2$ concentrates from the primary molybdenum porphyry ore deposits are typically lower in contaminant metals than those from copper mines. The Climax mine, for example, produces a special lubricant grade of concentrate containing a minimum of 98.2 per cent $MoS_2$ with essentially no copper. However, commercial-grade $MoS_2$ concentrates from molybdenum porphyry deposits usually analyse about 90 per cent $MoS_2$.

The $MoS_2$ concentrates are typically roasted in multiple-hearth furnaces to remove sulphur and convert the molybdenum to the trioxide. Hydrometallurgical dissolution is an alternative to roasting, but it accounts for only a minor part of concentrate processing.

Special grades of molybdic oxide, for example roasted higher grade concentrates or resublimed oxide, are the usual starting points for wet chemical manufacture of purified molybdenum chemicals such as sodium or ammonium molybdates or pure oxide.

Molybdenite was found by Scheele[3] in 1777 to yield a peculiar metallic acid when oxidised with nitric acid; this compound was subsequently reduced to the metal by Hielm[4] in 1782 and named molybdenum. It is suspected that although in earlier times no specific knowledge about molybdenum and its alloying properties existed, molybdenum may have been used unintentionally in the preparation of high quality steels. For example, chemical analysis of a fourteenth century Samurai sword revealed that the steel contained molybdenum.

Molybdenum can be described as an amenable element, that is to say, it often does things a little more easily than other elements around it in the periodic table. For example, although tungsten is harder and has a higher melting point, it is more difficult to work than molybdenum. Molybdenum is located vertically in Group VI of the periodic table, between chromium and tungsten, and horizontally in the 4d block between niobium and technetium (Table 1). It is one of the so-called refractory metals because of its hardness and high melting point. These properties reflect the high strength of interatomic bonding resulting from the efficient overlap of the 4d orbitals and the number of bonding electrons available.

TABLE 1

Elements surrounding molybdenum in the periodic table

| Group Va | Group VIa | Group VIIa |
|---|---|---|
| (6.1 Specific gravity)<br>Vanadium<br>1860°C (m. pt.) | 7.1<br>Chromium<br>1920°C | 7.4<br>Manganese<br>1260°C |
| 8.4<br>Niobium<br>2468°C | 10.4<br>Molybdenum<br>2610°C | 11.5<br>Technetium<br>2150°C |
| 16.6<br>Tantalum<br>2990°C | 19.3<br>Tungsten<br>3410°C | 20.5<br>Rhenium<br>3170°C |

TABLE 2

Polyion species of molybdenum and the corresponding pH values

| pH | Predominant species |
|---|---|
| 7 | $MoO_4^{2-}$ |
| 7 | $Mo_2O_7^{-2}$ |
| 6 | $Mo_7O_{24}^{6-}$ |
| 2-5 | $Mo_8O_{26}^{4-}$ |
| 1.6 | $Mo_{12}O_{37}^{2-}$ |
| 1.0 | $MoO_3/HMoO_4$ |

Molybdenum, in common with other members of the earlier transition metals, exhibits a wide range of oxidation states and is generally characterised by highly covalent bonding and the absence of simple ionic compounds. Its outstanding feature is extraordinary versatility, having formal oxidation states (-II) to (VI) and coordination numbers from 4 to 8, attributes which give rise to a very complex stereochemistry. The relatively large 4d radius of the molybdenum atom, and hence its diminished polarising power, is responsible for the stable quality of its higher oxidation states and is in fact the reason why we do not find any lower oxides or halides. In its higher oxidation states molybdenum has a particular affinity for the higher halogens, oxygen and oxygen-donor ligands. Unlike chromium it has an appreciable affinity for sulphur-donor ligands, a property which is important in a number of applications such as hydrodesulphurisation catalysts and in biochemistry when Mo-S bonds are present in enzymes.

These ligand environments found in additive groups or coordination compounds of molybdenum vary from relatively simple molecules such as water to the complex organic molecules which are to be found in enzymes and other biological systems. The bonding of these compounds is essentially covalent in character.

The aqueous chemistry of molybdenum is extremely complicated[5]. The variety of species in solution depends especially on the oxidation state of the molybdenum and the pH of the solution. During the formation of polyionic species in solution, an equilibrium state exists between the formation of oxo, hydroxo, and aquo-species. Because the molybdenum ions have relatively large radii - with a subsequent lowering of the molybdenum polarity - higher coordination numbers are usually observed as, for example, with $MoO_3$ where the molybdenum ions occupy octahedral sites. The degree of polymerisation has been discussed in terms of the low polarisability of the molybdenum ions arising from the greater extension of the 4d orbitals compared with the 3d orbitals of chromium. These polymeric species in solution may sometimes bear little resemblance to the complex oxo-anions isolated from solution. This may well be a significant factor with respect to the impregnation of catalyst bases using molybdenum compounds in solution.

In alkali solution Mo(VI) gives the simple tetrahedral oxo anion $MoO_4^{2-}$, but on lowering the pH, condensation occurs and polymeric units are formed containing the octahedrally coordinated metals and ions as discussed. The various polyion species which have been isolated and are used industrially are shown in Table 2. If condensation occurs in the presence of phosphate, silicate or certain other ions then these may be incorporated into the molybdenum-oxygen lattice to give heteropoly compounds[6] some of which find a use in the catalyst, pigments and inks industries as discussed later. The principal industrial chemical compounds of molybdenum and their properties are summarised in Table 3; a flow sheet of their production and end uses is given in Fig. 1.

The role of surface chemistry in the chemical applications of molybdenum and its compounds - the common factor which links them - has probably not been sufficiently emphasised. In fact, the extraction of molybdenum sulphide from the ore by flotation would be extremely difficult if it were not for the natural hydrophobic surface of the mined molybdenite. It is the different surface chemical effects of the basal and edge planes of $MoS_2$ (Fig 2) which largely determine the catalytic, lubricating and polymer reinforcing action of $MoS_2$. The edge plane appears to be a strong hydrogenating centre though the basal plane has a much lower level of hydrogenation activity. The basal planes of $MoS_2$ determine the degree of adhesion of an $MoS_2$ particle to a substrate and for good lubrication this adhesive force should be greater than the force required to shear the particles[7]. In crystalline thermoplastics the steps on the basal planes of the $MoS_2$ fillers acting as centres of nucleation probably determine compatibility with a particular polymer as well as the proportion of hydrophilic to hydrophobic centres[32]. The physicochemical properties of molybdenum and its compounds can be grouped under three headings, viz particle size, particle shape and specific surface area; these are related to applications given in Table 4.

TABLE 3

Pure molybdic oxide and ammonium dimolybdate*

| Element | Typical Analyses | |
|---|---|---|
| | Molybdic oxide | Ammonium dimolybdate |
| Molybdenum | 66.6 | 56.54 |
| Sodium and potassium | 0.0090 | 0.0080 |
| Aluminium | 0.0008 | 0.0007 |
| Calcium | 0.0012 | 0.0010 |
| Chromium | 0.0005 | 0.0004 |
| Copper | 0.0010 | 0.0008 |
| Iron | 0.0006 | 0.0005 |
| Lead | 0.0001 | 0.0001 |
| Magnesium | 0.0015 | 0.0015 |
| Nickel | 0.0005 | 0.0005 |
| Silicon | 0.0012 | 0.0010 |
| Tin | 0.0014 | 0.0012 |
| Titanium | 0.0005 | 0.0005 |

*As produced by Climax Molybdenum BV, Rotterdam

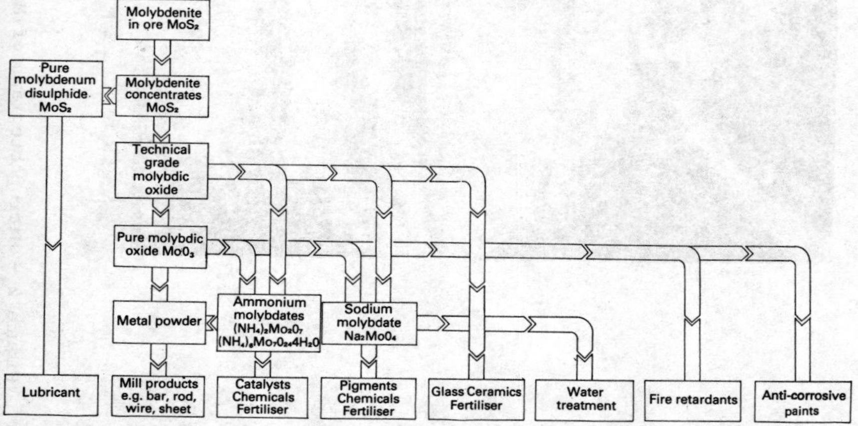


**Fig 1** Flow chart showing the production and uses of important molybdenum compounds

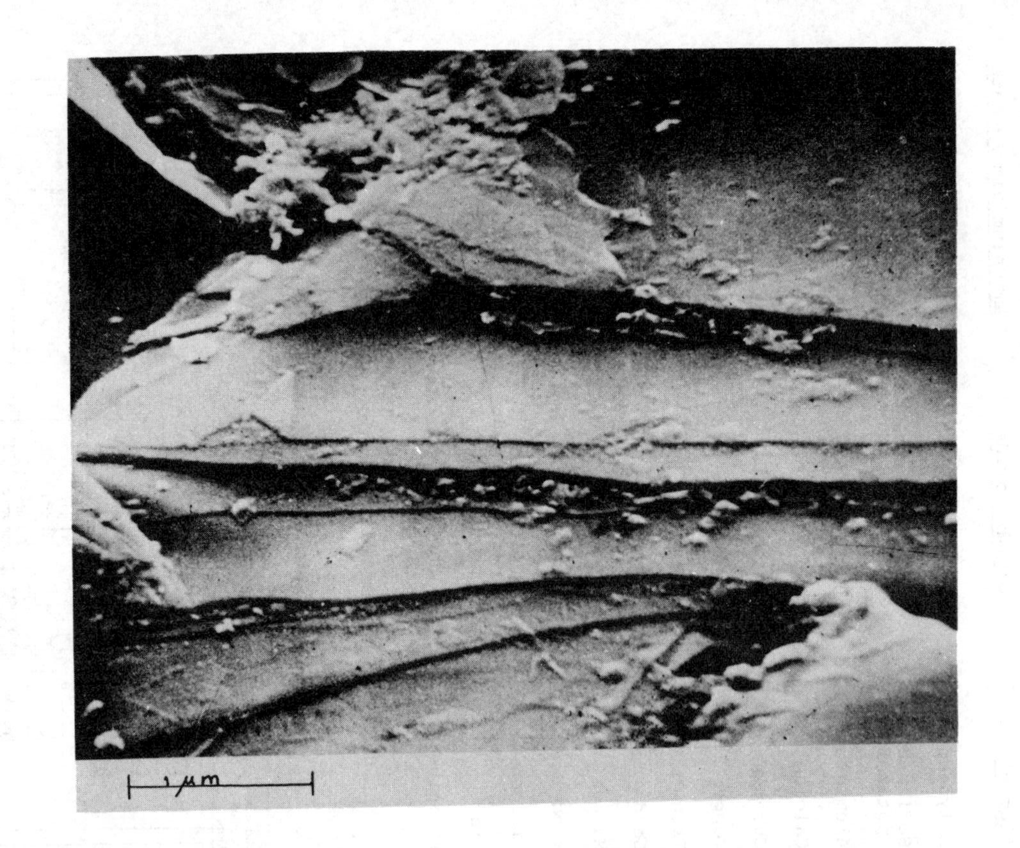


Figure 2 - Microphotograph of $MoS_2$ crystal showing plasticity & lamellae

TABLE 4

The physicochemical properties of molybdenum and its compounds and the significance of their relation to application

| | Particle property | | |
|---|---|---|---|
| | Size | Shape | Surface |
| Sintered molybdenum | *** | *** | *** |
| Catalysis | *** | * | *** |
| Pigments, inks | *** | ** | *** |
| Anticorrosion pigments | *** | ** | ** |
| Aqueous corrosion inhibitors | ** | - | ** |
| Flame retardants and smoke suppressants | *** | *** | *** |
| Lubrication ($MoS_2$) | *** | *** | *** |
| Oil soluble molybdenum compounds | - | - | - |
| Biochemical processes | *** | ** | *** |

(degree of significance increases with the number of stars)

The uses of catalysts, pigments, inks, and corrosion inhibitors are largely concerned with rates of dissolution of compounds and the final dispersed state of the molybdenum out of solution. These properties are influenced by size and surface area, shape probably being not to important. When the solid state properties of molybdenum are being fully utilised, as in the production of sintered molybdenum or in lubricating systems, then all three properties are interdependent and of vital importance. This is also the case with smoke suppressants and fire retardants, though with the latter the relevance of particle shape has not yet been evaluated.

## Sintered Molybdenum

This refers only to sintered molybdenum and not to the huge area of superalloys associated with steels. The reason for this is that sintered products are produced by the hydrogen reduction of ultra-high purity molybdic oxide or ammonium dimolybdate, which is so pure that its structure was determined by using directly the regular commercial material produced in the plant of Climax Molybdenum BV[8]. Molybdenum belongs to a group of refractory metals including tungsten, tantalum and niobium (Table 1) which are characterised by high melting point, hardness and high strength at elevated temperatures; it is mainly these properties which are exploited in the metallurgical industry.

The role of the refractory metals as alloying elements has increased since the Second World War. New advances in technology of the jet engine, nuclear power and small space vehicles require materials capable of resisting the highest temperatures and corrosive influences as well as having the optimum mechanical properties. Not only did these metals themselves fulfil these conditions but also, to a more limited extent, did their borides, silicides, nitrides and carbides. The accolade, however, goes to molybdenum because of its amenability, for although molybdenum and tungsten both have atoms with almost identical radii, the atomic weight and density of molybdenum is half of that of tungsten and, therefore, half the quantity of molybdenum is required to produce the same alloying effect as tungsten. An additional factor of practical significance is that molybdenum is easier to work than the other refractory metals because it can be made ductile at room temperature. Furthermore, it is easier to machine than tungsten and therefore causes less tool wear. As tungsten and molybdenum are readily oxidised at over 700°C, special precautions, such as working in the absence of air during sintering, must be taken during manufacture by the Coolidge and Fink (1910) pulvimetallurgical method of producing ductile bars.

The alloys of molybdenum are also important; by alloying small amounts of titanium, zirconium and carbon to molybdenum, it has been possible to increase the recrystallisation temperature strength from 1200 to 1400°C. According to Machenschalk[9] the potassium doped grades are also important - again due to improved recrystallisation behaviour. For example, in the case of fine molybdenum wire it is possible to maintain good ductility even after exposure at 2000°C. In order to prevent the grain growth caused by the water vapour evolved during the reduction, the trioxide is usually reduced in steps, giving $MoO_2$ at 600-700°C and metallic molybdenum in the second stage. The quality of the product depends mainly on the purity of the trioxide; the particle size, surface area and particle size distribution of the trioxide are also of critical importance[10].

One of the oldest applications of molybdenum metal is in the electric lamp, where, because of its low thermal expansion - which is only slightly greater than that of hard glass - molybdenum is highly suitable for gas-tight sealing into glass. It is also used for mandrels around which the tungsten filaments are wound into coils; molybdenum subsequently being dissolved by strong acids. Other exploitations of this thermal property include: current leads for electronic tubes and a wide range of electrodes and grids which are subject to high thermal stresses; the semi-conductor industry where molybdenum is used as supports for the semi-conducting discs of silicon and germanium; in special cases for electrical contacts because of molybdenum's high resistance to spark erosion and severe mechanical wear.

Use is made of molybdenum's high melting point in many applications including the components for high temperature furnaces operating in a vacuum or protective atmosphere. Because of its resistance to molten glass, molybdenum is widely used for electrodes, heat shields and various ancillary parts connected with glass furnaces. A more recent application is in halide lamps which formerly used nickel caps for two-filament car lamps and now use molybdenum because of its superior corrosion resistance.

## Catalysis

Molybdenum catalysts are used in a wide variety of applications including petroleum hydrotreating, selective oxidation, ammoxidation, polymerisation, disproportionation, the water-gas shift reaction and naphtha reforming. One of the oldest and most extensive uses of molybdenum is in petroleum hydrotreating where in the earlier thirties heterogeneous catalysts containing molybdenum and tungsten were introduced by I. G. Farben. This was followed in the early forties by the Co-Mo-$Al_2O_3$ system for catalytic hydrodesulphurisation.

According to Basila[11], Co-Mo catalysts enjoy very wide usage for the following reasons:

1. Co-Mo catalysts have the highest desulphurisation activities.

2. Co-Mo catalysts are more rugged with respect to process operating conditions, especially during the regeneration cycle. Higher regeneration temperatures may be used for Co-Mo (480-510°C) than for Ni-Mo (425-450°C). The resulting shorter regeneration cycles minimise the possibility of high temperature steam causing catalyst damage. The lower temperature requirement for Ni-Mo catalyst is necessary to avoid loss of activity caused by reduced dispersion of the metals.

3. The use of a Co-Mo catalyst makes possible minimum hydrogen consumption per barrel of feed processed because of its lower hydrogenation activity. This factor is important in many refineries where the quantity of hydrogen is limited. It is also important in certain processes such as the desulphurisation of naphthas when minimum olefin saturation for maximum octane retention is required.

4. Modern Co-Mo catalysts are effective at much lower hydrogen partial pressures than Ni-Mo catalysts.

Hydrodesulphurisation catalysts containing molybdenum are generally recognised as the best because they are less susceptible to poisoning and thus have a longer life. These catalysts have taken on increasing importance in view of the stricter legislation on the maximum allowable sulphur content of the oils. Typical catalysts contain 8-15 per cent $MoO_3$ and 2-4 per cent CoO on alumina; they are reacted with sulphide and are usually deposited onto carriers from solutions of oxide or molybdate. The carriers interact with the molybdenum to give altered structures and therefore a good hydrodesulphurisation catalyst represents a delicate balance between porosity, mechanical strength, pore radius and pore volume of carrier[12].

More than a decade ago a selective oxidation process based on bismuth-molybdenum catalysts was introduced for the industrial production of acrylonitrile - a very important intermediate for the synthetic fibres and rubber industry. The use of bismuth molybdates as catalysts for the oxidation of propylene to acrolein was first reported in patents[13]. They used a catalyst of composition: 50 per cent $Bi_9Mo_{12}O_{52}$/50 per cent $SiO_2$ silica gel acting as a support. This same catalyst also converts propylene and ammonia to acrylonitrile, and butene to butadiene. The most prevalent commercial use of the process has been the ammoxidation of propylene to acrylontrile.

The Oxirane process is based on a homogeneous catalytic reaction in the liquid phase using an organic molybdenum compound as catalyst - various by-products are obtained depending on the reactants used, e.g.

$$RO_2H \;+\; >C{=}C< \;\xrightarrow{Mo}\; >C\overset{O}{—}C< \;+\; ROH$$

According to Ference[14] the Oxirane Co uses isobutane as the hydrocarbon, producing tert-butyl alcohol as the co-product in some of its plants; other plants employ ethylbenzene hydroperoxide as the oxidant. Over $10^9$ lb $year^{-1}$ of styrene are produced in Texas by the dehydration of the co-product 2-phenylethyl alcohol. Major derivatives of propylene oxide include propylene glycol, polypropylene glycol and polyetherpolyols - end products include urethanes, polyesters and surfactants.

Industrial processes for the oxidation of methanol to formaldehyde over an unsupported non-stoichiometric mixed ferric oxide catalyst offer a number of advantages over the traditional silver catalysed process. Because of the low methanol feed concentration, yields of formaldehyde in excess of 90 per cent can be achieved, the catalyst gives longer cycle life and is resistant to poisoning. A process using lower methanol concentration is less subject to fire and explosion hazards, and operates at a temperature about 200°C lower (400°C) than the silver process. Although the technology has been known for almost 50 years[15] it was not applied commercially until the 1950s.

Substitute natural gas (SNG) has been produced outside the USA for years and the current American shortage of natural gas necessitates the USA relying on SNG in future as an alternative energy source. The first USA unit was completed in 1972; by 1980 some 7 per cent of US demand will probably be met by SNG. It is probable that the initial desulphurisation of the sulphur feedstocks will be done on cobalt-molybdenum or nickel-molybdenum catalysts. In this same area the energy crisis is intensifying interest in coal gasification as a route to gaseous fuels in the USA using sulphur-resistant, water-gas shift catalysts[16]. In general, the catalysts are based on Co-Mo supported on alumina. The formulations are somewhat different from typical hydrotreating catalysts in that the alumina base is required to be stable in high temperature steam.

Following the pioneering work of the Ann Arbor laboratories of the Climax Company of Michigan USA, it has been reported[17] that $Mo/MoO_2$ is of interest at a fundamental level because it offers the opportunity to investigate the catalytic properties of low valent molybdenum. The $Mo/MoO_2$ catalyst consists of an intimate mixture of Mo metal and $MoO_2$. They have high BET nitrogen surface areas ($> 100 m^2 g^{-1}$) and possess a fair degree of porosity, the pores ranging in size from very small micropores (diameter $< 20 Å$) to large mesopores (diameter $\simeq 200 Å$). The Mo/$MoO_2$ mixtures will catalyse the isomerisation of straight chain hydrocarbons to branched chain hydrocarbons, and the conversion of methylcyclopentane to cyclohexane and benzene.

Recently there has been interest in using molybdenum as a promoter in multicomponent catalysts. For instance, Mo has been used to promote Pt/Rh auto exhaust catalysts. In the presence of Mo nitrogen oxides are reduced to nitrogen rather than ammonia. The detailed mechanism of the process is not known, but it is possible that the role of Mo is to accelerate the conversion of ammonia into nitrogen rather than to affect directly the properties of the Platinum metals. Molybdenum has also been used to modify the properties of metallic nickel catalysts, and it is possible that these bimetallic catalysts could be used as sulphur-resistant methanation catalysts.

The most promising new application for Mo-containing catalysts is as hydrogenation catalysts in coal liquefaction processes, and in upgrading oil residuals. The great advantage of molybdenum over other hydrogenation catalysts is its high activity in the presence of sulphur. This is presently an area of intense Research and Development activity.

It is evident from the foregoing that molybdenum is a highly versatile ingredient in catalysts; there are few other elements which have been used in such diverse applications. It is likely that new uses for molybdenum catalysts will continue to be found as the need arises.

<u>Pigments and inks</u>

The use of molybdenum compounds as colouring agents was first published as early as 1818. In 1863 the first molybdate orange was prepared by Schulze, but this did not become a commercial preparation until the late 1930s. They are currently used on certain makes of cars and international airport markings; they are cheap and stable up to 180°C - an advantage over most organic pigments.

Molybdate orange is a mixture of white lead molybdate, yellow lead chromate and white lead sulphate. It is therefore of interest to consider briefly the formation of molybdate orange which is normally far redder than the light orange of pure lead chromate alone. This brilliant colour characteristic arises from the crystallisation of lead chromate in the abnormal tetragonal form, which normally exists only at high temperatures, instead of its normal rhombic or monoclinic form. In the three-component system, the lead chromate and lead sulphate build into a tetragonal lead molybdate structure. When a solution of lead nitrate is mixed with a solution of sodium chromate-molybdate-sulphate, the initial precipitate is yellow orthorhombic lead chromate and the reaction proceeds through deep orange to a red colour. This happens on stirring at pH $>$ 7 when tetragonal lead molybdate, being less soluble than lead chromate, acts as a seed and the lead chromate builds into a lead molybdate structure.

Basic dyes can be laked by precipitation with the soluble salts of organic acids such as tannic or 12-molybdophosphoric. At the beginning of the century, very fugitive dyestuffs had been used because of their beauty, strength and wide colour range. They were fixed to the fibre with tannic acid and tartar emetic and were therefore sensitive to light and washing. As a result of biochemical work in 1914 which demonstrated that the addition of amino compounds to complex inorganic acids gave insoluble compounds, the use of tungstophosphoric and molybdophosphoric acids with amino compounds commenced. Their commercial use, however, was delayed until about 1920 because molybdenum and tungsten were needed for the hardware of the First World War. Heteropoly-compounds of most transition metals will also precipitate dyes to give light-fast materials and additionally, due to the bulky size of anion and cation, they are able to fix onto the fabric. They are better u.v. absorbers, preventing the organic part from losing its natural colour.

Anticorrosion pigments

The corrosion inhibiting properties of molybdates are also put to good use in a commercial white pigment introduced as a potential replacement for chromium and lead-based inhibitive pigments in paint systems. These 'Molywhite' pigments[18] are basic zinc and calcium molybdate compositions combining non-toxic and corrosion inhibiting properties in a product particularly applicable to areas where is has not been possible to use corrosion resistant primers. Food processing machinery, transport containers and water tanks are likely places to use Molywhites. The mode of action of the molybdate in these products is similar to that described for aqueous corrosion inhibitors.

Lizlovs[19] has shown that molybdate anions in neutral or basic solution form a complex with $Fe^{2+}$ substrates, which can be oxidised by dissolved oxygen to $Fe^{3+}$ giving a tenacious layer of iron oxide incorporating Mo. Molybdate anions interact synergistically with sodium nitrite on steel whilst sodium nitrate is a better synergist on aluminium. These synergistic effects are exploited in cooling tower water treatments, anti-freeze formulations and cutting fluids, as well as in certain paints. There is also a phosphate molybdate synergism found in aqueous and solvent media[20]. New water dispersible primers employ one or both synergisms above, and though the resin component of the primers, properly coalesced, is itself quite resistant to water permeation, passivation of the substrate by molybdates anions is still required.

Aqueous corrosion inhibitors

For many years chromates have been among the most popular of the traditional aqueous corrosion inhibitors, particularly in cooling water treatments, but in low concentrations they are highly toxic to fish and can have other undesirable side effects. For example, the strong oxidising character of Cr(VI) would almost certainly poison sewage processing plants and consequently the amount of chromates that can be discharged is already strictly regulated, particularly in the USA. The water treatment industry has, therefore, to seek more acceptable, less toxic alternatives. Molybdates are reported[21][22] to be likely candidates because of their favourable ecological position; the mechanism of molybdate action, although not completely understood, is believed to be similar to that of chromates. The molybdate ion, in basic solution, is not itself an oxidising agent and cannot by itself polarise the corroding iron electrode into the passive potential region where a protective film can form. However, polarisation curves and electron micro-probe analyses clearly indicate that film formation does take place[19]. The chemical nature of such a film is the subject of much speculation but it has been suggested that the mechanism involves the formation of an iron-molybdate complex which acts as the passivating film itself or repairs discontinuties in the formation of an $Fe_2O_3$-type film.

Reports[23] indicate that soluble molybdates used as corrosion inhibitors have extremely low or negligible toxicity and that sodium molybdate is considerably less irritating than sodium dichromate, presenting no dermal or local hazard[24]. The sensitivity of tropical fish to sodium molybdate was the subject of recent acute toxicity tests[25] designed under static bioassay conditions and these results show that typical no-effect levels are 2400-7500ppm, 50ppm, and 0.01ppm for molybdates, chromates and DDT respectively.

## Flame retardants and smoke suppressants

As plastics find acceptance for more sophisticated applications and wider uses in the building, transport and consumer industries, so the requirements for the flammability of these products are becoming stricter and existing flame retardant treatments are sometimes unable to meet the new demands. In particular, the majority of existing flame retardants, though adequately controlling the rate of burning and flame-spread in most polymers, do little or nothing to reduce the amount of smoke generated by the burning polymer. Analyses of the statistics of fire casualties in the UK[26] and USA[27] show that the effects of smoke and toxic gases are increasing and now account for almost a half of all fire deaths.

Molybdenum compounds have long been known to possess some flame retardant properties, but until recently the fire performance requirements of polymers were easily met by the use of acceptable amounts of additives, such as antimony trioxide or halogen compounds. However, antimony trioxide, the traditional fire retardant additive, does not reduce the amount of smoke generated, whereas addition of molybdenum trioxide can reduce smoking by over a half. Investigations into the mechanism of the smoke suppressing action of molybdenum are at an early stage, but already it is known to differ from that of antimony trioxide which functions in the vapour phase in the presence of halogen, forming $SbCl_3$. Molybdenum trioxide, on the other hand, appears to work in the solid phase[28] (90 per cent or more of the molybdenum remaining in the char of burnt PVC compared to less than 10 per cent for an antimony containing sample) and by so doing it may prevent combustion of the carbon components of the polymer.

Skinner et al[29], using oxygen index measurements of a series of unsaturated polyesters containing $MoO_3$ and various halogenated compounds, have provided substantial evidence for some form of flame retardant synergistic effect between molybdenum and halogen. They further claim that the magnitude of the effect is greater in the presence of bromine but is dependent on the type of compound. It should also be noted (Table 4) that the present results of various workers appear to indicate that the particle size and specific surface of the molybdenum oxide are of primary importance. There is little doubt that, at least in Europe, as the legislation for smoke emission is tightened, molybdenum compounds will have an important part to play.

## Lubrication

Molybdenum compounds are used as solid lubricants in both the oil-insoluble and the oil-soluble forms. The latter are referred to as antiwear or EP additives and the former ($MoS_2$) as a lamellar solid lubricant. The essential difference in behaviour between them is that $MoS_2$ is attached mechanically, or at lest by physisorption, to a metal substrate and acts spontaneously; the soluble EP additive must first interact with the metal substrate of the rubbing surfaces to produce a new chemical compound which becomes the solid lubricant. In the absence of such decomposition oil-soluble compounds would be at best poor boundary lubricants.

### Oil-insoluble ($MoS_2$)

$MoS_2$ is classed as a solid lubricant, viz a solid which reduces the mechanical interaction between two moving surfaces and imparts antiwear properties to the system. If the latter is absent we may refer to the solid as a parting compound. It is not generally realised how much Concorde and Jumbo jets owe to $MoS_2$, or that the Apollo could still have been in its hangar, or that the moon probe might not have functioned - but for the unique contributions of $MoS_2$ to the solution of a lubrication problems under extreme environmental conditions. $MoS_2$ differs essentially from graphite in its superior performance at very high or low pressures and the effect of environmental conditions on lubricating behaviour.

The mechanism of lubrication by $MoS_2$ is somewhat complicated in that it is dependent on three factors, namely crystal structure, electronic configuration and surface properties. Of these, the first two are indigenous to the solid and are interdependent; surface properties arise out of both of these. The various hypotheses connected with crystal structure and electronic configuration have been summarised by Braithwaite and Rowe[30]. Following Bragg's work, it was long thought that $MoS_2$, like graphite, gets its lubricating qualities from the easy sliding of the lamellae over one another when under shear, but this is not the whole story; although several elements from layer structure sulphides[31], [32] only niobium, tungsten and molybdenum are truly lubricating sulphides. An interesting hypothesis that has been put forward by Jamison attempts to link crystal structure and lubricating properties of various disulphides and diselenides.

It proposes that an additional factor, viz the manner in which the chalcogenides surround the metal atoms, may be the most important parameter and that structure alone does not explain lubricating behaviour. Having then allowed for these somewhat specific structural and electronic parameters, attention should be focussed on the surface properties of $MoS_2$ without which the lubricating behaviour would be quite different.

In practice the environmental influences on the lubricating behaviour of graphite and $MoS_2$ are diametrically opposed, that is to say, $MoS_2$, unlike graphite, operates best under vacuum and non-oxidising conditions[33] largely due to the increased adhesion of the platelets to the metal substrate under these conditions. Other surface characteristics which influence its lubricating behaviour are its particle size and shape.

$MoS_2$ functions as a lubricant in two ways according to the type of system in which it is being used. Sacrificial lubrication occurs in one-pass systems such as extrusion or drawing where the $MoS_2$ is acting essentially as a lubricating parting compound.

Contact lubrication[34] occurs in systems where reciprocating motion is predominant, such as a bearing or gear, and the lubricant is mechanically worked into the sliding surfaces during running-in to form a sort of 'Beilby layer'. $MoS_2$ is usually supplied as an additive to oils, greases or aerosols and its uses have been well documented[37]. For a fuller appreciation of the mechanism of action of $MoS_2$ many inter-disciplinary boundaries have to be crossed.

Two industrial uses in particular are of topical interest as they are connected with energy conservation; these are as friction modifiers in motor vehicles and as lubricating additives to engineering plastics. With respect to the former it has been reported[35] that $MoS_2$, which is properly dispersed in oil at not less than 0.3 per cent concentration, can make a significant contribution to the efficiency of both the engine and rear axle of a motor vehicle; these authors stress the importance of the synergism that they have found to exist between Moly Van L* and $MoS_2$ in the rear axle assembly. It was concluded that the addition of 1 per cent $MoS_2$ to the crankcase oil of an internal combustion or compression ignition engine improves the efficiency up to 3 per cent - with a corresponding saving in fuel, and the addition of 1 per cent $MoS_2$ to the sump of a rear axle increases the efficiency 1.5 - 2.0 per cent. This figure can be improved to 4.0 per cent utilising the synergism between $MoS_2$ and Moly van L. It is calculated that an improvement of only 2 per cent in overall fuel consumption of one hundred million cars, each averaging 20mpg and only 10,000 miles per year running, would save 1000Mgal of fuel per year - equivalent to 50M barrels of crude.

*Registered by R T Vanderbilt Co Ltd

With respect to plastics it has long been known that the addition of $MoS_2$ to plastics or metal composites will usually reduce the wear rate and enable them to accept dry and boundary conditions without seizing. The main difference between the plastics and metal composites is in the amount of $MoS_2$ required. Plastics composites generally contain 1 - 5 per cent $MoS_2$, often coupled with a secondary filler such as Fibreglass; the metal composite contains at least 30 per cent $MoS_2$ and in some cases as much as 90 per cent. Obviously the role of $MoS_2$ is quite different in these two systems. In the case of the plastics composites the $MoS_2$ is an antiwear agent whereas in the metal composite its major role may be to combat friction. With such composites the sliding process continues to feed on the exposed film of $MoS_2$, even after prolonged rubbing.

In addition to these purely tribological effects, $MoS_2$ may also improve the mechanical and physical properties of thermoplastics where it may act as a nucleus for crystallisation[36]. A specific widely used example is $MoS_2$-filled Nylon (Nylatron GS) which has superior tensile strength, modulus of elasticity and flexural strength. Such enhanced mechanical properties allow the use of these reinforced plastics in bearings where weight saving (energy saving) is important.

Oil-soluble molybdenum compounds

The initial interest in these compounds stems largely from the erroneous assumption that molybdenum thiophosphates, etc. would decompose under the influence of the frictional heat between contacting asperities to produce $MoS_2$. It is now thought more likely that these compounds are able to react with the metal surface to produce iron sulphides which have antiwear and EP properties up to about 800°C. The author feels that the developing importance of molybdenum in this field is in its ability to provide the oxidation state required to form sulphur bridges in complex organo-metallic compounds containing phosphorus and sulphur. Such a contribution is important because the efficiency of an EP lubricant depends on the speed of film formation (reaction with the metal substrate) and yet if the emergency is too quick for film build up, higher activity may give too much corrosion.

Thus a delicate balance must be maintained between film performance and reactivity, and this is where the type of organic ligand coupled to the Mo can be influential. Not only is the type of compound important but also the type of system in which it is to be applied. For example, EP additives which function satisfactorily in predominantly sliding conditions - as with a hypoid gear which is relatively bulky and relies solely on the sump oil for cooling - may not function so well between the piston and cylinder wall of an internal combustion engine where the action is reciprocal and additional water cooling is available. The molybdenum compounds referred to earlier as the most widely used commercially are sulphurised oxymolybdenum organo-phosphorodithiolate[35] and variously substituted derivatives.

Biochemical processes

The biochemical importance of molybdenum was first noticed in connection with plants in the 1940s when it was observed that certain plants would not grow in molybdenum-deficient soils. It seems that molybdenum is an essential constituent of bacteria associated with legumes - especially those which catalyse the conversion of the elemental nitrogen in the air into nitrogen-containing compounds. In fact all plants benefit because they cannot utilise nitrogen in the elemental form. Atmospheric nitrogen is assimilated via symbiotic bacteria with the help of the nitrogenase enzyme complex made up of two protein components - the Fe protein and the Mo-Fe protein. Other bacteria use a different enzyme which also requires molybdenum to fix nitrogen from nitrate in the soil.

The agricultural industry uses 500,000 lb molybdenum per year at a level of 0.03ppm. On large seeded legumes such as soya beans and peas, seed treatment provides an economical and effective method of supplementing the crops' molybdenum requirements. Liquid or dry fertilisers can be formulated to contain molybdenum so that molybdenum can be added with the regular fertiliser programme. Foliar sprays are used to eliminate molybdenum deficiency in citrus crops and frequently these sprays are used on other crops where deficiencies are apparent or anticipated. For example, in parts of Australia, one ounce of molybdenum spread over sixteen acres of molybdenum-deficient mixed pasture land (containing a legume) was beneficial. The molybdenum is contained in a phosphate fertiliser and its first effect is to increase symbiotic nitrogen fixation which gives larger growth of the legume. Ultimately this increases the fertility of the soil and the yield of grasses and other non-legumes and hence considerably increases the stock carrying capacity of the pasture. In the Netherlands the occurrence of molybdenum deficiency during propagation of plants is mainly caused by insufficient molybdenum in the potting soil[38].

However, other factors can have an important influence. An especially low pH of the potting soil promotes the occurrence of molybdenum deficiency; symptons appear more quickly when the potting soil contains a good deal of nitrate nitrogen. Though deficiency symptoms in young plants can be controlled by spraying with a 0.05 - 0.1 per cent solution of ammonium molybdate, a dressing of 5g ammonium molybdate $cm^{-3}$ of potting soil is also effective. It has been calculated that dry plant material contains 400,000ppm carbon and that the C/Mo ratio is $100 \times 10^6$. Thus one begins to appreciate the enormous energies which man uses in supplying fixed nitrogen to plants when we consider that one molybdenum atom alone in nature can bring together $100 \times 10^6$ atoms to give carbohydrates and proteins.

In 1953 molybdenum was identified as a constituent of xanthine oxidase; as the human body requires approximately 12mg of cobalt and 18mg of molybdenum, we can calculate that 250,000 lb of molybdenum is locked up in the human race. Molybdenum is a cofactor of certain enzymes which catalyse many important reactions in both animals and plants. Its action is not fully understood though we are beginning to understand better the structure of the cofactor in nitrogenase due to the most recent work of Steifel[39] and his colleagues. They have shown on model systems using X-ray absorption spectroscopy that the molybdenum atom is surrounded by four to five sulphur atoms (at normal Mo-S distance) and two to three iron atoms (at Mo-Fe distance). This would indicate that the molybdenum may be sulphur bridged to the iron forming an Fe/Mo cluster. They have also demonstrated that unlike most aqueous molybdenum compounds this cofactor contains no terminal or bridging oxygen systems. It probably derives its reactivity from a strained environment - not unlike a lattice defect in an industrial catalyst. Being present in proteins in a particular complex it may also acquire conformational energy from the host molecule. Such enzymes, according to Bray[40], occur in mammals, birds, insects and bacteria and have been isolated from these sources. The main biological roles of the xanthine oxidases and dehydrogenases are purine degradation in mammals and birds and nitrogen assimilation from hypoxanthine in fungi and elsewhere.

The anticarcinogenic activity of molybdenum is not clear and the effect of molybdenum deficiency in animals is well known, although the effect of such deficiencies on oesophagal cancer was not appreciated until a little over a decade ago[41]. Although molybdenum is a heavy metal, it is generally classed as non-toxic and there are no recorded cases of poisoning in humans attributable to molybdenum - despite ample opportunity for contact during the long history of the industry. It is clear from experimental results on laboratory animals that massive dosages would be required to produce acute effects in humans.

In conclusion one can fairly ask - whither molybdenum? Indeed this question can apply to a greater or lesser extent to most of its sister elements which are also being used in large quantities relative to their natural availability and ease of extraction. For example, when one considers that the latest Amax mine at Henderson, Colorado has taken ten years to complete at a cost in excess of five hundred million dollars, and that future mines of this type will approximately double in price with each decade, then it will be appreciated that by the second millenium these important elements (Fig 1) may become exotic if not rare. Man has postponed his energy problem until it has probably reached an almost irreversible crisis; the same thing could and will happen to many other natural products if strategic planning is further delayed. Linking these thoughts then to the most vitally important problems facing mankind in the foreseeable future, one logically arrives at the overlapping and interrelated problems of feeding man and keeping him warm. It is perhaps strange that as technology advances it is just possible that the things we take for granted today may become luxuries of the future.

Returning then to the problem of heat and food we are dealing, in the biochemical sense, with a total energy problem. In this connection two approaches seem to present themselves which may rely heavily on the catalytic properties of molybdenum and its compounds - the one being the fast developing petrochemicals industry which will more sensibly utilise our fossil fuels and the other the biochemical routes to synthetic foods. This latter technology will serve the dual prupose of not only allowing us to copy nature's economic syntheses but also to release her abundant energy which is locked up in its organic species through photosynthesis.

At present the production of nitrogenous fertiliser is linked strongly to the energy problem; in fact it has been calculated that more than 80 per cent of all the hydrogen required for making ammonia is produced from fossil fuels by a reaction which is very energy consuming. Furthermore, it has been estimated that more than 2 per cent of the USA's total natural gas consumption is used in the production of nitrogeneous fertilisers which is so wasteful of energy compared to nature's route. It is, therefore, hoped that our knowledge of the coordination chemistry of molybdenum may help us understand nature better and as a result of this improve our knowledge of synthetic catalysts.

References

[1]J. W. Goth, Journal of Engineering & Mining, 1977, 173, 88.

[2]J. W. Vanderwilt and R. U. King, Mining & Metallurgy, 1946, 2, 274, 299.

[3]Scheele, Act Stockholm, 1777.

[4]Hielm, Journal de Physique, 1789.

[5]G. P. Haight, and D. R. Boston, Climax lst International Conference on the Chemistry & Uses of Molybdenum (Reading), September 1973, 48.

[6]G. A. Tsigdinos, Topics in Current Chemistry, 1978, 76, 1.

[7]E. R. Braithwaite and J. B. Peace, Proc. Ind. Process Heating Symposium, 1963, 3, D33.

[8]A. W. Armour, M. G. B. Drew, and P. C. H. Mitchell, J.C.S. Dalton, 1975, 1493.

[9]R. Machenschalk, Private Communication to Author, September 1976.

[10]M. J. Kennedy, PhD Thesis, Brunel University, 1972.

[11]M. C. Basila, Plenary Lecture, Climax 2nd International Conference on the Chemistry & Uses of Molybdenum (Oxford), September 1976.

[12]W. Ripperger and W. Saum, Climax 2nd International Conference on the Chemistry & Uses of Molybdenum (Oxford), September 1976, 175.

[13]US Patents, 2904580 (1959), 2941007 (1960).

[14]R. A. Ference, The Oil & Gas Journal, February 1978, 117.

[15]H. Adkins and W. R. Petersen, J. Am. Chem. Soc., 1931, 53, 1512.

[16]M. A. Segura, et al., US Patent 339406, 1976.

[17]R. Burch, Private Communication to Author, November 1976.

[18]Sherwin Williams Co. Technical Bulletin 343.

[19]E. A. Lizlovs, Corrosion, 1976, 32, 263.

[20]A. Marchese, A. Papo, G. Torriano, Anti-Corrosion, September 1976.

[21]M. S. Vukasovich and D. R. Robitaille, Proc. Climax 2nd International Conference on the Chemistry & Uses of Molybdenum (Oxford), September 1976, 225.

[22]A. W. Armour and D. R. Robitaille, J. Chem. Tech. Biotechnol., 1979, 29, 619.

[23]C. G. Farnsworth, Water Sewage Works, 1970, 117, 418.

[24]A. J. Lehman et al., Food Drug Cosmetic Journal, 1955, 10, 679.

[25]Bionics Inc., Wareham, Mass., Private Communications to Climax Molybdenum Company of Michigan, 1972.

[26]P. C. Bowes, Ann. Occup. Hyg., 1974, 17, 143.

[27]J. R. Gaskill, Journal Fire Flammability, 1973, 4, 279

[28]F. W. Moore and G. A. Tsigdinos, Proc. 1978 International Symposium on Flammability and Fire Retardants, Toronto, 160.

[29]G. A. Skinner, L. E. Parker, and P. J. Marshall, Fire & Materials, 1976, 1, 154.

[30]E. R. Braithwaite and G. W. Rowe, Scientific Lubrication, March 1963.

[31]W. E. Jamison and S. L. Cosgrove,ASLE Trans, 1971, 14, 62-72.

[32]W. E. Jamison, Report AL727008 SKF King of Prussia P.A.

[33]E. R. Braithwaite, Scientific Lubrication, April and May 1966.

[34]Idem, Ibid.

[35]E. R. Braithwaite and A. B. Greene, Wear, January 1978.

[36]E. R. Braithwaite, Wear, 1975, 34, 455-465.

[37]J. P. G. Farr, Wear, 1975, 35, 1-22.

[38]J. van den Ende and G. A. Boertje, Acta Horticulturae, 26, 1972.

[39]C. I. Steifel, Chas. Kettering Research Laboratories, Ohio, USA, Private Communication to author, June 1978.

[40]R. C. Bray, Proc. Climax 2nd International Conference on the Chemistry & Uses of Molybdenum (Oxford), September 1976, 271.

[41]I. J. T. Davis, Intake (Abbott Laboratories Limited) Issue 39, January 1976.

# The Chemical Uses of Nickel and Its Compounds

By E. R. Braithwaite
CLIMAX MOLYBDENUM COMPANY LTD., 1/3 GROSVENOR PLACE, LONDON SW1X 7DL, U.K.

## Occurrence

Nickel is not found in the native form and makes up 0.009 per cent of the earth's crust in the form of sulphides or oxides (laterites) and silicates (garnierites). The sulphides to date have been mainly exploited due to their relatively high concentration of nickel, ease of access and technical and economical feasibility.[1] This situation must change due to the fact the laterites and garnierites account for over 75% of the world's known reserves.

The sulphides are concentrated ores formed below the earth's surface in geological times and are found mainly in Canada, Australia, Finland and the USSR, and consist of three main types viz:

Pyrrhotite ($Fe_7S_3$)
Pentlandite $(NiFe)_9S_8$
Chalcopyrite ($CuFeS_2$)

and of these Pentlandite is the commonest and accounts for more than 60% of the current world nickel production.

The laterites are usually found in tropical or semitropical zones and are formed by the atmospheric weathering of rocks in hot climates - a process known as laterisation. During laterisation nickel is taken into solution by surface water which then filters through the permeable earth and eventually concentrates the nickel up to 1 per cent or more some way below the surface. The two principal ores formed in this way are the limonite type containing iron oxide and the silicate or serpentine ores where the nickel is mixed with silicate during laterisation.[2]

## Extraction

Most of the world's nickel is derived from the sulphide

ores though many companies are now seriously concerned with the extraction of nickel from the more abundant lateritic ores. The output of most mines is rock containing a low percentage of nickel sulphide and hence process of concentration is not unlike those of molybdenite in that the rock is subjected to a series of surface chemical treatments including comminution and flotation to get rid of unwanted gangue and hence obtain a nickel sulphide concentrate. Such concentrates additionally contain iron, cobalt, copper and silica as principal impurities which have to be separated at a later stage; the proportions of these impurities vary with each ore body.

The concentrates are further treated in two ways either pyrometallurgically or hydrothermally, the latter becoming more widely used due to its lower energy consumption. It should, however, be emphasised that each plant has its own expertise in conducting these operations and should the reader require more than the general outline given below he should consult[1]

The pyrometallurgical processing of nickel ores in concentrates commonly includes three distinct chemical operations:

(a) Roasting: to convert sulphur to sulphur dioxide and iron to iron oxide

(b) Smelting: to remove most of the iron as slag by heating the product if roasting with a silica flux, leaving behind a molten sulphide solution

(c) Converting: to remove more sulphur from molten sulphide and remaining iron as a slag

The product of these operations is termed MATTE and is a homogeneous solution of Ni, Cu, Co, Fe, and S containing up to about 80 per cent Ni.

The matte is further processed to separate nickel from the

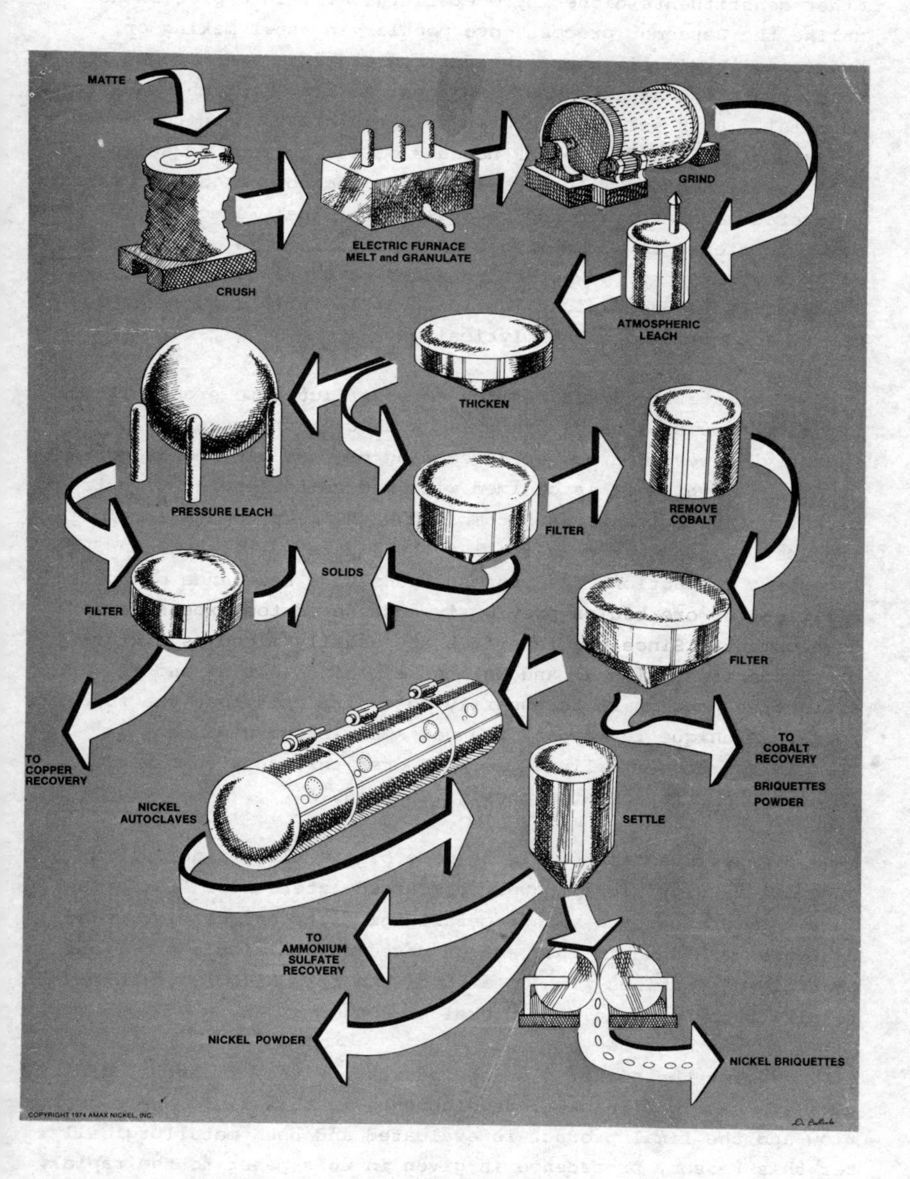


Flowsheet for products of nickel at the Amax Port Nickel
Louisiana plant.

other constituents either pyrometallurgically in a process not unlike the Bessemer process once popular in steel making or hydrometallurgically where the desired elements are leached selectively into the aqueous phase to separate them from the unwanted material. Both acidic and basic solvents are used though ammonia is probably the most popular because of its ability to selectively form stable ammines with nickel, cobalt and copper, and is easily recoverable for further use. Usually the final reduction to metallic nickel is accomplished in an autoclave in an atmosphere of hydrogen, nickel powder being added to the liquor to act as seed for the reduced nickel nuclei. Still further purification may be accomplished electrolytically.

The laterite ores are much more difficult to process; thus far many processes have been tried without much success. Quite recently, however, a new acid leaching process has been reported by Amax[3] which it is claimed will reduce the energy to extract nickel by 40 per cent, when compared with the pyrometallurgical process. Since energy is the largest single cost factor in lateritic extraction, the process opens up a new range of possibilities for ore bodies previously considered too expensive for development. Since the laterites are usually mined as mixtures or garnierites (high Ni and Mg) and limonite (low Ni and Mg) the extraction process is necessarily more difficult than the sulphide route. A unique feature of the new Amax leaching process is that it can economically treat both ores through a combination of high pressure and atmospheric leaching.

Finally mention should be made of the ultra fine, ultra pure low density nickel powder discussed later under "batteries". This is produced by a vapour metallurgical technique discovered by Langer and Mond in 1889 and is based on the simple principle that nickel combines with CO at 50$^{o}$C to form $Ni(CO)_4$ which readily decomposes to pure nickel at 200$^{o}$C.

Whilst the nickel plating industry is the biggest single user of 'chemical' nickel, developments in this field are rather slow and the final product is evaluated and used metallurgically. For this reason, precedence is given in this paper to the rapidly changing and expanding field of catalysis with its interesting and often incompletely understood chemical problem.

Pigments, ceramics and glazes are included for the sake of completeness, for whilst nickel has been used in these areas for a very long time, its contribution is on a relatively small scale. However, the underlying chemistry is well understood and it may be that the more stringent legislation on the use of toxic materials will see a re-emergence of nickel in some of these areas. Batteries and hydrogen storage devices are obviously important and fundamental studies in solid-state physics and chemistry with respect to the latter are attracting much attention.

## Chemistry

Nickel is similar in its chemical reactions to iron and cobalt but less reactive than iron. It is resistant to alkalis, including fused caustic soda, due to the presence of a highly resistant film of black nickel oxide which forms on its surface. However, the chemistry of nickel is rather simpler than the other elements of the first transition series as the only oxidation state presently of industrial importance is Ni(II) and its compounds are stable to oxidation by air and to reduction. Alkaline solutions of Ni(II) may be oxidised by bromine to Ni(III) hydroxide, NiO(OH), and also electrochemically. The redox equilibrium between Ni(II) and Ni(III) hydroxides is used in the Ni-Cd or Ni-Fe batteries. The chemical compounds and so-called simple salts of iron, cobalt, and nickel have metals in the +2 oxidation state; they differ in their redox chemistry. Aqueous solutions of Fe(II) salts are readily oxidised to Fe(III) whilst Co and Ni as aquo-ions are much more difficult to oxidise, although cobalt unlike nickel in the presence of ammonia can be oxidised aerobically to Co(III) ammine complexes.

The stereochemistry of Ni(II) is one of the most complex of the transition metals and may occur in six co-ordinate octahedral systems, e.g. $[Ni(H_2O)_6]^{2+}$, $[Ni(NH_3)_6]^{2+}$, as five coordinate, e.g. $[Ni(CN)_5]^{3-}$ and very often as four coordinate, where normally square planar $(R_3P)_2NiCl_2$ stereochemistry is found.

A marked difference between Ni(II) and the secondary oxidation states of iron and cobalt lies in its ability to form square-planar complexes; this is associated with the 8d electron configuration of Ni(II).

Palladium(II) and platinum(II), the vertical congeners of nickel, show an overwhelming tendency to form square-planar complexes. Compared with nickel, palladium and platinum have more class b character and their bonding preferences for the less electronegative donor atoms such as sulphur and phosphorus are greater. A consequence of the marked affinity of palladium and platinum for sulphur compounds is that platinum and palladium catalysts are very susceptible to poisoning by sulphur compounds, more so than nickel. The Ni-S bond is probably weaker than the Pd-S and Pt-S bonds and therefore bound sulphur is more readily stripped from nickel e.g. by hydrogen.

It is of interest to note that nickel never occurs in nature as the metal although palladium and platinum do so. Similarly palladium and platinum are much more resistant to corrosion than nickel; palladium and platinum are noble metals, nickel is not. The main reason for the difference is that the ionisation energies of nickel atoms are much less than those of palladium and platinum; also platinum, in particular, is stabilised by stronger metal binding. The lattice of the $Ni^{2+}$ compounds, especially the oxide, is also more stable as a consequence of the smaller size of the nickel.

Nickel, palladium, and platinum also differ in the relative stability of the higher oxidation states, so that Pd(IV) and Pt(IV) unlike Ni form a larger number of complexes e.g. $PtCl_6^{2-}$. This leads to a difference in the types of compounds used for catalyst preparation; Pd and Pt are utilised as the hexachloro-ions Pd(IV) and Pt(IV), whilst Ni is usually presented as the hexahydrated cation of Ni(II) e.g. nickel formate or nitrate. It is further relevant to the carbonyl process for the extraction of nickel that Pt and Pd do not form a simple carbonyl analogous to $Ni(CO)_4$ directly from metal.

Within recent years there has been developed, for example, by Wilke and others[4] [5] [6] an elegent and extensive chemistry of zero-valent organo-nickel compounds which have demonstrated remarkable catalytic properties. For example the trimerisation of butadiene to cyclododecatrienes by the so called 'naked' nickel compounds such as bis-cycloocta-1,5-diene nickel. This chemistry is, however, at an early stage of development with respect to industrial utilisation. However, it is clear that this field has very great potential especially in homogeneously catalysed reactions of unsaturated hydrocarbons. A recent example being used by Du Pont de Nemours Inc. is of zero valent trialkyl phosphates in the homogeneous catalytic addition of hydrogen cyanide to butadiene giving adiponitrile.

## Industrial Applications

### Catalysis

Metallic nickel or one of its alloys is usually in the highly dispersed state and is either distributed on a support by the precipitation or impregnation of one of its salts or used in the form of an evaporated metal film or metal cluster. Silica-supported nickel catalysts have been used almost since the discovery of olefin hydrogenation by Sabatier and Senderens in 1897. In fact it was Sabatier and Senderens who first reduced organic vapours over various heated metals in the presence of hydrogen and expressed their preference for nickel. Since this time there have been many theories of action of which the writer considers two to be relevant: the delocalised electron theory and the localised atomic theory. There has unquestionably been a gradual shift in popularity from the former to the latter.

### Supported nickel catalysts

There is little doubt that the supported nickel catalyst is becoming very popular as it may offer some advantages over the non-supported nickel, such as more metal surface being made available due to the dispersed state of the metal together with the reduction or prevention of sintering.

E.S.C.A. studies by Wu and Hercules (7) have shown that for low nickel content $\alpha$-alumina supported catalysts, nickel interacts strongly with the support and is not present as NiO, whereas at high concentrations most of it appears as NiO. This is due to lack of interaction with the substrate, which is also the case for nickel on a silica substrate.

Oxide and sulphide catalysts

Nickel oxide. This is a p-type semiconductor as it gains oxygen when heated:

$$4Ni^{2+} + O_2 \longrightarrow 4Ni^{3+} + 2O^{2-}$$

Thus for each new oxide ion formed, two $Ni^{3+}$ ions result, and each of these species, having excess positive charge, constitutes a positive hole. The metal in the p-type metal oxide is in the lowest oxidation state but additionally $Ni^{3+}$ is available. It is possible to change the concentration of positive holes by doping, i.e. by the introduction of small amounts of another oxide whose cation has a different valency.

For example, the $Ni^{2+}$ ion in nickel oxide can be replaced by $Li^{+}$ and the lithium ion occupies a normal nickel ion site, but is only singly charged. Each atom of lithium forming $Li^{+}$ contributes only one electron to the crystal instead of two. Therefore, each lithium ion must be accompanied by on electron hole which can be regarded as $Ni^{3+}$. The nickel oxide containing lithium ions can be prepared by heating lithium oxide and nickel together in oxygen.

$$xLiO + (1-x)NiO + (x/4)O_2 \rightleftharpoons (Li_x Ni^{2+}{}_{1-2x} Ni_x{}^{3+})O$$

Conversely the addition of trivalent-chromium ($Cr^{3+}$) to NiO decreases the number of positive holes and hence decreases the conductivity.

Nickel sulphide. Interest in catalysis by metal sulphides arose at the beginning of the century due to the poisoning of industrial catalysts by sulphur compounds during the hydrogenation of coal and the hydro-processing of crude oils. Nickel sulphide like the oxide differs from the nickel metal catalyst due to its semi-conducting properties. The sulphide is polyfunctional so that in addition to its oxidation-reduction properties it also has the properties of an acid catalyst particularly when used on a support for hydrocracking. It is typically a high-temperature catalyst and its more stable form is $Ni_3S_2$ formed by the hydrogenation of NiS.

When used as a hydrogenation catalyst for poly-ene, the degree of hydrogenation depends on the fact that the rate of reaction increases with the number of double bonds in the hydrocarbon chain. For metallic nickel the selectivity for partial hydrogenation is 40-50 per cent whilst for $Ni_3S_2$ it is greater than 80 per cent. The cis-trans isomerisation rate for metallic nickel is 40-50 per cent whilst the corresponding value for $Ni_3S_2$ is greater than 90 per cent[8] and it is the cis-trans isomerisation at the double bond which causes the hardening of fats since all trans-isomers have high melting point.

Bimetallic catalysts

This type of catalyst, supported and unsupported, has been developed over the last decade, the impetus in its rapid development springing from the success of the Pt-Re catalysts in reforming operations. An important alloy catalyst is Ni-Cu where not only are the surface properties different from the bulk properties but also the active sites on the surface are heterogeneous.

<u>Copper-nickel</u>. The unsupported Cu-Ni alloys can be prepared by the hydrogen reduction of oxides[9] formed via the precipitated hydroxide, carbonate and so on. Inhomogeneities can arise in alloy catalysts either because of segregation during preparation or in surface enrichment by one of the components in the reduced alloy. According to Anderson[10], surface enrichment usually occurs with the component of lowest surface energy, although even this behaviour will be exclusively modified by the presence of a chemisorbed gas which reacts preferentially with one of the components and so leads to surface enrichment of that particular component.

Copper nickel alloys supported on silica can be prepared by in situ calcination of supported nitrates followed by hydrogen reduction. Robertson et al.[11] have shown that due to the immiscibility of NiO and CuO before reduction, the behaviours of Cu and Ni after hydrogenation are entirely distinct. However, this may be avoided by more careful control of the preparative procedure.

Raney nickel. The Raney nickel catalysts are skeletal types made by leaching out Al from Al-Ni alloy[12]. The composition of a Raney nickel catalyst is somewhat variable depending on the mode of preparation. A residue of Al or $Al_2O_3$ is always present to a greater or lesser extent, the amount and proportion of which can greatly influence catalytic activity, life, and in particular sintering and mechanical strength and thermal stability.

According to Freel et al.[13] the alloy before leaching usually contains 40-50 per cent Ni principally as $NiAl_3$ and $Ni_2Al_3$ together with a smaller proportion of eutectic, which can influence the final pore structure. The majority of the metal particles are secondary aggregates (up to 100 $\mu$m) of very small particles. When dry, Raney nickel is pyrophoric and is preserved out of contact with air in either water, dilute alkali or ethanol. Dispersed multimetallic Raney catalysts are also available containing one or more transition metals other than nickel[14]. These not only modify their specific surfaces, but also their structures[15] can be changed to give crystallites of about 40Å by addition of molybdenum, the amount determining the specific activity and the concentration of $NiMo_5Al_{10}$.

Nickel-molybdenum. Several catalyst manufacturers are now actively concerned with the replacement of cobalt by nickel in cobalt-molybdenum catalysts. This is not only due to the recent world shortage of cobalt caused by political problems in Africa, but also it is considered that Ni-Mo has the edge over Co-Mo in treating feedstocks containing high levels of aromatics plus nitrogen and oxygen compounds, i.e. nickel catalysts have a great future in coal conversion.

The most important of these is the nickel-molybdenum alumina which was developed for hydrotreating following the success of the Co.Mo.$Al_2O_3$ family of catalysts. At a relatively early state the greater hydrogenating activity of Ni.Mo.$Al_2O_3$ catalysts was recognised, particularly in treating feedstocks containing nitrogen compounds and unsaturated compounds. However, an increasing acceptance of Ni-Mo catalysts, for uses in which hydrodesulphurisation is not the major objective, arose from the following[16] main developments:

1. Improvements in technique of activation, particularly the recognition that sulphiding at temperatures below about 220°C is essential. Higher temperatures lead to reduction to metal which can rapidly sinter and cause a permanent loss of activity.

2. Improvements in the technique of catalyst preparation, particularly the realisation that Ni-Mo catalysts unlike Co-Mo catalysts are more susceptible to promoter influence e.g. by phosphoric acid, which appears to increase its hydrodesulphurisation activity.

The oxides of Ni and Mo supported on alumina, silica-alumina, boria alumina are used in a variety of industrial processes including hydrogenation, hydrocracking, hydrodesulphurisation, hydrodenitrification and hydroisomerisation. These catalysts are usually sulphided before use and are particularly effective in up-grading hydrocarbon feedstocks containing sulphur and nitrogen. One of the reasons for presulphiding is to suppress the formation of metallic nickel which would have been formed during hydrogenation. Metallic nickel leads to excessive hydrocracking of hydrocarbon molecules and the deposition of coke. Fivansanker et al.[17] have shown that the formation of metallic nickel decreases in the order NiWAl $>$ NiWSiAl $>$ NiWBAl; hence the underlying support exerts appreciable influence.

Recently Burch[18] has suggested that the addition of a second metal, such as molybdenum, can materially alter the catalytic properties of the nickel. For example, catalysts prepared by the coprecipitation and reduction of hydrated oxides of Ni and Mo do not have Ni-type properties unless the Ni content is greater than 75 per cent. The influence of the Mo appears to be electronic rather than geometric (in contrast to the effect of adding Cu to Ni) because the relative surface energies make it unlikely that there would be a sufficient number of Mo atoms present in the Ni surface to significantly affect the number and distribution of Ni atoms.

There are four main catalytic processes involving Ni catalysts:

(a) Methanation. As oil stocks become more depleted, the gasification of coal is becoming more important. One of the most promising ways of providing high b.t.u. gas is the hydrogenation of carbon oxides to methane.

Whilst the methanation of CO and $CO_2$ is catalysed by a number of metals as originally demonstrated by Fischer et al.[19] there is little doubt that at temperatures up to 800°C, nickel is the cheapest, most active and most selective and still forms the core of most commercial methanation catalysts. In common with other catalysts however, it is susceptible to sulphur poisoning, though with improved plant design and control this is not considered by some companies to be a major problem. The nickel content varies from 25-77 per cent by weight and is usually dispersed on a high surface area refractory support such as $Al_2O_3$.

Araki and Ponec[20] have recently demonstrated that carbon deposits on the surface of the nickel, or carbide may be important to the hydrocarbon synthesis, which is confirmation of the earlier work of Kummer et al.[21] who first showed the importance of surface carbon in methanation. Rare earth-nickel metal catalysts have also been reported. One rare earth atom to five nickel atoms may give optimum combination of specific activity and surface area[22]. These catalysts have low surface areas and interest in them arises from the fact that $LaNi_5$ reacts with hydrogen to give $LaNi_5H_6$ which is a reservoir of absorbed hydrogen.

(b) Steam reforming of hydrocarbons. The reaction may be

summarised by the following equations:

$$C_nH_{2n+2} + nH_2O \longrightarrow nCO + (2n + 1)H \quad \text{(i)}$$

$$CO + H_2O \longrightarrow CO_2 + H_2 \quad \text{(ii)}$$

$$CO + 3H_2 \longrightarrow CH_4 + H_2O \quad \text{(iii)}$$

The product distribution is determined by (ii) and (iii) viz. the water gas shift and methanation. The production of CO, $CO_2$ and $H_2$ favour a temperature in excess of 700°C for the manufacture of $H_2$ or CO-$H_2$ mixtures, whereas the production of $CH_4$ to be used in town gas or substitute natural gas is carried on at lower temperatures ($<$550°C). Temperatures above this tend to favour the formation of $CO_2$ with the added risk of coking. The principal catalyst for steam reforming is almost invariably nickel; this is supported on a high-area refractory oxide such as alumina or magnesia plus an alkali metal to reducing coking at higher processing temperatures.

(c) Hydrogenation. The conversion of fats to edible products involves the selective hydrogenation of polyene fatty acids in the presence of nickel catalysts to achieve the required hardness. Isomerisation may occur and has to be suppressed. Catalytic hardening dates back to Sabatier (1897) who discovered that unsaturated compounds can be hydrogenated in the gaseous phase by finely divided nickel. In the hydrogenation of fatty oils using supported nickel there is an empirical relationship between activity and the ratio of surface area to pore size of the support. Furthermore, it is claimed that if the pore size is greater than 20-25Å then this has a depressing effect on the selectivity of the hydrogenation at the nickel surface.

The supported catalysts are produced in two ways: dry reduction, where the oxide or hydroxide is first prepared by the precipitation of a nickel salt with caustic soda or the ignition of the nitrate followed by hydrogen reduction of the oxide at 300-350°C; wet reduction, where nickel formate or carbonate together with diatomaceous earth is suspended in a fatty oil and hydrogen reduced. The finely divided nickel thus produced is in the form of a hard fat flake containing about 20 per cent Ni. More recently Ni/Cu and Ni/Ag catalysts have been used.

Whilst a nickel catalyst is marginally less selective than copper, nickel is preferred because traces of copper left behind in the final product can lead to autooxidation and rancidity. However, certain of the advantages of Cu as a Cu-Ni catalyst are to be found in a Ag-Ni catalyst which has recently been introduced in a commercial scale[23]. Contrary to many sources of information, Raney nickel catalysts are not now a major factor in this industry due to their lower selectivity, activity and process difficulties associated with their filterability, although of the commercially available fat hardening catalysts at least 75 per cent of the nickel is reduced to the metallic form.

(d) Hydrocracking. This is used extensively in petroleum refining to produce high quality gasoline, jet fuel and lubricants. Many hydrocracking catalysts contain both a hydrogenation and dehydrogenation component and an acid component. The nickel catalysts can either be $Ni_3S_2$ or Ni metal on a non-acidic or weakly acid support. With Ni on a non-acid support hydrogenolysis is the only observed cracking reaction of the paraffins - no isomerisation occurs.

With Ni on a acid support, isomerisation is the most important reaction. Cracking probably occurs by hydrogenolysis at the Ni-metal surface, and also at the acidic oxide sites. Nickel helps to prevent polymerisation of olefines formed in cracking by hydrogenating them. During the impregnation with the soluble Ni-salt, the acidic protons of the catalytic sites on the $SiO_2/Al_2O_3$ support are exchanged with Ni to give salts of the $SiO_2/Al_2O_3$ sites. One of the effects of sulphiding Ni on $SiO_2/Al_2O_3$ is that $H_2S$ reacts with the salts and regenerates new active sites which are cracking centres.

## Batteries

### Operations and advantages

The first nickel storage (Ni-Cd) battery was introduced

by Edison in 1900 and this was followed some years later by the commercial Ni-Fe alkaline cell. In the discharge state the active materials are nickel hydroxide and cadmium hydroxide or Fe(II) hydroxide; the electrolyte is a 21 per cent aqueous solution of KOH to which a small amount of LiOH is added. During charging the cadmium or iron hydroxide is reduced to cadmium or iron; the nickel is oxidised to NI(III) forming the positive plate. Current research centres on the correlation between electrode behaviour and the structures and composition of the hydroxide and oxyhydroxides involved.

The addition of LiOH to the electrolyte dates from early manufacturing technology. It is now accepted[(24) (25) (26)] that small additions of LiOH favour a higher degree of oxidation of the $Ni(OH)_2$ during charging and an over-all increase in the storage capacity. Larger amounts of LiOH may be harmful to the active material, altering the characteristic flat discharge potential time curves that are a valuable feature of this electrode.

The charge and discharge characteristics of the negative plate are also subject to research[(27)], the capacity being limited by the diminution in surface area of active cadmium due to growing of crystallites by rearrangment and passivation or choking of the porous structure with $Cd(OH)_2$.

The KOH is not consumed in the cell reactions, so that its volume can be much smaller than it is, for example, in the lead-acid cell. However, KOH will dissolve atmospheric $CO_2$ which impairs the performance. Carefully vented cells are, therefore, necessary; it has also been possible to provide an hermetically sealed cell. In sealed cells an excess of active material is included at the negative (Cd) electrode. This prevents the evolution of hydrogen at the negative plate during overcharge. Overcharge does produce oxygen at the $Ni(OH)_2$ positive plate, but the cell is designed so that the oxygen liberated can pass easily to the negative electrode where it is reduced to hydroxyl ions.

Perhaps the greatest advantage of the Ni-Cd cell over the lead-acid system is the possibility to have complete discharge and has twice the life of the Pb-acid cell. Ni-Cd cells retain their capacity for long periods and do not deteriorate when left uncycled or in a discharged state. The Ni-Fe cell can be similar in performance to Ni-Cd, but its discharge characteristic is not so flat and the charge retention is poorer. Ni-Fe cells are made in larger sizes than Ni-Cd and they tend to be used for heavy traction applications.

The Ni-Fe is the most rugged cell and lasts longest. It is not damaged by short circuit, overcharge or overdischarge. Its cost is intermediate between lead-acid and Ni-Cd. Its capacity varies from 5Ah to over 1,000Ah. The Ni-Cd can replace almost any other battery, its use being limited solely by cost and size.

Nickel powder is the starting material for the production of some secondary electrochemical power sources and for some fuel-cell electrodes. The two most common ways of currently preparing such powders are the carbonyl method and the products of reaction of the ammonium complexes of metal salts by hydrogen at high temperatures or pressure, though alternative routes are the subject of intense studies.

The batteries are made up of two types of nickel plates viz. sintered and pressed or pocket plate.

Sintered plates

The development of the sintered plate was a major step in achieving a storage cell with good life characteristics and capable of operating at much higher charge/discharge rates than other plate types. The sintered nickel-plaque which serves as a current collector for this plate involves moulding a quantity of nickel powder on to a supporting matrix. Perforated sheet metal (Ni plated steel) is popular as it is relatively inexpensive and allows continuous sintering. The forming takes place in a reducing atmosphere at 900-1,000$^{\circ}$C followed by slow cooling to ambient temperature during which time the $H_2/N_2$ mix is replaced by air. Nickel or cadmium hydroxides are deposited in situ within the pores resulting in a highly conductive but compact electrode system. The physical properties of the sintered nickel plate before impregnation are critical. The structure should be continuous, pure and highly ($>$80 per cent) porous.

Pressed plates

Are formed in situ by impregnation of an aqueous bath containing $NiNO_3$ or $CdNO_3$ with sodium hydroxide followed by a water rinse to remove the nitrate and to give a porosity of 90 per cent.

According to Tofield et al.[28] nickel batteries will play an important role in the second generation of advanced batteries connected with energy economics. The industrial development of traditional alkaline electrolytic batteries (i.e. Ni-Zn, Ni-Fe) should produce improved power systems for transport vehicles in the not-too-distant future. It is forecast that the realisable energy densities of such batteries will be about twice that of the lead-acid cells. The final choice of these sources of energy is probably a decade away; nevertheless one can confidently predict that nickel will be part of the system whatever the ultimate form. Now that a possible nickel shortage is not unlikely it is well to recall the work of Binkley and Watson[29] who have detailed techniques for making the optimum use of nickel. Such design modifications include the correct separation of anode and cathode and the use of bipolar anodes.

Finally, mention should be made of the lithium-nickel halide battery which represents a potentially high energy-density source of electric power. Not only is lithium weight saving but also has one of the highest electrochemical potentials of any known electrode and material. Gulton has devised a 'dispersed' lithium anode which consists of a dispersion of lithium particles, nickel powder and graphite in mineral oil. This mixture is bonded in an inert atmosphere to a nickel grid with an organic binder and then the electrode is hermetically sealed into the battery. Organic liquids such as propylene carbonate are used as electrolytes due to the reactivity of lithium with water. The theoretical capacity of such a battery is about 280kW $kg^{-1}$ cf. 16kW $kg^{-1}$ for Ni-Cd.

## Nickel in Hydrogen Storage

Currently there is much discussion of future 'hydrogen economy' i.e. the use of hydrogen to store and carry energy derived from nuclear or solar sources. In the former, the large enthalpy of hydride formation is utilised. In the latter, hydrogen is absorbed (condensed) by hydride formation. Fundamental to such a strategy will be the electrolysis of water, the storage of hydrogen and the development of fuel cells. Nickel promises to play an important part in each of these technologies. Nickel boride and nickel oxide have been used as catalytic electrolysis fuel-cell electrodes. There is also considerable interest in nickel-based alloys that store hydrogen to densities higher than liquid hydrogen because the hydrides are stable below their dissociation temperatures and pressures[30] [31].

The attractiveness of these compounds is their use as rechargeable stores; the prototype is $LaNi_5H_6$, but many others are possible and are being explored. These metallic hydrides include a series of Ni-mischmetal-calcium alloys of the type $Ca_xM_{1-x}Ni_5$, where M = mischmetal and $x \leqslant 1$. Whilst this material is economically attractive, its hydrogen storage capacity is much less than that of $LaNi_5$. Other examples include $Ti_2$-Ni and Ti-Ni[32], which absorb large amounts of hydrogen at temperatures below 121°C and pressures below one atmosphere; Ni-Nb and Ni-V have also been extensively studied. Introduction of some $LaNi_5$ into metallic magnesium also makes the lighter element suitable for hydrogen storage.

According to Videm[30] even heavier metal hydrides ($LaNi_5$ and $TiNi_2$ and Ti-Ni hydrides) contain sufficient hydrogen for a storage/fuel cell system to have an energy-to-weight ratio about double that of the lead-acid cell. A storage capacity greater than that of $LaNi_5$ seems unlikely. Moreover, the practical applications at the moment are limited by the chemical stability required in contact with aqueous electrolytes, the pressure at which the hydrogen is desorbed, and the effect of cycling on the systems. Quite recently Wallace et al.[33] have proposed a mechanism for the adsorption of hydrogen by $LaNi_5$. This mechanism depends primarily on the nature of the surface of $LaNi_5$, the top 50-100Å consisting of lanthanum oxide and nickel; the oxide becomes hydroxide in the presence of atomic hydrogen. The oxide can be formed either from minute traces of oxygen impurity in the alloy or from atmospheric oxygen to form a heterogeneous $Ni/La_2O_3$ surface layer. It is the removal of oxygen by La in the surface which protects the surface nickel from impurities. Since neither $La_2O_3$ nor Ni is permeable to hydrogen it is suggested that hydrogen reaches the underlying $LaNi_5$ by migration along the interface between Ni and $Li(OH)_3$. It can leave by the same route and then reassociate to molecular hydrogen.

Pigments

Whilst it is appreciated that the industrial use of nickel as a constituent of a pigment or dye has been on a relatively restricted scale, nevertheless the writer feels that due to increasing legislation on toxicity and pollution it may become more important.

The best known pigment is nickel titanate which is a mixed-phase pigment consisting of the oxides of titanium, nickel and antimony which are calcined at about $800^{o}C$, the chromophoric activity being supplied by the nickel and antimony. The principal colouring element is nickel which replaces the titanium randomly in some of the cation sites thus producing a defect structure. The antimony acts as a stabiliser by preserving the neutrality of $Ti^{4+}$ and $Ni^{2+}$ and additionally changes the yellow colour of nickel titanate to yellow-orange due to the formation of nickel antimonate. The rutile or spinel structures are basically favoured for the matrix due to their stability. Nickel titanate is a 'substrate' pigment having nickel titanate oxide core. It has excellent resistance to acids and alkalis and is heat stable up to over $400^{o}C$; it is also completely light stable. Whilst its hiding power is high, its tinting strength is low. It also has the advantage of chalking without colour change which makes it ideal for outside applications.

Nickel-cobalt aluminate mixtures provide good blue pigments of spinel structure which consists of a solid solution of 15-20 per cent nickel aluminate in cobalt aluminate and retains a good blue colour when diluted with $TiO_2$[34]. Such a pigment has higher tinctorial strength than either Ni or Co aluminate alone. The colour development of a $ZnO$-$SnO_2$ spinel by $Ni^{2+}$ in the octahedral or tetrahedral form has also been demonstrated[35], whilst the effect of nickel monoxide on Willemite pigments has also been studied. The replacement of zinc oxide in blue ceramic pigments by NiO gives a green colour which increases in intensity with increasing concentration of NiO due to the formation of $Ni_2SO_4$[36]. Nickel phosphate systems have also useful pigmenting properties[37]. A bright-yellow pigment is formed by calcining basic nickel carbonate and ammonium phosphate at 1,000°C to give $Ni_2(PO_4)_3$ and $Ni_2P_2O_7$. If the $NiCO_3$ is present in excess the colour proceeds to a dark yellowish green.

Nickel forms stable co-ordination compounds with azo dyes containing certain functional groups ortho to the azo-bridge. For example Ciba-geigy describes such a dyestuff[38] which is an azo-methine nickel complex. This is a yellowish-green pigment which is light and migration fast in PVC and is also used in green printing units. Azo-nickel dyestuffs are generally fast to washing and wet processing - the wet processing being superior to the azo dye without nickel. These nickel co-ordination compounds have the advantage of being non-toxic and their outstanding property is their light fastness. According to Roger and Amick[39] the use of certain surface-active agents in the application of nickel co-ordination compounds of azo dyes gives dyeings on wool of increased strength, uniformity and brilliance.

## Ceramics

Nickel oxides and other nickel compounds are used for the colouration of ceramics and to promote adherance between porcelain and iron. In the latter applications, the iron shapes are given a pre-coat of nickel by dipping in a $NiSO_4$ bath and then adding a 1-1.5 per cent NiO to the ground-glass frit. According to Harman and King[40], the stability of the nickel oxides or sulphates in the ceramic industry is based on a few specific properties, viz. the ability of NiO to absorb light selectively and the ease with which nickel can be replaced by iron in a plating solution.

The colouring action of nickel in glass has been described by Weyl[41] who showed that the oxidation state of nickel oxide is not affected by a wide variation in the oxygen pressure over a molten glass and that nickel is present in solution in the glass as Ni(II) only. However, the colour in the glass may vary from purple to yellow depending on the co-ordination of the nickel ion and the composition of the glass. For glasses in which nickel is surrounded by four oxygen atoms the colour is purplish whilst a sheath of more than four oxygen atoms gives a yellowish colour. In glasses which contain potash rather than soda, nickel oxide gives a brilliant purple colour. A useful property of NiO-glasses is that they only weakly absorb u.v. light but strongly absorb visible light, an effect which can be accentuated by the addition of traces of cobalt oxide.

Ferrites

Ferrites are one of the relatively new classes of ceramic materials that have been developed for high-frequency use. In the broad sense a 'ferrite' is a material containing iron oxide which exhibits ferrimagnetic and semiconducting properties. Another important characteristic of the ferrites is that they possess in general a spinel structure.

Of the two types, soft and hard, nickel ferrite belongs to the former class, which are only magnetically active in a magnetic field. Such Ni-Zn ferrites are used for transformer cores, cup cores and radio antennas. Almost all of the soft materials contain zinc oxide as one of the bivalent constituents. One of the advantages of solid solutions of zinc ferrite with another compound such as nickel ferrite is that the zinc preferentially goes into one type of lattice site causing the remaining ions to rearrange themselves in the structure. The effect of this is to give a larger net magnetic moment and so lower the anisotropy of the ferrite with increase in permeability.

Ferrites can be fabricated to high densitites by means of hot pressing at relatively low temperatures. The small average grain size of a hot-pressed Ni-Zn ferrite results in high-frequency properties which differ from those of normally sintered coarse grained ferrites. According to de Lau[42] grain size can influence the various magnetisation processes in magnetic fields of small amplitudes. This substantiates the early findings of Blum[43] who showed that the microstructure of $Ni_2Fe_4O$ significantly effects the magnetic and electrical properties of ferrites.

The chemical composition of Ni-Zn ferrites is also of critical importance. For example it has been shown by Brockman and Matteson[44] that by varying the mole per cent of $Fe_3O_4$ in a Ni-Zn ferrite in very small increments an optimum composition exists. Deviations from this composition as small as 0.1 per cent can cause a decrease in resistivity of up to three orders of magnitude.

Nickel in Electroplating

Electroplating is the application of the displacement of elements in the electromotive series. Nickel is typical of metals which can be electrodeposited to provide durable surface coatings on steel and other substances. Some 15 per cent of the free-market world production of nickel is used in the electroplating industry. Although nickel electroplate was first produced 130 years ago, modern plating is based on the formulation invented by Professor O. P. Watts of the University of Wisconsin in 1916. This solution contains nickel sulphate $NiSO_2.7H_4O$ (240g $l^{-1}$), nickel chloride $NiCl_2.6H_2O$ (20g $l^{-1}$), boric acid $H_3BO_3$ (20g $l^{-1}$). Each ingredient of the electrolyte has a definite role. $NiSO_4$ is used to provide the major part of the nickel-ion content on grounds of economics, whilst the chloride reduces polarisation and at higher concentrations increases the deposition rate and forms finer and more homogeneous coatings. Boric acid acts as a buffering agent to control pH at the cathode. Success of this process depends initially on the pH of the bath and the purpose of the boric acid is to buffer the electrolyte to about pH 5.5. This permits the electrodeposition at useful rates (e.g. 40mÅ $cm^{-2}$) of satisfactory metal.

The benefit of nickel plating is not simply that the nickel surface resists corrosion much better than does steel. The nickel plate can be obtained with a range of engineering properties such as hardness and ductility. The electrolyte can also be made to smooth ('levelling') roughness in the substrate surface and to give brightness. The presence of specific organic agents (e.g. Coumarine) at very low concentrations (e.g. 0.01-0.1g $l^{-1}$) in the Watts' bath brings about levelling; these substances are thought to be absorbed on asperities, thus locally inhibiting the deposition of the nickel. Brighteners are other organic compounds which are strongly adsorbed and their effect is observed on the submicroscopic topography. Fundamental electrochemical mechanisms of brightening and levelling are obscure but there is no doubt that these effects are enormously important to industry. Levelling can be obtained without significantly impairing the good corrosion resistance of Watts' nickel but bright plate is not so resistant. Therefore, current practice is to apply a 'duplex' coating where a thick deposit of levelling nickel is overlayed with hard, more decorative bright nickel; corrosion is then limited to the bright layer and is not penetrative. Comparatively recently baths based on nickel sulphamate have been developed industrially; nickel fluoroborate is also used. These solutions contain very much higher concentrations of nickel so that correspondingly high plating rates can be obtained. The metal produced is comparatively stress free; the reason for this is unknown and these baths are particularly suitable for electroforming nickel.

### Electroless nickel plating

There is another technique for obtaining a coherent, adherent nickel coating in electroless plating; here a solution of nickel ions also contains a metastable reducing agent, usually a hypophosphite or a borohydride. Under controlled conditions of pH and temperature and in the presence of catalytic surfaces the reducing agents decompose giving nascent hydrogen. The substrate surface need not be conductive; it is frequently activated by the addition of minute quantities of palladium from a dilute aqueous solution. On transferring the activated surface to the electroless plating baths, nascent hydrogen is generated and this reduces nickel ions at the active centres. Fortunately nickel metal is also catalytic and so the process once started is self-maintaining. The metal is locked into pits in the previously deeply etched plastic surface and therefore chemical bonding is not involved. However, with proper etching and activation very satisfactory results are obtained and the enhancement in durability of plastics that results allows many small components to be made more cheaply than they could from solid metal.

A new black nickel coating has been developed[45] and due to its very selective light absorption characteristics this makes it a potentially low-cost coating for flat-plate solar collectors.

Biochemical Systems

Given the important biological roles of the 3d-transition metal ions Mn, Fe, Co, Cu and Zn, which are present some or all of the time in the functional systems as the 2+ cation, it is remarkable that $Ni^{2+}$ does not seem, at the present time, to be an essential requirement for any biological process. The element cannot be classed as unavailable; it is present in the earth's crust at a level of ca. 75ppm and in sea water at a concentration of $2x10^{-3}$mg $l^{-1}$. The absence of a facile redox change for nickel would appear to rule out its participation as a redox centre in enzymes or electron-transfer sequences. However, it is surprising that nickel apparently does not have a significant role as a Lewis acid centre in, for example, hydrolytic-enzymes, since its co-ordination chemistry has some resemblance to Zn and Mn. (Perhaps, though, it is early days to make such a statement).

It has been shown by Dixon et al.[46] that highly purified urease from jack beans (Canavalia ensiformis) contains stoichiometric amounts of nickel and, in an attempt to understand the role of nickel in urease, the above workers[47] turned to the study of other enzymes. It was concluded that the nickel metalloenzymes $Ni^{2+}$, carboxypeptidate, and $Ni^{2+}$, phosphoglucomutase, which are all comparable in catalytic activity to the natural metalloenzymes as well as jack beans urease have electrical absorption spectra which are similar to many octahedral $Ni^{2+}$ complexes.

A flexible rather than an entactic site is suggested which may act as a Lewis acid site but this is, as yet non-proven.

Nickel has, however, been shown(48) to bind some metallo-proteins and, in a number of instances substitute for the normal $M^{2+}$ cation with the retention of biochemical activity. Galactose oxidase, normally a Cu(II)-containing enzyme, may be reactivated from the apoprotein and a solution of Ni(II). Carboxypeptidase A retains its peptidase and esterase activities, if anything slightly enhanced, when Ni(II) is present in place of the normal Zn(II) centre(49). Although Ni(II) will substitute for Zn(II) in carbonic anhydrase, the activity of the system is then lost;(50) Ni(II) will replace Mn(II) in concanavalin(51).

The major study of nickel in biological systems has been the classification of plants, shrubs and trees which are able to assimilate nickel(52) (and/or other metals) in concentrations greater than 8,000ppm (dry weight of leaf). These metals seem to have no function essential to the metabolism of the plant. Rather the latter, when growing on soils rich in the element, has developed the ability to store and tolerate large quantities of the metal, in particular, by binding to the carbohydrates in the cell walls. The nickel appears to be transported in the plant as a complex with citrate. The use of such plants to extract nickel from very low-grade ores remains a speculative possibility, since it is possible to extract the metal from the leaves by washing, e.g. with a solution of ammonium oxalate.

References

[1]J. R. Boldt, "The Winning of Nickel, Methuen, 1967.

[2]International Laterite Symposium, A.S.M.E., February 1979.

[3]E.M.J., November 1978.

[4]G. Wilke, et al., Agnew. Chem. Int. Edn., 1963, 2, 105.

[5]W. Reppe, N. von Kutepow, and A. Magin, Agnew. Chem. Int. Edn., 1969, 8, 727.

[6]C. Manfredi, "Aspects of Homogeneous Catalysis", Ugo, R. (ed.) Milano, 1970.

[7]D. M. Hercules and M. Wu, J. Phys. Chem., 1979, 83.15, 2008.

[8]T. N. Nalibaev, A. B. Farman, and N. Inalyatov, Russ. J. Phys. Chem., 1971, 45, 211.

[9]D. H. Cadenhead and N. J. Wagner, J. Catal., 1971, 21, 312

[10]J. R. Anderson, "Structure of Metallic Catalysts", Academic Press, 1978.

[11]J. D. Robertson, B. D. Merricoz, J. H. de Baas, and S. C. Kloet, J. Catal, 1975, 39, 234.

[12]H. Raney, U.S. Patent 1563787, 1925.

[13]J. Freel, W. J. M. Peelers, and R. B. Anderson, J. Catal., 1970, 10, 281.

[14]G. A. Pushkareva, A. B. Fasman, Yu. F. Kluchnikov, and I. A. Sapukov, Russ. J. Chem. Phys., 1972, 46, 843.

[15]J. Batesm, B. Cornils, and C. D. Frohning, Chem. Ing. Tech., 1975, 47, 522.

[16]T. Beecroft, Private Communication, February 20, 1979.

[17]S. Fivansanker, A. V. Ramaswamy, S. Vishnoi, and P. Ratnasamy, J. App. Chem. Biotechnol., 1978, 28, 387.

[18]R. Burch, Private Communication, March 16, 1978.

[19]F. Fischer, H. Tropsch, and P. Dalthey, Brennst. Chem., 1925, 6, 265.

[20]M. Araki and V. Pontec, J. Catal., 1976, 44, 439.

[21]J. T. Kummer, T. W. Dewitt, and P. H. Emmett, J. Am. Chem. Soc., 1948, 70, 3632.

[22]P. J. Denny and D. A. Wham, Chem. Soc. Spec. Periodical Rep., Catalysis, 1978, 2, 71.

[23]J. Lefebre and J. Baltes, Fett Seifen Anstrichmittel, 1975, 77, 125.

[24] G. W. D. Briggs, Chem. Soc. Spec. Periodical Rep., Electrochemistry, 1974, 4, 33.

[25] N. Y. Uflyand, S. V. Mendeleva, and S. A. Rosentveig, Sov. Electrochem., 1970, 6, 1268.

[26] E. J. Rulin and R. Baboin, J. Electrochem. Soc., 1971, 118, 428.

[27] R. D. Armstrong, K. Edmonson, and G. D. West, Chem. Soc. Spec. Periodical Rep., Electrochemistry, 1974, 4, 18.

[28] B. C. Tofield, R. M. Dell, and J. Jensen, Nature (London), November 1978, 276.

[29] Binkley and Watson,"South of Scotland Electricity Board Ann. Rep.", London HMSO 1967/68.

[30] A. F. Anderson and A. J. Macland, "Hydrogen for Energy Storage", Pergamon, Oxford, 1978.

[31] Ed. G. Alefield, and J. Volkl, "Topics of Applied Chemistry", Springer-Verlag, 1978.

[32] D. P. 2, 307, 177.

[33] W. E. Wallace, R. F. Karlicek, and H. Imamura, J. Phys. Chem., 1979, 83, 13, 1708.

[34] B. Hill, U.S. Patent 3748165, 1973.

[35] Lee, Eung-Sang, Hwang Sung-Yun, Yo Op, Hoe, Chi, 1977, 14(3), 187.

[36] G. N. Masternikova, Stekko Kerum, 1974, 8, 23.

[37] Tatum Tositako, Skitizai Kyoaishi, 1975, 48(8), 498.

[38] D. P. 2, 443,017, 1975.

[39] G. L. Roger and C. A. Amick, U.S. Patent 2,520,106, August 1950.

[40] C. G. Harman and B. W. King, Ind. Eng. Chem., 1952, 1015.

[41] W. A. Weyl, J. Soc. Glass. Technol., 1944, 28, 128.

[42] J. C. M. De Lau, Proc. Brit. Ceram. Soc., 1968, 10, 275.

[43] S. L. Blum, J. Am. Ceram Soc., 1958, 11, 489.

[44]F. G. Brockman and K. E. Matteson, J. Am. Ceram. Soc., 1973 53(9), 517.

[45]R. E. Peterson and J. H. Lin, N.A.S.A. Contract Rep., 1976.

[46]N. E. Dixon, C. Grazzola, R. L. Blakeley, and B. Zernur, J. Am. Chem. Soc., 1975, 97, 4131.

[47]N. E. Dixon, C. Grazzola, R. L. Blakeley, and B. Zernur, Science, 1976, 191, 1144.

[48]D. Amaral, L. Bernstein, D. Morose, and B. L. Hozecker, J. Biol. Chem., 1963, 238, 2281.

[49]J. E. Coleman and B. L. Vallee, J. Biol. Chem., 1961, 236, 2244.

[50]A. Thursland and S. Lindskog, Eur. J. Biochem., 1967, 3, 117.

[51]K. D. Hardman and C. F. Answorth, Biochemistry, 1972, 11, 4910.

[52]M. E. Farago, A. J. Clark, and M. J. Pitt, Coord. Chem. Rev., 1975, 16, 1.

# Rare Earth Speciality Inorganic Chemicals

By K. A. Gschneidner, Jr.
RARE EARTH INFORMATION CENTER, AMES LABORATORY,* and DEPARTMENT OF MATERIALS SCIENCE AND ENGINEERING, IOWA STATE UNIVERSITY, AMES, IOWA 50011, U.S.A.

## Introduction

The rare earth elements comprise about 20% of the naturally occurring elements of the periodic table. The rare earth elements (as defined by the International Union of Pure and Applied Chemistry) include the Group IIIA elements Sc, Y and the lanthanide elements La, Ce, Pr, Nd, Pm, Sm, Eu, Gd, Tb, Dy, Ho, Er, Tm, Yb and Lu. This definition will be used throughout this paper (although it is noted that many scientists and engineers use the term "rare earths" to mean the 15 lanthanide elements and do not consider Sc and Y to be "rare earths").

Many introductory chemistry textbooks view the rare earth elements as being so chemically similar to one another that collectively they can be considered as one element. To a certain degree this is correct — many uses are based on this close similarity — but as one examines these elements more closely vast differences in their behaviors and properties become apparent. An obvious example is the difference in the melting points of the lanthanide elements which vary by almost a factor of two for the end members La - 918°C and Lu - 1663°C, and the remaining trivalent lanthanide melting points lie between these two values. This difference is much larger than that found in many of the groups of the periodic table, e.g. the melting points of Cu, Ag and Au vary by about 100°C (1083, 961 and 1063°C, respectively).

*Operated for the U.S. Department of Energy by Iowa State University under contract No. W-7405-ENG-82. This research was supported by the Director of Energy Research, Office of Basic Energy Sciences, WPAS-KC-02-01.

In addition to differences in properties and behaviors for the normal 3+ valence state, several of the lanthanide elements also exhibit a second valence state. This can also account for vast differences in properties, e.g. the vapor pressure of La (trivalent) at 1300 K is $1.2 \times 10^{-11}$ atm while that of Yb (divalent) is 0.254 atm, a difference greater than <u>one billion.</u>

<u>Abundances.</u> The term "rare" implies that these elements are scarce and limited in availability, but in fact the rare earths are quite abundant and exist in many viable deposits throughout the world. Of the 83 naturally occurring elements[1] the sixteen naturally occurring rare earth elements fall into the 50th percentile of the elemental abundances. Ce, which is the most abundant, ranks 28th and Tm, the least abundant, ranks 63rd. Collectively, the rare earths rank as the 22nd most abundant "element" (at the 68th percentile mark).

The light lanthanides (La through Eu) are more abundant than the heavy lanthanides (Gd through Lu). Thus the individual light lanthanide elements are generally less expensive than the heavy lanthanide elements. Furthermore, the even atomic number metals (Ce, Nd, Sm, Gd, Dy, Er and Yb) are more abundant than their respective odd atomic number neighbors (La, Pr, Pm [which is radioactive], Eu, Tb, Ho, Tm and Lu).

Considering some of the other elements discussed in this volume, Zr (20th), Cr (21st), Ni (25th) and Li (27th) are slightly more abundant than Ce, while Sn (46th) and Mo (tied for 50th with B and Tl) are just as abundant, while all the platinum group and noble metals, Ag (tied for 67th with In), Os (70th), Pd (72), Ru, Pt and Au (all tied for 74th) and Rh and Ir (tied for 78th with Re) are less abundant.

<u>Classification of Compounds Based on Their Usage.</u> The chemical usage of the rare earth speciality compounds is best discussed by dividing the usage and/or compounds into two major classes and treating each separately. Compounds which consist of a mixture of rare earths in proportion to their occurrence in the ores are defined as Class IA compounds. Class IB compounds are also mixtures of the rare earth elements, but in which one or several of the rare earth elements have been removed. Compounds which contain an individual (at least 90%) rare earth element are members of Class II. The

costs and volumes utilized differ significantly, and in the case of some Class II applications, i.e. the refined (>98% pure) elements, the chemistry involved is much different.

Class IA compounds account for about 66% of the total tonnage consumed and about 25% of the dollar value realized by the world-wide industry. The tonnage and monetary value realized from Class IB applications are much smaller. The Class I applications are based on the fact that the chemical behavior of the rare earth elements are quite similar and typical of a trivalent, large size atom (or ion). The major applications involving the mixed rare earths are: as gasoline cracking catalysts, the starting material for making mischmetal and rare earth silicides for various metallurgical uses, mixed rare earth oxide polishing compounds, and carbon arcs.

The Class II applications are further divided into three subcategories based on: (A) the presence or absence of 4*f* electrons, (B) valence states different than 3+, and (C) nuclear properties. Examples of Class IIA applications are: hosts and activators in phosphors, magnetic and electronic materials, optical devices, and basic and applied research. The uses of $CeO_2$ as a decolorizing agent and polishing compounds are the most important Class IIB applications, while the utilization of rare earth elements with high neutron capture cross sections accounts for the major Class IIC application.

## Properties of the Rare Earths

In order to have a better appreciation of the chemical behaviors of the rare earth elements the more important properties will be discussed first. These include the electronic configurations, ionic radii, 4*f* energy levels and thermodynamic properties. The physical properties of the pure metallic forms have been thoroughly reviewed[2] and will not be included here.

Electronic Structure. The electronic structures of the rare earth elements in various states of matter are given in Table 1. These include the free gaseous atom (the ground state), the metallic form and the $R^{2+}$, $R^{3+}$ and $R^{4+}$ ionic species. The first two states are of indirect interest to the chemist, but they can be important when making certain thermochemical cycle calculations — those which involve the heat of sublimation, i.e. R (metal) → R (gas) at 298 K. This will be discussed momentarily.

Table 1. Electronic Structures of the Rare Earths

| Element | Neutral Atom Configuration | 4f Configuration for Known Oxidation States $M^{+2}$ | $M^{+3}$ | $M^{+4}$ | Metallic State No. of Electrons Valence | 4f |
|---|---|---|---|---|---|---|
| Sc | $3d4s^2$ | – | 0 | – | 3 | 0 |
| Y | $4d5s^2$ | – | 0 | – | 3 | – |
| La | $5d6s^2$ | – | 0 | – | 3 | 0 |
| Ce | $4f\ 5d6s^2$ | – | 1 | 0 | 3 | 1 |
| Pr | $4f^3\ 6s^2$ | – | 2 | 1 | 3 | 2 |
| Nd | $4f^4\ 6s^2$ | – | 3 | – | 3 | 3 |
| Pm | $4f^5\ 6s^2$ | – | 4 | – | 3 | 4 |
| Sm | $4f^6\ 6s^2$ | 6 | 5 | – | 3 | 5 |
| Eu | $4f^7\ 6s^2$ | 7 | 6 | – | 2 | 7 |
| Gd | $4f^7\ 5d6s^2$ | – | 7 | – | 3 | 7 |
| Tb | $4f^9\ 6s^2$ | – | 8 | 7 | 3 | 8 |
| Dy | $4f^{10}\ 6s^2$ | – | 9 | – | 3 | 9 |
| Ho | $4f^{11}\ 6s^2$ | – | 10 | – | 3 | 10 |
| Er | $4f^{12}\ 6s^2$ | – | 11 | – | 3 | 11 |
| Tm | $4f^{13}\ 6s^2$ | – | 12 | – | 3 | 12 |
| Yb | $4f^{14}\ 6s^2$ | 14 | 13 | – | 2 | 14 |
| Lu | $4f^{14}\ 5d6s^2$ | – | 14 | – | 3 | 14 |

If one considers the number of $4f$ electrons in the $R^{3+}$ state as the normal one, then the lanthanide series as a whole has $4f^n$ electrons, where n = 0 for La, n = 1 for Ce, n = 2 for Pr ..... to n = 14 for Lu. However, in the gaseous state the most common configuration is $4f^{n+1}\ 6s^2$ while only La, Ce, Gd and Lu have the normal $4f^n$ configuration ($4f^n 5d^1 6s^2$). Examination of the metallic state reveals that the normal configuration is $4f^n\ (5d6s)^3$ with only Eu and Yb having a $4f^{n+1}\ (5d6s)^2$ configuration. For the elements which do not change the number of $4f$ electrons upon sublimation (La, Ce, Gd and Lu — $4f^n$ to $4f^n$ — and Eu and Yb — $4f^{n+1}$ to $4f^{n+1}$) the heat of sublimation is ~420 kJ/mole for the four trivalent metals and ~148 kJ/mole for the two divalent metals.[3] For the remaining metals which gain one $4f$ electron upon vaporizing the heats of sublimation lie between these two limits.[3]

As seen in Table 1 the $R^{3+}$ state is common to all of these elements but a few of the lanthanides exhibit another valence state, $R^{4+}$ for R = Ce, Pr and Tb, and $R^{2+}$ for R = Sm, Eu and Yb. These additional valence states are a striking example of Hund's rule which states that empty, half-filled and completely-filled levels tend to be more stable states: $Ce^{4+}$ and $Tb^{4+}$ give up an f electron to have an empty and half-filled 4f level, respectively; and $Eu^{2+}$ and $Yb^{2+}$ gain an f electron to have a half-filled or completely-filled 4f level, respectively. $Pr^{4+}$ and $Sm^{2+}$ by giving up or gaining an f electron, respectively, can in rare instances gain extra stability. In these two cases they tend towards but do not reach the respective empty or half-filled level.

The $Ce^{4+}$, $Eu^{2+}$ and $Yb^{2+}$ states are quite common and are utilized by industry to separate these three rare earths (and also Sm) from the remaining rare earth elements by relatively cheap chemical methods. The other rare earths can also be separated but the more expensive liquid solvent extraction or ion exchange techniques are required. Furthermore, several important applications rely on the $Ce^{4+}$ and $Eu^{2+}$ valence states.

The overall chemical properties of the rare earth elements are due to their outer electrons 5d6s (3d4s for Sc and 4d5s for Y). There is some variation in the chemical properties of the lanthanides due to the lanthanide contraction (see next section) and hybridization or mixing of the 4f electrons with the valence electrons. For most major uses it is not cost effective to separate the lanthanides in order to use the element which gives the best properties. However, it may be worthwhile to separate the lanthanides into a group of lights or heavies to optimize the chemical behaviors for some limited applications, e.g. a Ce-free La-Pr-Nd concentrate.

Although some hybridization of the 4f electrons occurs, primarily at the beginning of the lanthanide series, the amount is small and the 4f electrons can be considered to be localized and part of the ion core. For uses which depend upon the number of 4f electrons, i.e. magnetic and optical behaviors, separated, and sometimes quite pure, individual rare earth elements are generally required.

Ionic Radii. The ionic radii of the rare earth elements are given in Table 2. The systematic and smooth decrease from La to Lu is known as the lanthanide contraction. It is due to the increase in the nuclear charge, which is not completely screened by the additional 4f electron as one goes from one lanthanide to the next. This increased effective charge draws the electrons (both core and

Table 2. Effective Ionic Radii for a Cordination Number of 6, after Shannon and Prewitt.[4]

| | $R^{2+}$ | $R^{3+}$ | $R^{4+}$ |
|---|---|---|---|
| Sc | - | 0.745 | - |
| Y | - | 0.900 | - |
| La | - | 1.045 | - |
| Ce | - | 1.01 | 0.80 |
| Pr | - | 0.997 | 0.78 |
| Nd | - | 0.983 | - |
| Pm | -- | 0.97 | - |
| Sm | 1.19[a] | 0.958 | - |
| Eu | 1.17 | 0.947 | - |
| Gd | - | 0.938 | - |
| Tb | - | 0.923 | 0.76 |
| Dy | - | 0.912 | - |
| Ho | - | 0.901 | - |
| Er | - | 0.890 | - |
| Tm | - | 0.880 | - |
| Yb | 1.00[a] | 0.868 | - |
| Lu | - | 0.861 | - |

[a]Estimated

outer valence electrons) closer to the nucleus thus accounting for the smaller radius of the higher atomic number lanthanides. The lanthanide contraction also accounts for the decreased basicity on going from La to Lu, and is the basis of various separation techniques.

The ionic radii of the rare earths are quite large with respect to the transition metal ions, but the radii are generally smaller than the alkali and alkaline earth metal ions. $Na^+$ and $Ca^{2+}$ are comparable in size to the rare earth elements, and substitution for $Na^+$ and $Ca^{2+}$ by $R^{3+}$ (and the reverse) is sometimes possible provided the electronic charge is compensated for by the addition (or removal) of anions, or by some other mechanism.

Energy Levels. The energy levels of the trivalent lanthanide ions, which primarily involve the 4*f* electrons, are shown in Figure 1.[5]

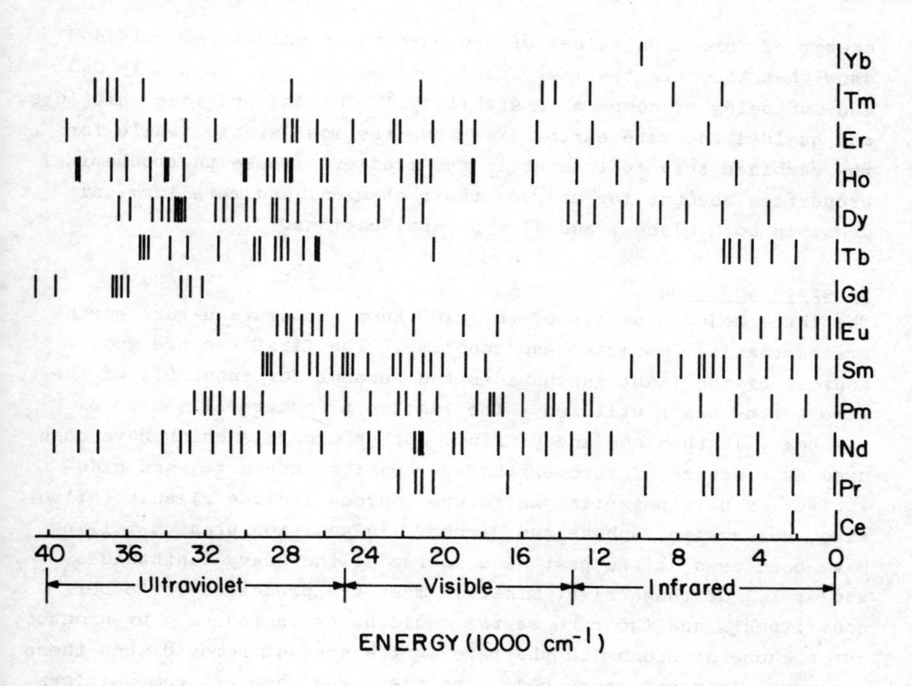


Figure 1. Energy levels of the lanthanide ions which contain unpaired 4f electrons.[5]

These energy levels are utilized in many ways - in methods of identifying the rare earths and quantitatively determining their concentrations in materials, and in countless optical applications as phosphors and laser materials.

The lower energy levels (those below about 7000 cm$^{-1}$ for the lights, and below about 14,000 cm$^{-1}$ for the heavies) are associated with the ground state multiplet for the 4f electrons. These levels (especially those close to 0 cm$^{-1}$) are the most important ones with respect to the thermal, magnetic and electrical properties of rare earth materials.

Thermodynamic Properties. The rare earths are among the most electropositive elements in the periodic table, exceeded only by the alkali and alkaline earth metals. Thus it is not surprising that the rare earths form stable compounds with the strongly electronegative elements, H, C, N, O, the halides, etc. Indeed the free energy of formation values of the rare earth oxides and sulfides show that they are the most stable oxides and sulfides, only CaO and CaS being of comparable stability.[6,7] Of the hydrides, nitrides and halides the rare earths are among the most stable, while for the carbides they fall about in the middle.[6] These thermochemical properties account for many of their chemical and metallurgical uses, in both Class I and II type applications.

## Minerals and Ores

The three major minerals of the ∿160 known to contain rare earths are bastnasite, monazite and xenotime. The first two are good sources of the light lanthanides and account for about 95% of the rare earths being utilized. The last is a commercial source of the heavy lanthanides and yttrium. Other minerals which have been used as a source of rare earths are apatite, euxenite, and gadolinite; or have potential as future sources include allanite, fluorite, perovskite, sphene and zircon. In addition uranium tailings have been used in the past as a source of the heavy lanthanides and/or Y. Of these eight minerals most are processed for other constituents and the rare earths could be extracted as a by-product, but because of economics the rare earths are not removed when these ores are mined and processed. The idealized chemical compositions of these eleven minerals are given in Table 3.

Table 3. Composition of Rare Earth Minerals

| Name | Idealized Composition | Primary Rare Earth Content |
|---|---|---|
| allanite | $(Ca,Fe^{2+})_2(R,Al,Fe^{3+})_3Si_3O_{13}H$ | R = lights |
| apatite | $Ca_5(PO_4)_3F$ | R = lights |
| bastnasite | $RCO_3F$ | R = lights (60-70%) |
| euxenite | $R(Nb,Ta)TiO_6 \cdot xH_2O$ | R = heavies(15-43%) |
| fluorite | $CaF_2$ | R = heavies |
| gadolinite | $R_2(Fe^{2+},Be)_3Si_2O_{10}$ | R = heavies(34-65%) |
| monazite | $RPO_4$ | R = lights (50-78%) |
| perovskite | $CaTiO_4$ | R = lights |
| sphene | $CaTiSiO_4X_2(X=\frac{1}{2}O^{2-},OH^- \text{ or } F^-)$ | R = lights |
| xenotime | $RPO_4$ | R = heavies(54-65%) |
| zircon | $ZrSiO_2$ | R = either |

<u>Bastnasite.</u> Bastnasite, a fluorocarbonate, is the principal source of rare earths today and about 55% of the rare earths utilized in the free world comes from the Molycorp's open pit mine at Mountain Pass, California; see upper portion of Plate 1. Mainland China is reported to have vast bastnasite deposits which are thought to be equal to or greater than all the other bastnasite deposits in the remainder of the world. Other significant deposits are found in Burundi, Sweden and New Mexico in the U.S.A.

The rare earth distribution in the California and Chinese bastnasite is given in Table 4, where it is seen that the Chinese deposit is slightly richer in the heavy lanthanides and Y, primarily at the expense of the La content, which is 5% greater in the California bastnasite.

<u>Monazite.</u> Monazite is the second most important ore source for the rare earths and accounts for about 40% of the free-world utilization.

Table 4. Rare Earth Content of Source Minerals

| Rare Earth Element | Bastnasite | | Monazite | | | Xenotime |
|---|---|---|---|---|---|---|
| | Calif. (%) | China (%) | S. Carolina (%) | Australia (%) | India (%) | Malaysia (%) |
| La | 32.0 | 27 | 19.5 | 20.2 | 23 | 0.5 |
| Ce | 49.0 | 50 | 44.0 | 45.3 | 46 | 5.0 |
| Pr | 4.4 | 5 | 5.8 | 5.4 | 5.5 | 0.7 |
| Nd | 13.5 | 15 | 19.2 | 18.3 | 20 | 2.2 |
| Sm | 0.5 | 1.1 | 4.0 | 4.6 | 4.0 | 1.9 |
| Eu | 0.1 | 0.23 | 0.17 | 0.1 | | 0.2 |
| Gd | 0.3 | 0.4 | 2.0 | 2.0 | | 4.0 |
| Tb | | | 0.2 | | | 1.0 |
| Dy | | | 1.3 | | | 8.7 |
| Ho | | | 0.1 | | 1.5 | 2.1 |
| Er | 0.1 | 1.0 | 0.5 | 2.0 | | 5.4 |
| Tm | | | - | | | 0.9 |
| Yb | | | 0.2 | | | 6.2 |
| Lu | | | - | | | 0.4 |
| Y | 0.1 | 0.25 | 3.0 | 2.1 | | 60.8 |

Plate 1. Molycorp's Mountain Pass, California facility.
Top — bastnasite open pit mine showing three levels of old workings and the floor of the current level.
Bottom — multi-cell froth flotation lines.
Courtesy Molycorp, Inc., Union Oil Company.

Monazite is a rare earth phosphate containing 5 to 10% $ThO_2$ which presents some interesting problems associated with its processing and utilization. Because Th is a fissionable material several countries, notably Brazil and India, remove and stockpile the Th as a strategic item, and the Th-free monazite is exported. Australia and Malaysia have no such export restrictions. Because of Th's natural radioactivity several countries, notably the U.S.A., have stringent environmental requirements on the Th content of the monazite processed and utilized therein. The four major free world countries producing monazite are (the number in parentheses after each country is their percentage of total world monazite production in 1979): Australia (65%), India (12%), Brazil (10%), and Malaysia (8%).[8] In addition, the U.S.S.R. produces about as much monazite as India or Brazil. Other countries mining monazite are the U.S.A., Thailand, Sri Lanka, Zaire and the Republic of Korea, but all are believed to have produced less than 900 metric tons each in 1979.

The rare earth distribution in three different monazites is given in Table 4, where it is noted that the monazites have a significantly larger concentration of heavy lanthanides plus Y than the bastnasites. The U.S.A. monazite, as typified by the South Carolina variety, has a higher concentration of heavy lanthanides plus Y than does the Australian monazite, which in turn is greater in these rare earths than the Indian monazite. The La and Ce content is higher in the later two.

Most commercial monazite sources are alluvial and beach sand deposits. Monazite is a minor component (1 to 20%) of the heavy mineral sands whose primary constituents are rutile, ilmenite, cassiterite and zircon. In the past these sands were primarily processed for their Ti, Zr or Sn content, and thus the monazite production varied as the markets for Ti, Zr and Sn have gone up or down. However, the demand for rare earths has grown sufficiently in the last five years that some of the heavy mineral sands can be processed economically for their monazite and the excess Ti or Zr or Sn is stored when these markets are down.

Xenotime. Xenotime is the major ore source for the heavy lanthanides and Y. It is a phosphate containing small amounts of $U_3O_8$ and $ThO_2$ (up to ~3% each). The rare earth distribution is shown in Table 4, and the contrasts between xenotime and the bastnasites and monazites are quite evident — the light lanthanide concentrations are smaller by a factor of about ten, while the heavy lanthanide and Y concentrations are up by a factor of ten to several hundred. Because of high Y and heavy lanthanide concentrations most of the xenotime is used as a source material for the individual rare earth elements rather than being used as a mixture of heavy rare earths. The major producer of xenotime is Malaysia, while deposits are reported to exist in Norway and Brazil.

Beneficiation of Ores

The processes described below are those utilized by the largest rare earth producers, namely the Molycorp process for bastnasite[9] and the Rhone-Poulenc process for monazite.[10] Other companies use different methods for upgrading these ores, but these will not be discussed here because of space limitations. However, some of the alternate processes are described in the monograph by Callow on rare earth industrial processes.[11]

Bastnasite. The California bastnasite is mined by open pit methods. The ore, which contains 7 to 10% rare earth oxide (REO), is ground to a powder and is upgraded to a product containing 60% REO by a hot froth flotation technique; see lower portion of Plate 1. The heavier products, barite ($BaSO_4$) and celestite ($SrSO_4$), settle out while the bastnasite and other minerals are floated-off. The 60% REO concentrate is treated with 10% HCl to dissolve the calcite ($CaCO_3$). The insoluble residue now contains 70% REO, which is either sold as a mineral concentrate, or further processed at the Molycorp Mountain Pass mine site.

The 70% REO concentrate is roasted to oxidize the $Ce^{3+}$ to the $Ce^{4+}$ state. After cooling, the material is leached with HCl dissolving the trivalent rare earths (La, Pr, Nd, Sm, Eu and Gd) leaving behind the Ce concentrate which is refined to various grades and marketed. The dissolved lanthanides are separated into two groups, La-Pr-Nd and Sm-Eu-Gd, by liquid-liquid organic solvent extraction (these solvents will be discussed later in the section on separation chemistry). A typical bank of solvent extraction cells is shown in Plate 2. The La, Pr, Nd concentrate may be marketed as a Ce-free rare earth product or further processed for the individual elements by continued solvent extraction processes, as shown in Figure 2.

Plate 2. Liquid-liquid solvent extraction cells at the Molycorp Mountain Pass, California facility. Courtesy Molycorp, Inc., Union Oil Company.

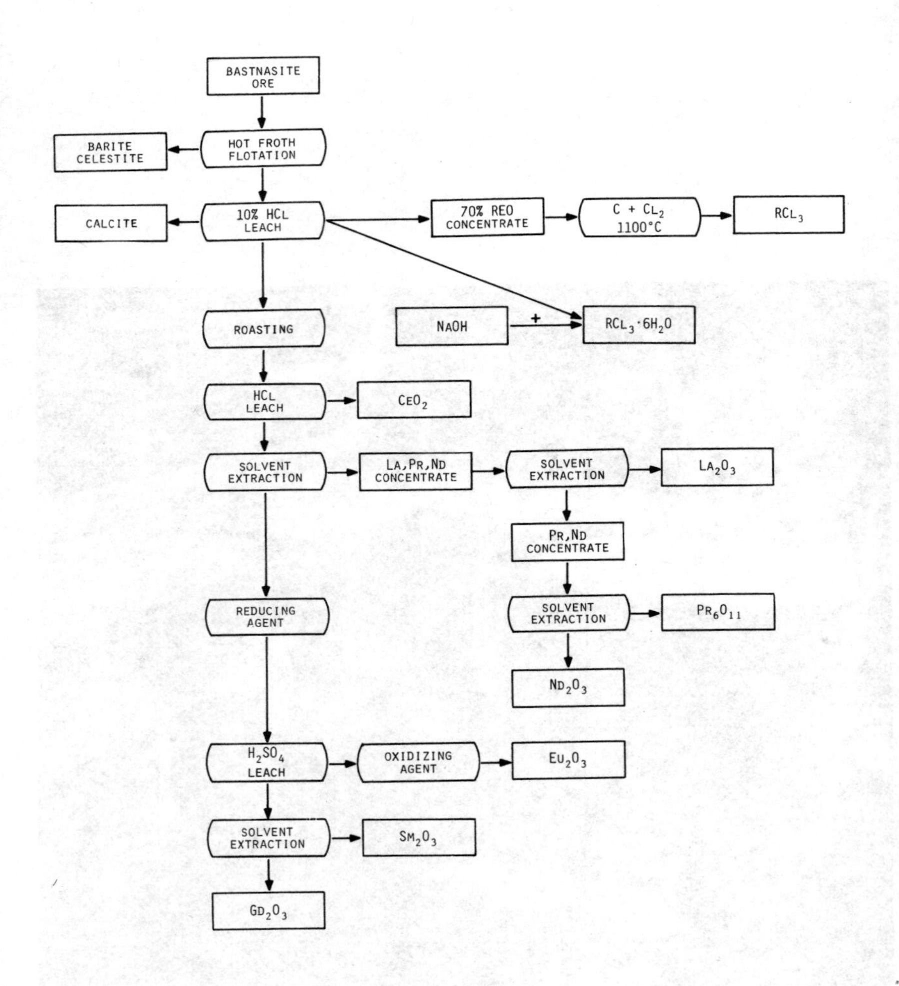


Figure 2. Flow chart of the Molycorp process for the beneficiation of bastnasite and the separation and purification of the rare earth elements.

The Sm-Eu-Gd fraction is treated with a proprietary reducing agent to reduce Eu to the divalent state ($Eu^{3+}$ can easily be reduced by Zn metal to $Eu^{2+}$, i.e. Jones reducer, while the $Sm^{3+}$ and the $Gd^{3+}$ remain in the higher oxidation states). Divalent europium is precipitated as the insoluble $EuSO_4$, oxidized to $Eu^{3+}$, recovered and sold as $Eu_2O_3$. The Sm and Gd are then separated by a solvent extraction step.

The 70% REO can be converted directly into anhydrous chloride by mixing the REO with carbon powder and briquetting the mixture. The briquettes are heated to 1000 - 1200°C in a $Cl_2$ gas atmosphere to give the molten chloride which is tapped-off from the bottom of the reaction chamber and the CO gas is pumped off and oxidized to $CO_2$ before releasing to the atmosphere.[12] The anhydrous rare earth chloride is primarily used to prepare mischmetal by electrowinning (mischmetal is the metallic form of the naturally occurring mixture of the rare earth elements).

The hydrated rare earth chloride ($RCl_3 \cdot 6H_2O$) can be prepared by reacting the bastnasite, after it has been leached with 10% HCl to remove the calcite, with caustic soda (NaOH) in the presence of HCl. This material is produced in large quantities for the production of mischmetal, in gasoline cracking catalysts and as a feedstock for solvent extraction and ion-exchange separations. Before the hydrated chloride can be used to prepare mischmetal it must be dehydrated in such a manner that minimizes the amount of oxychloride that might be formed. This is accomplished by heating $RCl_3 \cdot 6H_2O$ in rotary vacuum drum driers at 350°C in a vacuum of 700 torr.[13]

Monazite (and Xenotime). The beach sands which contain monazite and other heavy minerals are recovered by using dredges, shaking tables, jigs and spirals. The bulk of this material consists mainly of Ti and Zr minerals and from 1 to 20% monazite (and/or xenotime). The monazite (xenotime) is separated from the other minerals by a combination of gravity, electromagnetic and electrostatic techniques and then is cracked by one of two methods — the acid process or the basic process. Below, the Rhone-Poulenc basic process for cracking monazite is described. For more information on the acid process and other variations of the basic process the reader is referred to Callow's treatise.[11] Xenotime, which is also rare earth phosphate, can be treated in the same manner as the monazite and so no more will be said explicitly about xenotime.

The finely ground monazite is mixed with a 70% NaOH solution and heated in an autoclave at 140 - 150°C for several hours; see top portion of Plate 3. After the addition of $H_2O$ the soluble $Na_3PO_4$ is recovered as a by-product from the insoluble $R(OH)_3$, which still contains the 5 - 10% Th. Two different processes are used by Rhone-Poulenc to remove the Th. One process is to dissolve the hydroxide in HCl or $HNO_3$ and then selectively precipitate the $Th(OH)_4$ by addition of NaOH and/or $NH_4OH$. The other process is to add HCl slowly and carefully to the hydroxide to lower the pH to 3 - 4, and to remove the soluble $RCl_3$ from the insoluble $Th(OH)_4$ by filtration. These alternate processes are shown in Figure 3.

The Th-free rare earth solution may be converted to a usable form of the mixed elements (e.g. the hydrated chloride, anhydrous chloride, fluoride, carbonate or hydroxide), or calcined to form the oxide (see lower portion of Plate 3), or further processed to obtain other mixed concentrates or the individual elements (Figure 3). The separation of the rare earth elements in the Rhone-Poulenc process is carried out entirely by the liquid-liquid solvent extraction technique, rather than a combination of chemical and solvent extraction steps like that employed by Molycorp for bastnasite (see above). In the Rhone-Poulenc process the La is extracted first, followed by Ce and successively the other lanthanides according to their increasing atomic numbers. Yttrium is removed within the heavy lanthanide group and its location with respect to these elements can vary according to the choice of organic extractants. More details concerning the separation of the rare earth elements by solvent extraction are given below in the Section on "Separation Chemistry".

## Mixed Rare Earth Chemicals

About 99% of the world-wide rare earth market on a volume basis consists of mixed rare earth chemicals and chemical concentrates of one of the elements (75 to 98%) while the individual separated elements (>98% pure) make up the remaining 1%. On a monetary basis, however, the latter accounts for about 33% of the sales. The applications and markets of the mixed rare earths and chemical concentrates will be discussed here, while those of the separated elements will be discussed in the last section of this paper. Additional

Plate 3. Rhone-Poulenc's La Rochelle, France facility.
Top — autoclave for attacking monazite by NaOH.
Bottom — kiln for calcining the rare earths.
Courtesy Rhone-Poulenc, Inc.

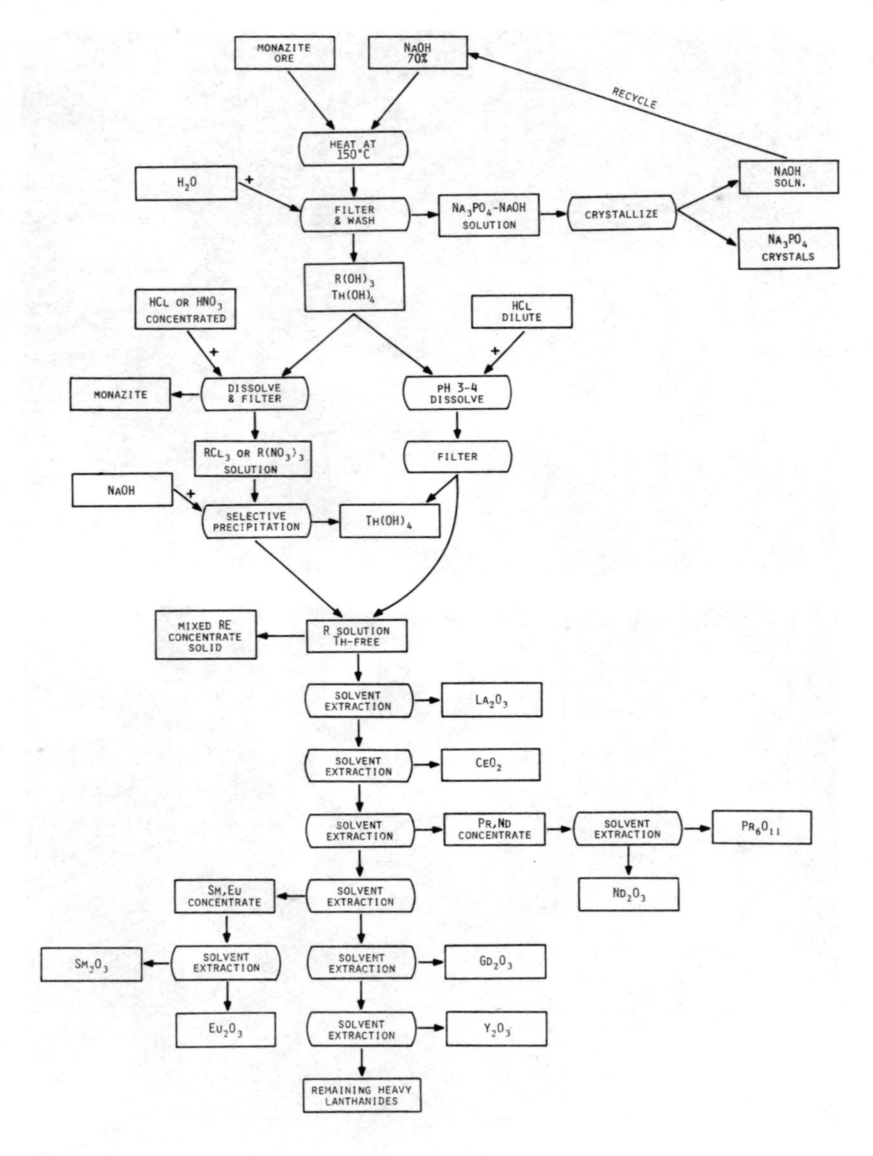


Figure 3. Flow chart of the Rhone-Poulenc basic process for the beneficiation of monazite and the separation and purification of the rare earth elements.

information on the uses, applications and potential applications can be found in the volumes by Michelsen[14] and Gschneidner,[15] the quarterly publication RIC News[16] and the irregularly published Over-View.[17]

Mineral Concentrates. The theoretical rare earth content of basnasite ($RFCO_3$) is 75% REO (rare earth oxide equivalent), while that of monazite ($RPO_4$) is 70% REO and that of xenotime ($R'PO_4$) is 67% REO, where R = light lanthanides and R'= heavy lanthanides plus Y. These mineral concentrates are sold by the major ore producers and/or processors to other companies which convert these materials to the desired chemical form(s) as required in their product(s) and/or process(es). These mineral concentrates account for a major share of the mixed rare earth market, but the percentage is difficult to determine. The major ore producers and/or processors also crack the mineral concentrates to prepare other mixed rare earth chemicals and enriched mixtures (enriched in a given rare earth). These are described next.

Chemical Concentrates. As seen from the flow charts for bastnasite (Figure 2) and monazite (Figure 3) the respective mineral concentrates can be converted to mixed rare earth chemical concentrates or to various concentrates which have been selectively processed to increase the content of one particular element or combinations of two or more rare earth elements. In general the single or binary elemental concentrations contain anywhere from 75 to 98% of the desired element(s), but these can be purer if there is a market for such products. These higher purity materials (>98% pure) are discussed in the next two sections of the paper.

The major chemical forms of the mixed concentrates are the acid soluble oxides, carbonates and fluorides, and the water soluble chlorides and nitrates. In the case of bastnasite, as noted earlier, the mineral concentrate is converted into the anhydrous mixed rare earth chloride for the preparation of mischmetal by electrolysis. The mischmetal is used to nodularize the carbon in ductile irons, to desulfurize steels, and as lighter flints. The bastnasite mineral concentrate can also be converted into a hydrated chloride, which is used in the manufacture of zeolite gasoline cracking catalysts or for making mischmetal, but in this case the chloride needs to be dehydrated first. From 1 to 5% mixed rare earths are added to $Na_2O \cdot Al_2O_3 \cdot nSiO_2 \cdot xH_2O$ zeolites to increase the catalytic efficiency. Today practically all petroleum cracking units in the world use rare earth zeolite catalysts. The manufacture of mischmetal for use in the iron and steel industries and the use of rare earths in cracking catalysts account for ∿60% of the world-wide use of the rare earth elements.

For monazite the mixed rare earth solution after the Th has been removed is ready to be converted into any of the chemical forms noted above, the major one being the hydrated chloride for the mischmetal and catalyst industries. Another important compound is the mixed rare earth fluoride which is used in carbon arc cores. The fluoride is easily prepared by adding a concentrated HF solution to Th-free mixed rare earth solution and the insoluble $RF_3$ precipitates out quantitatively. The mixed rare earth fluorides are added to carbon arc cores to increase the arc intensity by a factor of ten and to change the yellowish color of the carbon arcs to a light quality nearly identical to sunlight. It is fortunate that monazite has the proper distribution of rare earths, each with its own electronic energy level scheme (Figure 1), to produce this balanced white sunlight. Bastnasite does not have the proper rare earth distribution and a mixture of fluorides derived thereof does not work. <u>This is the only major application of rare earths which requires 4f electrons and in which they are used as a mixture.</u>

The mixed rare earth concentrations are further processed (as previously discussed) to obtain enriched elemental concentrates, usually in the form of the rare earth oxide or hydrated oxide, for a variety of large scale uses. Since the presence of other rare earth elements is not detrimental in the end use, further purification is not carried out and the source material can be either bastnasite or monazite. The major use of $La_2O_3$ is in optical lenses, whereby it increases the index of refraction of the lens. Up to 40% $La_2O_3$ may be added to some glasses. All high quality camera lenses contain $La_2O_3$. Another use includes the addition of $La_2O_3$ to barium titanate capacitors to improve the temperature dependence of the dielectric properties.

The two most important uses of $CeO_2$ are as polishing compounds and as a glass additive to decolorize flint glass beverage containers. The polishing compounds have $CeO_2$ contents which vary from 50% to 90%, with the remainder being the other light lanthanide oxides. The higher the $CeO_2$ content the faster the rate of polishing. The $CeO_2$ polishing compounds have replaced the cheaper polishing compounds (rouge [$Fe_2O_3$], $SiO_2$ and $ZrO_2$) because they are cleaner, faster, longer lasting and give a superior finish. The decolorizing application depends upon the $Ce^{4+}/Ce^{3+}$ oxidation-reduction reaction to maintain iron in the $Fe^{3+}$ state since the $Fe^{2+}$ state imparts a bluish color to the glass. In addition the presence of Ce (1) reduces the amount of color complement needed to eliminate the faint yellow color due to $Fe^{3+}$, and (2) absorbs ultraviolet light, which is beneficial since it reduces the rate of deterioration of the product in the bottle. Another important application involves the addition of $CeO_2$ to glasses undergoing solar, electron, x-ray or neutron radiation (e.g. TV tubes) to prevent browning effects induced by the radiation. $CeO_2$ is also used to color glass, but this is a minor application.

In the processing of bastnasite a Ce-free, La-Pr-Nd concentrate results as a natural consequence of the separation process because the Ce is removed first. This concentrate is not generally available in the monazite process, unless one mixes the La and Pr-Nd concentrates, or modifies the process to remove the Ce first. The La-Pr-Nd concentrate is used to prepare a Ce-free mischmetal which is utilized as an additive to some non-ferrous alloys. Another use of this concentrate is as a temperature compensating additive in barium titanate capacitors. Another concentrate, which enjoys a larger demand than the La-Pr-Nd concentrate, is the Pr-Nd concentrate which is commonly called "didymium". Didymium is converted to a metallic form and is added to some Mg-base alloys to improve the high temperature strength and creep resistance of the alloy. Didymium is also used in welding and glass blower goggles to protect the person's eyes from the intense light. The Pr-Nd mixture is especially effective in absorbing the strong yellow sodium light which is emitted during the glass blowing operation.

The major use of the praseodymium concentrate is as the coloring agent in zircon ($ZrSiO_4$) base yellow ceramic tiles. $Pr_6O_{11}$ is used as a thin vacuum-deposited film in fine optical glass components to prevent unwanted reflection. Furthermore, it has great potential as a $PrCo_5$ permanent magnet. In theory it should be almost as good as the current standard $SmCo_5$ magnet (see below) but in practice this potential has yet to be realized.

Although several of the lanthanide inorganic compounds exhibit pleasing colors, Nd is the most commercially important one as a glass coloring agent; others are Ce, Pr and Er. $Nd_2O_3$ is added to glass to give it a violet color, which can be shifted to pink by the addition of Se. $Nd_2O_3$ is also added to barium titanate capacitors to provide temperature compensation.

The remaining elemental concentrate to be considered here is Sm. Although the bastnasite (Figure 2) and monazite (Figure 3) flow charts indicate other elemental concentrates — $Eu_2O_3$, $Gd_2O_3$ and $Y_2O_3$, these concentrates are generally of a much higher purity (>98%) when utilized. The major use of $Sm_2O_3$ is as a starting material for the preparation of the $SmCo_5$ permanent magnets, which are the most powerful magnets known. Today all the Sm that can be prepared by processing bastnasite and monazite (∿200 tons/year) is utilized in the manufacture of $SmCo_5$ magnets. Additional Sm, interestingly enough, comes as a by-product from the manufacture of mischmetal. When the mixed rare earth chloride is reduced electrolytically to the metal the $Sm^{3+}$ is only reduced to its $Sm^{2+}$ state and it remains in the slag along with the $Eu^{2+}$. This slag is then easily chemically processed to give the Sm/Eu concentrate, which is separated by a process similar to that described above in the section on beneficiation of bastnasite to yield the two elemental concentrates. The amount of Sm obtained by this route is comparable to that obtained by the processing of bastnasite and monazite and depends greatly upon the mischmetal market. The Sm concentrate used to prepare magnets is about 96% pure, and the presence of small amounts of other rare earths does not greatly effect the magnet performance.

Markets and Statistics. Statistical data for the rare earth markets are difficult to obtain except for broad categories such as the total production. The two main sources for this information appear annually — in the U.S. Bureau of Mines Yearbook chapter on rare earths[8] and the March issue of Engineering and Mineral Journal.[18] Other information appears in print occasionally.[15,19] Another problem is that each author includes different products in his or her market categories and intercomparisons are difficult. The author, however, has attempted to present a more detailed view of the rare earth markets by using all of the available information and by making some judicious estimates.

The three major markets and how they have changed since 1975 are presented in Table 5 for both the U.S.A. and the total free world. It is seen that the U.S.A. fraction of the world markets, although still substantial, has been decreasing from about 75% in the mid-1970's to about 60% towards the end of the decade. This trend is expected to continue into the 1980's, but the U.S.A. share is still expected to be 50% or greater. The overall growth pattern of the rare earth markets is still excellent, averaging about 10% over the last half of the 1970's, but this growth rate is down considerably from the 15% rate in the 1960's.

A more detailed break down of the 1978 markets is given in Table 6, where it is seen that the use of the rare earths is much larger in the entire metallurgical and catalyst/chemical markets in the U.S.A. than in the rest of the free world. In the glass/ceramics category the use of rare earth polishing compounds is much larger in the rest

Table 5. Major Markets for the Rare Earths (in 1000's of metric tons)

| United States of America | | | | | |
|---|---|---|---|---|---|
| Market | 1975 | 1976 | 1977 | 1978 | 1979 |
| metallurgical | 4.0 | 4.5 | 6.4 | 5.4 | a |
| cat./chem. | 5.4 | 5.4 | 7.0 | 6.2 | a |
| glass/cer. | 2.0 | 2.0 | 3.2 | 3.5 | a |
| misc. | 0.2 | 0.2 | 0.2 | 0.2 | a |
| total | 11.6 | 12.1 | 16.8 | 15.3 | a |

| Free World (including U.S.A.) | | | | | |
|---|---|---|---|---|---|
| Market | 1975 | 1976 | 1977 | 1978 | 1979 |
| metallurgical | 7.2 | 4.9 | 7.7 | 7.8 | 11.0 |
| cat./chem. | 5.8 | 5.9 | 8.8 | 7.8 | 6.8 |
| glass/cer. | 2.8 | 4.3 | 5.9 | 8.6 | 8.0 |
| misc. | 0.3 | 0.3 | 0.3 | 0.3 | 0.3 |
| total | 16.1 | 15.4 | 22.7 | 24.5 | 26.1 |

[a] not available

Table 6. The 1978 Rare Earth Markets (in metric tons)

| Market | U.S.A. | Rest of Free World | Total Free World |
|---|---|---|---|
| Metallurgical | | | |
| nodular iron | 1800 | 850 | 2650 |
| steels | 2900 | 1100 | 4000 |
| flints/misc. | 750 | 450 | 1200 |
| catalyst/chemical | | | |
| cracking catalyst | 4800 | 1200 | 6000 |
| chem./other cat. | 1400 | 400 | 1800 |
| glass/ceramics | | | |
| polishing | 1600 | 2800 | 4400 |
| glass add./cer. | 1900 | 2300 | 4200 |
| refined (>98% pure) | 150[a] | 100 | 250 |
| total | 15300 | 9200 | 24500 |

[a]see Table 8 for more details

of the free world than in the U.S.A., but the use of rare earths as glass additives and in ceramics is about equal. The metallurgical subcategory "flints/misc." includes the applications of rare earth cobalt permanent magnets (primarily $SmCo_5$), non-ferrous alloy additives and lighter flints. The "refined" category includes the higher purity rare earth materials (>98%) and these will be treated in the next two sections.

In Tables 5 and 6 there is no distinction between the unseparated or mixed rare earth markets and those which rely on some of the rare earth chemical concentrates. In the metallurgical and catalyst/chemical markets mixed rare earths are used almost exclusively. But in the glass/ceramics markets most applications involve the rare earth chemical concentrates - $CeO_2$ in the polishing area, and $La_2O_3$, $CeO_2$, $Nd_2O_3$, $Pr_6O_{11}$, etc. concentrates as additives to glass (to increase the index of refraction, and for decolorizing and coloring) and ceramics (to give the yellow color to zircon base tile and for temperature compensated capacitors). On a world-wide basis the chemical concentrates account for about a third of the total volume of rare earths consumed and about 40% of the monetary value, the mixed rare earths account for two-thirds of the volume consumed and about 25% of the value, and the refined materials about 1% by volume and 33% of the monetary value.

## Separation Chemistry

As far as the rare earth scientific, technological and industrial community is concerned, two of the most important spin-offs which resulted from research on the atomic bomb in World War II and the years immediately following, were the development of the ion exchange and the liquid-liquid solvent extraction techniques for separating the rare earth elements. Work on the ion exchange process was carried out by G. E. Boyd and co-workers at the Oak Ridge National Laboratory (Oak Ridge, Tennessee) and by F. H. Spedding and co-workers at the Ames Laboratory (Ames, Iowa). Both groups simultaneously published their results which showed that the ion exchange process would work at least on a small scale for separating rare earths. In the early 1950's Spedding, J. E. Powell and co-workers showed that it was possible to separate large amounts of high purity (>99.99%) individual rare earth elements. This was then the beginning of the modern rare earth industry in which large quantities of high purity rare earth elements became available for many specialized uses, especially in the electronic, phosphor and optical industries.

Other researchers, notably D. F. Peppard and his colleagues at the Argonne National Laboratory (near Chicago, Illinois) and B. Weaver and his collaborators at the Oak Ridge National Laboratory developed the liquid-liquid solvent extraction method for separating rare earths in the early to mid-1950's. Today the liquid-liquid solvent extraction method is used by the two major rare earth producers of separated rare earths to separate the natural mixtures into the individual elements. This technique is capable of preparing 99.9% pure elements for all of the rare earths and, in the case of Y, a 5 nines pure material. On the other hand the ion-exchange technique has been used to prepare <u>all</u> of the rare earth elements at the 5 nines purity level. That the liquid-liquid solvent extraction method is capable of doing the same for <u>all</u> of the naturally occurring rare earths remains to be proven to the author's satisfaction.

In a process for separating two elements, A and B, which are intimately mixed together the separation factor, α, is given as

$$\alpha = \frac{C_A^f \; / \; C_B^f}{C_A^o \; / \; C_B^o} = \frac{C_A^f \; C_B^o}{C_B^f \; C_A^o} \qquad (1)$$

where C is the concentration of A or B, and f and o signify the final and original states, respectively. If $\alpha = 1$, then no separation is possible. Usually the choice of A and B is made such that $\alpha \geq 1.0$.

Thus the more α exceeds 1 the easier it becomes to separate A and B from one another. In general separation factors for ion exchange are greater than the corresponding ones for liquid-liquid extraction for the same pair of elements. The latter, however, is a quicker and easier process to manipulate and change.

Liquid-Liquid Solvent Extraction.[10,20-22] In this process two immiscible or partially immiscible solvents containing dissolved rare earths are mixed, the solutes are allowed to distribute between the two phases until equilibrium is established, and then the two liquids are separated. The concentrations of the solutes in the two phases depend upon the relative affinities for the two solvents. According to convention the product (liquid) which contains the desired solute is called the "extract", while the residue left behind in the other phase is called the "raffinate". At equilibrium the concentration of the solute in the extract is $C^o$, and in the raffinate $C^f$, and the distribution coefficient, λ, is given by

$$\lambda = \frac{C^o}{C^f} \qquad (2)$$

For the separation of two rare earths the separation factor is given as

$$\alpha = \frac{\lambda_B}{\lambda_A} = \left(\frac{C_B^o}{C_B^f}\right)\left(\frac{C_A^f}{C_A^o}\right) \qquad (3)$$

which is identical to Eq. (1). The best way to affect the separation of A and B is to use a multistage counter-current extractor on a continuous flow basis (see the upper portion of Plate 4). For the case in which A has a greater affinity for the organic phase, and B the greater affinity for the aqueous phase, the organic phase be-

Plate 4. Rhone-Poulenc's La Rochelle, France facility.
Top — liquid-liquid solvent extraction cells for separating the rare earth elements.
Bottom — control room for the liquid-liquid extraction separation process.
Courtesy Rhone-Poulenc, Inc.

comes enriched in A and the aqueous phase enriched in B. However, since the process is not perfect, the extract (enriched A) contains impurities (primarily B) which should be removed. This is done by a second stage operation in which the extract is "scrubbed" by another solvent to remove B without any significant loss of A in the extract.

The solvent or non-aqueous phase is a multicomponent mixture of organic compounds, generally consisting of an extracting agent, a diluent and a modifier. The solvent is the most important element in the extraction process — it controls the efficiency and thus the economics of the process.

There are three kinds of extracting agents — solvating, acidic and ionic extractants. The solvating agent is so-called because of its ability to dissolve the material to be extracted. As far as rare earth separations are concerned, tributylphosphate (TBP) is the most useful solvating extractant because it extracts the rare earth nitrate from the aqueous solution to form $R(NO_3)_3(TBP)_3$. The aqueous medium, however, must be strongly acidic (>8 M in $HNO_3$) for successful extractions. Extraction into the organic phase increases with increasing atomic number, but separation beyond Tb is difficult because the separation factors between adjacent lanthanides are small (see Table 7). For La, Pr and Nd high purities can be readily obtained by using TBP to extract the lanthanides from 13-14 M $HNO_3$ in just a small number of stages (10-15).

The acidic extractants exchange acidic hydrogen for the rare earth metal to form the metal salt in the organic phase. One of the more successful acid extractants is di-2-ethylhexyl orthophosphoric acid (HDEHP). This extractant because of its viscosity must be employed with a diluent. The aqueous medium is generally an HCl or $HClO_4$ solution, and separation factors average rather uniformly 1.9 and 2.4, respectively, for the lanthanide series La through Lu; see Table 7. Other acid extractants are discussed by Kaczmarek,[10] Hulet and Bode,[21] and Powell.[22]

The ionic extractants are long-chained amines (primary, secondary, tertiary and quaternary), which have a cation or an anion that readily exchanges with the appropriate metal species in the aqueous phase. The ammonium salts of the tertiary and quaternary alkyl amines appear to be the most useful in separating the rare earth elements.

Table 7. Separation Factors for Several Extracting Agents and Some Ion Exchange Complexing Agents.

| | Solvent Extraction | | | Ion Exchange | | |
|---|---|---|---|---|---|---|
| Pair | TBP 15.5M $HNO_3$ | HDEHP HCl | HDEHP $HClO_4$ | EDTA | HEDTA (25°C) | HEDTA (92°C) |
| La-Ce | 1.6 (La-Ce to Eu-Gd) | 2.4 | 3.0 | 3.3 | 5.0 | - |
| Ce-Pr | | 2.8 | 2.1 | 2.4 | 2.8 | - |
| Pr-Nd | | 1.7 | 1.4 | 2.0 | 1.8 | 2.1 |
| Nd-Pm | | 2.1 | 2.2 | 1.9 | 1.6 | 1.8 |
| Pm-Sm | | 2.4 | 3.1 | 1.8 | 1.6 | 1.8 |
| Sm-Eu | | 2.2 | 1.9 | 1.5 | 1.0 | 1.5 |
| Eu-Gd | | 1.6 | 1.4 | 1.1 | 0.7 | 1.0 |
| Gd-Tb | 1.3 (Gd-Tb to Yb-Lu) | 3.2 | 5.0 | 3.5 | 1.0 | 2.0 |
| Tb-Dy | | 2.0 | 2.1 | 2.7 | 1.0 | 1.9 |
| Dy-Ho | | 2.1 | 1.9 | 2.0 | 1.0 | 1.8 |
| Ho-Er | | 2.1 | 2.3 | 2.0 | 1.2 | 1.7 |
| Er-Tm | | 2.5 | 2.5 | 2.0 | 2.0 | 1.6 |
| Tm-Yb | | 1.8 | 3.1 | 1.8 | 1.6 | 1.7 |
| Yb-Lu | | 2.2 | 1.9 | 1.6 | 1.3 | 1.6 |

Diluents are added to the organic phase to lower the viscosity and/or facilitate contact between the two phases. These are usually aromatic compounds, such as toluene, or commercial mixtures or blends of several aromatic compounds which are sold under a variety of trade names. More details can be found in Kaczmarek's review about the physical and chemical property requirements of the diluents.[10]

Modifiers are chemicals which are added to the solvent to prevent the formation of a third phase, or modify the rate of extraction, or the equilibrium position of the given rare earth element relative to the others. Long chain alcohols or TBP have been used as modifiers.[10]

The only liquid-liquid solvent extraction plant which separates all the rare earth elements on a continuous basis is the Rhone-Poulenc plant in La Rochelle, France; see Plate 4. In order to do this the output of the counter current solvent extraction cells (see upper portion of Plate 4) is continuously monitored by on-line chemical analyses of the extract and raffinate. This information is fed to a computer, which then accordingly modifies the composition of the solvent and/or the aqueous medium. (The control room of this plant is shown in the bottom portion of Plate 4). There are a large number of variables which can be changed: in the solvent - the extractant, diluent and modifier, in the aqueous medium - the anion and the pH, and the temperature. The La Rochelle plant can process 5,000 tons of REO equivalent per year using monazite as the initial ore source. A second plant also capable of processing 5,000 tons is being built in Texas on the Gulf of Mexico by Rhone-Poulenc and is expected to be on stream in 1982.

Ion Exchange.[22,23] In the ion exchange process a metal ion, $R^{3+}$ in solution exchanges with protons on a solid ion exchanger — a natural zeolite or a synthetic resin — which is normally called the resin. The tenacity with which the cation is held by the resin depends upon the size of the ion and its charge. However, no separation of the rare earth is possible because the resin is not selective enough. By introducing a complexing agent, A*, separation is possible, if the equilibrium constant, K (also called the stability constant) for the reaction

$$R + 3A^* \rightleftharpoons RA_3^* \qquad (4)$$

varies sufficiently from one rare earth to another for the separation to take place. Furthermore in the R-A* system there should only be one complexing species. If there are several, as is the case for citric acid and the rare earths, there will be competition between the various complexing species and this will tend to cause the bands of the individual rare earths to overlap strongly and give poor separations. It can be shown[23] that the separation factor for two rare earths, A and B, is given by

$$\alpha = \frac{\lambda_A}{\lambda_B}\frac{K_B}{K_A} \qquad (5)$$

where $\lambda_A$ and $\lambda_B$ are the distribution factors for the rare earths between the aqueous solution and the ion exchange resin when no complexing agent is present. Since no separation is possible in this situation, i.e. $\lambda_A/\lambda_B \cong 1$ Eq. (5) can be rewritten as

$$\alpha \cong K_B/K_A \qquad (6)$$

The stability constants for three complexing agents are plotted in Figure 4; two of these materials are useful for separating the rare earths (i.e. EDTA and HEDTA) and one is not (i.e. DTPA). More information on other complexing agents can be found in the

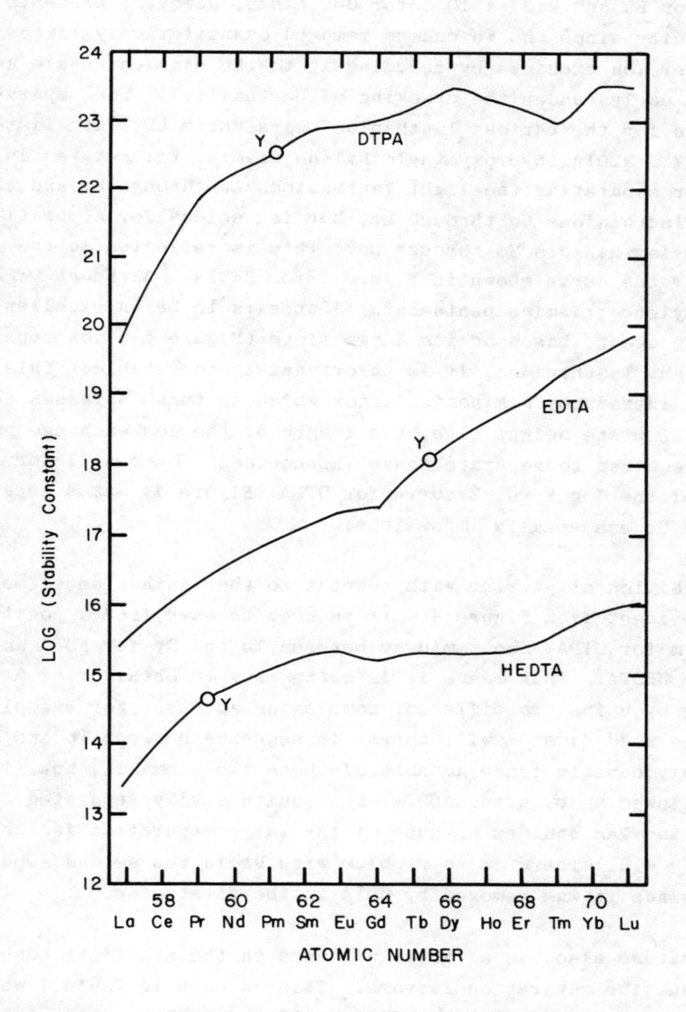


Figure 4. Stability constants for some selected polyaminopoly-carboxylate chelates at 25°C. DTPA - diethylene triamine pentaacetate, EDTA - ethylene diamine tetraacetate; and HEDTA - hydroxyethylethylene diamine triacetate. The stability constant for Y is shown for the respective complexing agent by an open circle on the three curves.

review by Powell.[22] EDTA (ethylene diamine tetraacetate) is the most useful since all neighboring pairs can be easily separated along the entire series, except for Eu-Gd where $\alpha$ = 1.05 because their equilibrium (stability) constants are nearly equal — 2.24 x $10^{17}$ for Eu and 2.34 x $10^{17}$ for Gd. This, however, presents no difficulty since the Eu can be removed quantitatively either before or after ion exchange by reducing it to the divalent state as discussed early (under the cracking of bastnasite). The separation factors for the various lanthanide pairs using EDTA are listed in Table 7. HEDTA (hydroxyethylethylene diamine triacetate) is useful for separating the light lanthanides La through Sm and the heavy lanthanides Ho through Lu, but is useless for separating the middle lanthanides Sm through Ho. This is reflected in the flatness of the curve shown in Figure 4 for HEDTA. Although DTPA (diethylene triamine pentaacetate) appears to be an excellent complexing agent, based on its large slope (Figure 4), for separating the light lanthanides, it is unfortunately too stable. This stability introduces a kinetic factor which in turn increases the theoretical plate height (i.e. the length of the ion exchange resin bed) required to separate these lanthanides. The oscillatory nature of the log K vs. Z curve for DTPA (Figure 4) makes separation beyond Sm essentially impossible.

The position of yttrium with respect to the lanthanide elements is also evident from Figure 4. It is seen to vary from a position near Pm for DTPA, about midway between Tb and Dy for EDTA and near Pr for HEDTA. This means it is quite easy to obtain Y of 6 nines purity by using two different complexing agents. For example, if EDTA is used first Y will appear in sequence between Tb and Dy and probably contain trace amounts of these two elements, but if this is followed up by using HEDTA, Y is quite easily separated from these two lanthanides because of the large separation factor, e.g. $\alpha^{HEDTA}_{Y/Tb} = 4.5$. There is no problem with Pr in the second separation step since it was removed by EDTA in the first step.

Temperature also has a large influence on the stability constants and thus the separation factors. This is seen in Table 7 where the separation factors for HEDTA at 25 and 92°C are listed. At 25°C separation is not possible for Ho-Dy, Dy-Tb, and Tb-Gd and Eu-Sm, but is possible at 92°C. Indeed many separation factors for adjacent lanthanide pairs have been improved by raising the temperature.

In the ion exchange process the resin bed is prepared by passing an acid through the column. Then the resin bed is loaded up with a mixed rare earth acid solution which contains the complexing agent and a retaining ion, such as $Cu^{2+}$ or $Zn^{2+}$. The retaining ion is needed to prevent the first rare earth ion from spreading out and being lost during the separation process. An eluant, $NH_4^+$, is needed to push the rare earths through the ion exchange columns. The most stable complex comes out first, i.e. the Cu or Zn complex, followed by Lu, Yb ....Y.... and finally La. With the correct choice of complexing agent the elution bands containing the individual rare earth elements are rectangular in shape and there is a minimum of overlap between adjacent bands. From the amount of various elements in the initial mixture one knows how much solution needs to be collected before starting to collect the next rare earth. Because of uncertainties in the exact chemical compositions and slight overlap of bands, an intermediate fraction of eluant is collected which contains both of the neighboring elements. This mixture is kept and recycled at a later date. The rare earth is then precipitated out of solution by using oxalic acid or methyoxalate. After filtering the rare earth oxalate is converted to the oxide by heating in air to 800 - 1000°C. The oxide is the most common chemical form for the individual elements sold by the rare earth producer.

To date the ion exchange process is the best method for producing the highest purity rare earth element. Only for Y is the liquid-liquid extraction process capable of reaching the same purity level obtainable with ion exchange. The major disadvantage of ion exchange is the length of time it takes to purify a given amount. It can take months before the last remaining lanthanide from a mixture which has been loaded on the columns is finally eluted from the resin bed. A number of companies throughout the world still use the ion exchange method for separating and purifying the rare earths. Some use it exclusively and others use it in combination with liquid-liquid solvent extraction, whereby the initial separations are made by using the latter technique and the final purification by using ion exchange.

## Separated (Individual) Rare Earth Chemicals

The use of pure (>98%) rare earth compounds is primarily dictated by the role of the 4*f* electron in the particular application, with the one exception noted earlier regarding the use of the mixed rare earth fluorides in carbon arc lighting. Although a number of uses discussed heretofore were based on the absence or presence of the 4*f* electron, lower purity (90 to 98%) materials could be used in many of these applications without adversely affecting the desired property. The uses discussed here require high purity elements, and in some cases certain impurities need to be held to a 1 ppm level or lower. The categories listed below are given in order of decreasing purity requirements: optical, electronic and magnetic, special uses of $Y_2O_3$ and miscellaneous.

Optical Applications. The unique spectral features of the various rare earth elements account for their use in many optical applications. Many of the elements are used as activators (in phosphors) and lasing ions (in lasers), but by far the largest volume (and monetary) usage is for the rare earths to serve as a host to hold these optically active ions. The hosts must not exhibit any optical activities in the electromagnetic region of interest and be able to dissolve the desired rare earth ion in its lattice. The rare earths without 4*f* electrons ($Y_2O_3$ and $La_2O_3$) or a completely filled 4*f* level ($Lu_2O_3$) or a half filled 4*f* level ($Gd_2O_3$) meet these criteria. Because of expense $Lu_2O_3$ is not used.

By far the largest use of the rare earths in the optical field is that of $Eu^{3+}$ activators as the primary red color in colored TV. The major host materials are Y-based oxides, such as $Y_2O_3$, $YVO_4$ and $Y_2O_2S$. $La_2O_2S$ and $Gd_2O_2S$ are also used as host materials. Up to 25 at.% Eu is dissolved in the host material in this application. Other cathode ray tube (CRT) activators are $Ce^{3+}$, $Eu^{2+}$ and $Tb^{3+}$. More details on this subject can be found in the recent review by McColl and Palilla.[24]

The use of rare earths, primarily $Eu^{3+}$ and $Tb^{3+}$ as activators and Y, La or Gd as hosts, in fluorescent lamps is a minor one today, but the potential is great considering the volume of lamps involved. Newly developed fluorescent tubes which use only three fairly narrow spectral colors, instead of all wavelengths of the visible region, offer much promise since one can obtain the same amount of light from three tubes as from four of the standard tubes. This represents both a 25% savings in electrical operating costs and capital investment in the fixtures. The $Eu^{3+}$ provides the red wavelength component and the $Tb^{3+}$ green color.[25]

One of the important newer uses involves the rare earth phosphors in medical x-ray intensifying screens. The x-rays activate the phosphor causing it to emit visible light which strikes a photographic film. The advantage of the rare earth-based intensifying screen is that the radiation exposure to patients is reduced by a factor of two to four, and the resultant films are sharper. The activators are $Tb^{3+}$, $Tm^{3+}$ and $Eu^{2+}$. The host materials are LaOBr or $Gd_2O_2S$ for the trivalent lanthanide activators and BaFCl for $Eu^{2+}$. Rabatin[26] has recently discussed this application in more detail.

Neodymium is one of the most important lasers in use today. Host materials for Nd include glass and YAG (yttrium aluminum garnet). YAG's have also been used as artificial diamonds, but in recent years this material has been replaced by $Y_2O_3$-stabilized $ZrO_2$, which is harder than YAG and has an index of refraction closer to that of diamond. The unique properties of PLZT (lead lanthanum zirconate titanate) — a ferroelectric which has good optical transparency — accounts for its use as optical shutters and modulators, eye protective devices, color filters, displays and image storage devices.[27]

<u>Electronic and Magnetic Applications.</u> In the electronics and magnetics area the most important rare earth compounds from an applications viewpoint are garnet based materials — yttrium iron garnet (YIG) and gadolinium gallium garnet (GGG). Both compounds are prepared by reacting the corresponding rare earth oxide with the appropriate transition metal oxide in the correct proportions to give the $R_3M_5O_{12}$ formula, where M = a trivalent metal ion and R = a trivalent rare earth metal ion. The reaction is carried out at elevated temperatures (1000 - 1500°C) which are below the melting point of the metal oxide and significantly lower than the $R_2O_3$ melting point.

The YIG based materials are used in the polycrystalline form in a variety of microwave devices. These include attenuators, circulators, isolators, phase shifters, power limiters and switches. These materials are also being used in microwave integrated circuits in which thin films are placed on garnet substrates. Properties of these materials are varied by substitution of Gd for Y and Al for Fe.

The use of GGG as bubble devices for domain memory storage is a fairly recent industrial application (∿3 years old).[28] A schematic representation of a typical bubble memory chip is shown in Figure 5. The non-magnetic single crystal GGG substrate is grown by the Czochralski technique, i.e. pulling GGG from a melt held just above its melting point (1740°C). The melt is contained in an Ir crucible under a $N_2$ atmosphere containing 2% $O_2$. Thin wafers are precision cut and polished (using rare earth polishing compounds) to the dimensions shown in Figure 5. Then a complex magnetic film is grown on the GGG substrate from a molten $PbO-B_2O_3$ flux at 1000°C. Finally a configuration of "tees" and "bars" of permalloy is overlaid on the magnetic film by electron beam deposition. Magnetic bubbles (polarities) induced on the ends of the tees and bars are about 1/20 the diameter of a human hair and are moved from one location by a rotating magnetic field, whose axis of rotation is normal to the face of the disk. Presently these bubble memories are

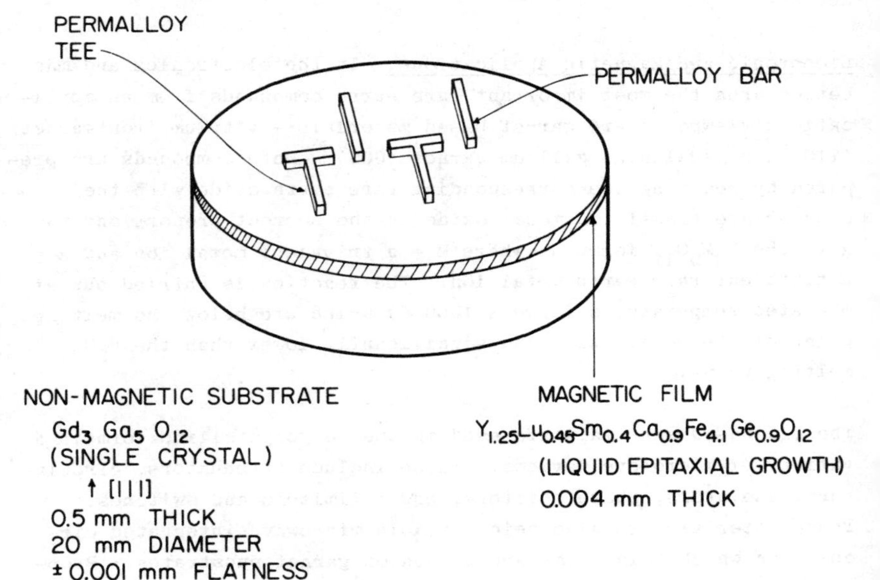


Figure 5. A schematic representation of a quarter million bit bubble memory chip.

used to record voice messages — about 10 seconds of voice communication can be stored on the one quarter million bit chip shown in Figure 5. The major problem to be solved, if any accelerated growth of this technology is to be realized, is the need to increase the density of bubbles on a chip.

Special Uses of $Y_2O_3$. The newest growth market for $Y_2O_3$ involves the stabilization of the high temperature cubic form of $ZrO_2$ — $Y_2O_3$-stabilized $ZrO_2$ (or YSZ). A 5 mol % $Y_2O_3$ addition will give a partially stabilized $ZrO_2$ product, while 8 mol % $Y_2O_3$ is required to give a fully stabilized material. YSZ has been used for many years as a ceramic material for thermal insulation. It has also been known for many years that because of the oxygen defects in this material the electrical resistivity varies linearly over several orders of oxygen pressure. Recently scientists and engineers have put this property to use in many applications, for example: to measure the oxygen partial pressure in automobile exhaust gases, the carbon potential in a CO-O-$CO_2$ atmosphere carburizing furnace, and the oxygen content in molten steel; and as a pH meter for use at high temperatures and high pressures. In the first two applications, because the response time is fast, these electrodes are also used to control the process: in the first to give as environmentally clean exhaust as possible, and in the second to keep the carbon potential constant to ±0.05% during the heat treating process. In 1980 the General Motors cars alone required 5 million sensors.

Another recent use of YSZ is as simulated diamonds. The properties of this material so closely resemble those of diamond that in the mounted condition even experts have a difficult time distinguishing between the two.

A recent development which shows considerable promise is the addition of $Y_2O_3$ particles to Ni-based superalloys which are used in critical parts of gas turbine engines. The presence of the dispersed oxides is to improve the high temperature creep resistance of the superalloys. $Y_2O_3$ particles, 100 to 500 Å in diameter, are mechanically alloyed with the superalloy metal powders before the parts are compressed and sintered to the desired shape. The interparticle spacing is 500 to 3000 Å, giving an alloy with an oxide volume fraction ranging from 0.4 to 3.0%. The $Y_2O_3$ dispersion strengthened superalloy has allowed the turbine operating temperatures to be increased from 1100°C to 1150°C. $ThO_2$ had been used previously, but it has been replaced by $Y_2O_3$ because $Y_2O_3$ has two important advantages over $ThO_2$: $Y_2O_3$ is a more stable oxide (thermodynamically) and it is not radioactive. The current market is small but it is expected to grow over the next few years.

Miscellaneous. Other important applications are the use of rare earth materials in the nuclear industry, as alloying additives to superalloys, as hydrogen storage materials, and in applied and basic research.

In the nuclear field the rare earths are usually used in the form of oxides. The major application of Gd, because of its extremely large nuclear cross section, is as a burnable neutron absorber for regulating the neutron level and criticality in nuclear reactors. Sm, Eu and Dy also have large nuclear cross sections and are used as control rods.

The rare earth metals, primarily La and Y, are added to a number of superalloys and other specialty alloys to help getter the nonmetallic and tramp elements, and to improve the high temperature oxidation and corrosion resistance. Additions range from 0.01 to 0.2%. The rare earth metals are prepared by converting the oxide to the fluoride and then reducing the fluoride by Ca to the metal.[3] This application has been in use for many years, but it continues to grow at ∿20% a year.[29]

A fairly recent application is the use of $RNi_5$ base compounds as hydrogen storage materials. The amount of rare earth, primarily La, used today is small, but the future is quite promising. These hydrides are also used as hydrogenation catalysts, for hydrogen purification, thermal compression, waste heat engines, heat pumps and refrigeration.[30]

Markets. From the above discussion one notes that the use of the pure individual rare earth elements is diverse and is generally concerned with highly technical and advanced applications. In these applications the specific property or behavior of the rare earth element is critical and the cost is of a secondary consideration. The markets for these pure materials are summarized in

Table 8. 1978 Markets for the High Purity Individual Rare Earth Elements (in metric tons).

| Market | $Y_2O_3$ | $La_2O_3$ | $Eu_2O_3$ | $Gd_2O_3$ | Other $R_2O_3$ |
|---|---|---|---|---|---|
| Phosphors | 70 | 5 | 7 | 9 | 0.2 |
| Bubble devices | - | - | - | 20 | 0.2 |
| Nuclear | - | - | - | 2 | 0.1 |
| YSZ[a] | 20 | - | - | - | 0.1 |
| Miscellaneous | 11 | 9 | - | - | 0.4 |
| Total | 101 | 14 | 7 | 31 | 1.0 |

[a] yttrium-stabilized $ZrO_2$

Table 8. In addition to the items shown, some applications which might be included in Table 8 are: $Sm_2O_3$ in the $RCo_5$ permanent magnets and high purity oxides (>98%) in optical grade glasses. These items have been included in the data presented in Table 6.

Acknowledgments

The author wishes to acknowledge the assistance and/or helpful discussions concerning some aspects of this paper with his friends from several industrial concerns: G. A. Barlow, and T. A. Wilson (Molycorp, Inc., Union Oil Company), I. S. Hirschhorn (Ronson Metals Corp.), J. Kaczmarek (Rhone-Poulenc) and H. W. Smith, Jr. (Technical Information Programs). Thanks are also extended to the author's co-workers for their critical comments and assistance: B. J. Beaudry, B. L. Evans and O. D. McMasters. Finally, the author is grateful to Molycorp, Inc., Union Oil Company and Rhone-Poulenc, Inc. for supplying photographs of their facilities and for permission to include them in this chapter.

References

[1]V. V. Cherdyntsev, "Abundances of Chemical Elements", University of Chicago Press, Chicago, 1961.

[2]K. A. Gschneidner, Jr. and L. Eyring, eds., "Handbook on the Physics and Chemistry of Rare Earths", North-Holland Publishing Co., Amsterdam, 1978, Vol. 1.

[3]B. J. Beaudry and K. A. Gschneidner, Jr., "Handbook on the Physics and Chemistry of Rare Earths", K. A. Gschneidner, Jr. and L. Eyring, eds., North-Holland Publishing Co., Amsterdam, 1978, Vol. 1, Chapter 2, p. 173.

[4]R. D. Shannon and C. T. Prewitt, Acta Cryst. 1969, B25, 925 and 1970, B26, 1046.

[5]G. H. Dieke, "Spectra and Energy Levels of Rare Earth Ions in Crystals", H. M. Crosswhite and H. Crosswhite, eds., Interscience Publishers (Wiley) New York, 1968.

[6]K. A. Gschneidner, Jr. and N. Kippenhan, "Thermochemistry of the Rare Earth Carbides, Nitrides and Sulfides for Steelmaking", IS-RIC-5, Rare-Earth Information Center, Iowa State University, Ames, Iowa, 1971.

[7]K. A. Gschneidner, Jr., N. Kippenhan, O. D. McMasters, "Thermochemistry of the Rare Earths. Part 1. Rare Earth Oxides, Part 2. Rare Earth Oxysulfides, Part 3. Rare Earth Compounds with B, Sn, Pb, P, As, Sb, Bi, Cu and Ag", IS-RIC-6, Rare-Earth Information Center, Iowa State University, Ames, Iowa, 1973.

[8]C. M. Moore, Chapter on Rare-Earth Minerals and Metals in "1978-79 Bureau of Mines Minerals Yearbook", U.S. Dept. Interior, U.S. Printing Office, Superintendent of Documents, Washington, D.C.

[9]J. G. Cannon, "Van Nostrand's Scientific Encyclopedia", 5th Ed., D. M. Considine, ed., Van Nostrand Reinhold Co., New York, 1976, p. 1886.

[10]J. Kaczmarek, Chapter 8, Discovery and Commercial Separation in "Industrial Applications of the Rare Earth Elements", K. A. Gschneidner, Jr., ed., ACS Symposium Series, Amer. Chem. Soc., Washington, D.C. (to be published 1981).

[11]R. J. Callow, "The Industrial Chemistry of the Lanthanons, Yttrium, Thorium and Uranium", Pergamon Press, London (1967).

[12]W. Brugger and E. Greinacher, J. Metals, 1967, 19, [12], 32.

[13]I. S. Hirschhorn, J. Metals, 1968, 20 [3], 19.

[14]O. B. Michelsen, ed., "Analysis and Application of Rare Earth Materials", Universitetsforlaget, Oslo, Norway (1973).

[15]K. A. Gschneidner, Jr., ed., "Industrial Applications of the Rare Earth Elements", ACS Symposium Series, Amer. Chem. Soc., Washington D.C. (to be published 1981).

[16]RIC News is published quarterly by the Rare-Earth Information Center, Energy and Mineral Resources Research Institute, Iowa State University, Ames, IA 50011, U.S.A.

[17] Overview is published on an irregular basis by Molycorp, Inc., Union Oil Center, P.O. Box 54945, Los Angeles, CA 90054, U.S.A.

[18] J. G. Cannon, Eng. Mining J. 1980, [3], 178 (the latest issue available).

[19] Anoymous, Industrial Minerals, 1979, [3], 21.

[20] D. F. Peppard, "Progress in the Science and Technology of the Rare Earths", L. Eyring, ed., Pergamon Press, London, 1964, Vol. 1, p. 89.

[21] E. K. Hulet and D. D. Bode, "MTP International Review of Science, Inorganic Chemistry, Series One, Vol. 7, Lanthanides and Actinides", K. W. Bagnall, ed., Butterworths, London, 1972, p. 1.

[22] J. E. Powell, "Handbook on the Physics and Chemistry of Rare Earths", K. A. Gschneidner, Jr. and L. Eyring, eds., North-Holland Publishing Co., Amsterdam, 1979, Vol. 3, Chapter 22, p. 81.

[23] T. Moeller, "The Chemistry of the Lanthanides", Reinhold Publishing Corp., New York (1963).

[24] J. R. McColl and F. C. Palilla, Chapter 10, Use of Rare Earths in TV and Cathode Ray Phosphors in "Industrial Applications of the Rare Earth Elements", K. A. Gschneidner, Jr., ed., ACS Symposium Series, Amer. Chem. Soc., Washington, D.C. (to be published 1981).

[25] W. A. Thornton, Chapter 11, Lamp Phosphors in "Industrial Applications of the Rare Earth Elements", K. A. Gschneidner, Jr., ed., ACS Symposium Series, Amer. Chem. Soc., Washington, D.C. (to be published 1981).

[26] J. G. Rabatin, Chapter 12, Rare Earth X-ray Phosphors for Medical Radiography in "Industrial Applications of the Rare Earth Elements", K. A. Gschneidner, Jr. ed., ACS Symposium Series, Amer. Chem. Soc., Washington, D.C. (to be published 1981).

[27] G. H. Haertling, Chapter 16, PLZT Electrooptic Ceramics and Devices in "Industrial Applications of the Rare Earth Elements", K. A. Gschneidner, Jr., ed., ACS Symposium Series, Amer. Chem. Soc., Washington, D.C. (to be published 1981).

[28] J. W. Nielsen, Chapter 13, Bubble Domain Memory Materials in "Industrial Applications of the Rare Earth Elements", K. A. Gschneidner, Jr., ed., ACS Symposium Series, Amer. Chem. Soc., Washington, D.C. (to be published 1981).

[29] K. E. Davies, Chapter 9, Industrial Applications of Pure Rare Earth Metals and Related Alloys in "Industrial Applications of the Rare Earth Elements", K. A. Gschneidner, Jr., ed., ACS Symposium Series, Amer. Chem. Soc., Washington, D.C. (to be published 1981).

[30] E. L. Huston and J. J. Sheridan, III, Chapter 14, Application for Rechargeable Metal Hydrides in "Industrial Applications of the Rare Earth Elements", K. A. Gschneidner, Jr., ed., ACS Symposium Series, Amer. Chem. Soc., Washington, D.C. (to be published 1981).

# Precious Metal Chemicals

By P. E. Skinner and L. P. Vergnano
JOHNSON MATTHEY CHEMICALS LTD., ORCHARD ROAD, ROYSTON, HERTFORDSHIRE SG8 5HE, U.K.

## Introduction

Before discussing the main chemicals of the precious metals it is best to say a little about the metals themselves. The precious metals may be divided into two classes. The first are called 'platinum group metals' (P.G.M.'s) which are found in the second and third row transition series of Group VIII of the periodic table according to Mendeleef. They are, of course, ruthenium and osmium from the iron group, rhodium and iridium from the cobalt group and palladium and platinum from the nickel group.

The second class consists of silver and gold from Group IB.

## The P.G.M.'s

The P.G.M.'s are generally found together in the Earth's crust although the ratio of each varies with the source. The principal sources are from mines in the Merensky Reef in South Africa where the P.G.M.'s are the primary products and the Canadian and Russian deposits where P.G.M.'s are separated as by products from extensive nickel refinery operations.

Traditional placer deposits are still worked but only account for around 2% of the world output of newly mined metals which was 200 tons in 1978.

Separation of the P.G.M.'s is a lengthy and complex operation but procedures are constantly being improved and modern techniques such as solvent extraction, ion exchange and complex formation are exploited.

Silver and Gold

Principal sources of silver are Mexico, North America and Australia with world production at 6000 tons per annum.

South Africa and Russia are principal sources of gold with world production estimated at over 1500 tons per annum.

Secondary Refining

The high intrinsic value of materials containing the precious metals gives rise to an extensive recycling industry known as secondary refining. Examples of materials regularly handled are spent cyanide electrolytes, X-ray film, used watch batteries, gilt picture frames, spent homogeneous and heterogeneous catalysts, printed circuit board scrap, jeweller's findings and even floor sweepings from precious metal workshops.

Relative Values

The value of precious metals must always be appreciated when handling them as was shown by Dr. K. Turner in his Sir Robert Horne Memorial Lecture 'Precious Metals' published in Chemistry and Industry on 19th June 1980.

Thus rhodium is priced at over £10,000,000 per tonne and even humble silver has been priced at over £500,000 per tonne. The latter price has been very variable and during the past year was at one time seven times its previously considered 'base' price. Other metals which are available in tonnage quantities and are of high price for comparison are rhenium at £1,200,000 per tonne, and indium at £254,000 per tonne.

Prices are dependent on speculation and hoarding in some cases, but availability, rarity, and the complex refining procedures coupled with strong industrial demand because of their unique properties play a part in determining the prices of the metals.

## The Precious Metal Chemist

Chemists working in the precious metal field very soon become aware of special disciplines corresponding to an unwritten code of practice which have been built up. At first these are somewhat awesome but, once adopted, the chemist can derive a great deal of personal satisfaction from working in this field.

Financial risks resulting from low yields, losses or analytical errors can be large and thus special procedures for checking all work are carried out. An analytical error in a second decimal place could easily be measured in thousands of pounds depending on the lot size!

Any discrepancies must be investigated immediately and certain questions answered e.g.

1. have any volatile products formed?
2. have any spillages occurred?
3. has the highest yield been obtained?
4. have any precious metal salts entered the drainage system?
5. has anything been stolen?
6. can any improvements be made?

The chemist soon learns, for example, to:-

1. save all filter papers and cleaning rags for precious metal recovery. Nothing must be thrown away unless free of precious metal.
2. avoid mixing together materials from different preparations.
3. make use of specialised equipment to absorb volatile species and to prevent materials entering the drainage system, etc.
4. keep materials as secure as possible by making use of safes and security protected laboratories.

To sum up, the old fashioned terms integrity, reliability and honesty are major requirements for a P.G.M. chemist, who should also keep up to date by background reading and liaison with colleagues.

Nomenclature

Several systems of nomenclature are used in this paper. The I.U.P.A.C. system is universal but gives rise to some names not generally understood by the trade and can be unwieldy in certain circumstances. It is anticipated that the names used will allow the reader to obtain a feeling for the chemistry as they are still commonly used in reference material and text books.

Chemistry

The chemistry of the elements will be considered in turn. Emphasis will be placed on those materials which are supplied for industrial use to a wide range of industries. Many of the materials mentioned are key materials from which an extensive chemistry has been developed.

New uses for precious metal compounds are constantly being sought and research activity is high, particularly in such fields as metal deposition, new drugs, new catalysts, pollution and environmental control and energy related materials.

Gold, Au, Atomic Weight, 196.97

The most common starting material for gold compounds is the well known chloroauric acid ($HAuCl_4$). This is readily formed by dissolving gold in grain form in aqua regia and then 'freeing off' the excess acids by boiling.

Alkali metal salts of chloroauric acid are readily formed by addition of the corresponding halide and evaporation. The hydroxide can be precipitated from chloroauric acid by addition of sodium hydroxide followed by washing to remove sodium salts. It provides a useful intermediate for a range of compounds.

A material known as gold fulminate is obtained by addition of ammonia to chloroauric acid. It is a brown solid which is retained in the damp state for safety reasons and can be used to prepare gold potassium cyanide for example as it is soluble in potassium cyanide. The dry form is explosive. Gold potassium cyanide is prepared on a large scale with world usage probably in excess of 100 tonnes /annum.

Chloroauric acid also forms the starting point in the preparation of materials known as liquid golds and for materials used in chemotherapy.

Industrial uses of gold

The largest industrial use of gold is in the electroplating of electronic components such as printed circuit boards, button cells, connectors, etc. A range of proprietary plating solutions based on gold potassium cyanide have been developed over many years operating under alkaline or mildly acid conditions to obtain different properties of gold deposits.

Decorative uses have also emerged and white and pink golds, for example, can be obtained by alloy plating and are favoured for such items as spectacle frames.

More recently a range of non-cyanide plating baths have been developed based on sulphite complexes. These can be used to build up large thicknesses which are stress free. A large variety of objects can be plated which faithfully reproduce the detail of the formers.

Not all the cyanide systems require the use of electricity as some work by galvanic action caused by a zinc wire attached to the workpiece.

Sodium chloroaurate is used in Africa to deposit gold onto copper bangles and jewellery by a displacement process.

Gold has been used since ancient times for pottery decoration in the form of organo metallic intermediates. A typical preparation may contain 10 or more ingredients to give a desired effect. Some of the materials can be incorporated into transfers which are positioned and fired on the ware to give an infinite choice of designs and effects. Others can be brushed or painted on and then fired.

Again stemming from ancient times gold has applications in the medical field. Originally used as a 'cure' for all ills it is now particularly important in treating arthritis. The simple drugs of the 1930's such as gold sodium thiomalate are being replaced by more advanced forms today with a corresponding reduction in side effects.

## Silver, Ag, Atomic weight, 107.87

The most important silver chemical is silver nitrate which is easily obtained by dissolution of silver in nitric acid followed by evaporation and crystallisation.

Treatment of the nitrate with halide ions gives the silver halides which are widely used in various forms.

Silver I oxide may be made by addition of potassium hydroxide to silver nitrate, washing out the potassium nitrate formed. Oxidation of this material in alkaline persulphate gives the so called divalent silver II oxide.

Silver I oxide may be dissolved in long chain fatty acids to give the corresponding salt. These materials may be heat sensitive as well as light sensitive and these properties are exploited industrially.

Silver plating makes wide use of silver cyanide, AgCN, and the derivative silver potassium cyanide $KAg(CN)_2$.

Amminated species may be readily prepared from the nitrate or halide and a wide range of silver salts can be made by treatment of silver nitrate with the sodium or potassium salt of the desired acid radical.

Industrial uses

Silver salts are widely used in many industries. The halides are particularly important as the chloride and bromide form the basis of the photographic industry albeit prepared from silver nitrate under carefully controlled proprietary conditions. Although efforts are made to substitute for silver salts in this business they are still used on a large scale.

Silver iodide is used for seeding clouds either to prevent the formation of large ice crystals which can damage crops and property or to allow rain to fall. The crystal structure of AgI is similar to that of Ice h with approximately 1% mismatch. The material is administered either by burning an acetone suspension on the ground or by specially equipped aircraft.

Silver difluoride is a useful fluorinating agent.

The battery industry absorbs multiple tonnages of silver I oxide, silver II oxide and silver chloride, the latter being used in the so called 'seawater battery' in the fused state.

The mirror industry makes use of amminated silver nitrate which is reduced by alkaline glucose to form the mirror. Certain explosion hazards may be encountered with this process which is best left to companies with expertise in this field.

Photochromic glasses for use in buildings or in sun glasses make use of silver salts in their composition.

Thermal copying, which is rapidly growing, makes use of such materials as silver behenate or stearate and numerous patents are registered in this field.

Silver electrodeposition onto nickel steels makes use almost exclusively of cyanide based electroplating baths.

Platinum, Pt, Atomic Weight 195.09

The best known platinum compound which is used to prepare most of the industrially important platinum compounds is chloroplatinic acid, $H_2Pt\ Cl_6.6H_2O$ (C.P.A.). This material may be prepared in solution by dissolution of platinum powder in aqua-regia, removing residual acids by evaporation, or as a red deliquescent solid.

Treatment of this material with ammonium or potassium chloride gives the yellow, insoluble, chloroplatinates. The sodium salt is similarly formed but is more soluble and can be obtained by crystallisation.

Treatment of a solution of C.P.A. with a base, typically sodium hydroxide, leads to the formation of the strongly basic hexahydroxy species from which hexahydroxy platinic acid can be precipitated and a useful series of salts derived. Careful drying of the acid gives a group of oxides with catalytic properties.

A well known oxide, 'Adams oxide', can be obtained by fusing a mixture of C.P.A. and sodium nitrate and quenching the melt to obtain the oxide in finely divided catalytic form.

The materials described above are of platinum in the +4 oxidation state, however, platinum also forms many important compounds in the +2 oxidation state.

The commonest chloroplatinite is potassium chloroplatinite made by evaporation of a slurry of potassium chloroplatinate which has been reduced with hydrazine hydrate.

A wide range of ammines may be prepared by additions of ammonium hydroxide to a potassium chloroplatinite solution for example. One that is sometimes encountered is known as Magnus' green salt ($Pt\ (NH_3)_4 .\ Pt\ Cl_4$).

Use of Platinum Salts

Adam's oxide is used in many organic hydrogenation reactions.

C.P.A. is used as the basis for many supported platinum catalysts and platinum black. The latter is platinum in a finely divided form with a trace of oxygen present and is popular in fuel cells and for making inks used in thick film electronics. C.P.A. may be used in solutions for cladding e.g. Titanium electrodes.

Platinum diammine dinitrate ('P' salt), hexahydroxy platinum salts and dinitrosulphato platinous acid (D.N.S.) and some cyano derivatives find applications in electroplating of such things as turbine blades.

A new field of great promise is the use of platinum salts in chemotherapy as anti-cancer agents with cis-diamminedichloroplatinum (II) as the original, and still the most widely used.

Palladium, Pd, Atomic Weight 106.4

Palladium (II) chloride, $Pd\ Cl_2$, provides the most accessible chemical for preparing the range of palladium salts. It can be made by dissolving palladium metal grain or sponge in hydrochloric acid and chlorine. After removal of the surplus acid and chlorine the material can be obtained by evaporation as a red-brown polymeric solid.

The most useful series of salts of palladium are those with the +2 oxidation state. A series of 'oxides' can be obtained, for example, by treating chloropalladous acid, $H_2Pd\ Cl_4$, which is simply palladous chloride dissolved in hydrochloric acid, with a base, washing out the surplus sodium chloride, and then drying by a variety of methods.

Salts of chloropalladous acid, known as chloropalladites, are obtained by treatment with an alkali halide and crystallisation.

Palladium is soluble in nitric acid to form a series of nitrates.

Ammine compounds of palladium are easily prepared by treating chloropalladous acid with ammonium hydroxide. The diammine which is precipitated with hydrochloric acid, can be redissolved in excess ammonium hydroxide to form the corresponding tetrammines.

The +4 oxidation state of palladium is not often encountered, but ammonium chloropalladate is stable and is made by oxidation of the corresponding chloropalladite with chlorine. The alkali chloropalladates are unstable releasing chlorine.

Uses of Palladium Salts

Palladium chloride is used as the basis for making supported catalysts such as Pd on alumina, charcoal, etc. It is also used as the basis of materials for sensitising plastics prior to electrodeposition and to form a range of palladium blacks and powders for use in the preparations of electronic inks and components.

Tetrammine palladium nitrate, palladium diammine dinitrite ('P' salt) and dinitrosulphato palladous acid (D.N.S.) are used in electroplating applications, the deposit being generally white and bright but not as hard or corrosion resistant as other P.G.M. deposits. A variety of organo palladium salts are also used to make proprietary plating solutions.

Palladium acetate and acetylacetonate are useful homogeneous catalysts and a wide range of organo palladium compounds are available for industrial and research applications.

Ammonium chloropalladate is used in the photographic industry.

### Rhodium, Rh, Atomic weight 102.91

Rhodium salts are generally prepared via the trichloride which may be obtained by treatment of sodium rhodium chloride with concentrated hydrochloric acid.

The sodium rhodium chloride itself may be obtained by fusion of rhodium powder in sodium chloride with chlorine passing through the melt.

Rhodium salts are normally encountered in the +3 oxidation state, but the ligand stabilised +1 state is important industrially.

The trichloride gives rise to several materials made in appreciable quantities, such as the well known Claus salt, chloropentammine rhodium chloride $[Rh(NH_3)_5Cl]Cl_2$.

A wide range of rhodium organometallic compounds have been developed and are available for industrial use. Typical are Wilkinson's catalyst, chloro (tris-triphenylphosphine) rhodium (I) and hydridocarbonyltris-triphenylphosphine rhodium (I).

### Uses of rhodium salts

Rhodium sulphate solutions are widely used in decorative (say up to 0.5 microns thickness) and industrial (say up to 10 microns) electroplating applications. The decorative market is shared with rhodium phosphate and gives hard, bright deposits used for jewellery, spectacle-frames, watch cases, etc. The finish is tarnish resistant. One of the main industrial uses is in electroplating microswitches for the electronics industry. Rhodium deposits are hard, do not form oxide coatings and can operate in corrosive environments for many million operations.

Catalytic applications for rhodium salts are many. A major industrial process known as hydroformylation to make normal aldehydes, which are plastics precursors, had until the 1970's made use of cobalt catalysts. Substitution of rhodium homogeneous catalysts has allowed the reaction to take place at much lower temperature and pressure; this has resulted in energy and plant costs savings. The process gives rise to a higher normal to iso ratio for the product aldehyde.

$$CH_3-CH=CH_2 + H_2 + CO \xrightarrow[\text{Triphenyl phosphine}]{\text{Rh homogeneous catalyst}} CH_3-CH_2-CH_2-CHO$$

propylene principally n - butyraldehyde

Example of hydroformylation process

The LPO (low pressure oxo) process which is Rh catalysed operates at $80^{\circ}C$-$120^{\circ}C$ whereas the corresponding cobalt process requires $180^{\circ}C$-$200^{\circ}C$. The low pressure is 200-400 psi compared to 800-1,500 psi or higher for the cobalt based versions. Hence considerable capital and operating savings are possible with the rhodium based process.

Rhodium iodide may be used as a catalyst for the preparation of acetic acid from methanol, and rhodium chloride can be used to prepare supported catalysts.

In organic solvents rhodium trichloride can be used for electrode coating and pottery decoration applications.

Iridium, Ir, Atomic weight 192.22

The most accessible iridium chemical is ammonium chloroiridate. Iridium powder may be fused in sodium chloride with chlorine passing through the melt to give sodium chloroiridate. Ammonium chloroiridate may be precipitated from sodium chloroiridate by addition of ammonium chloride. The ammonium chloroiridate may be converted to chloroiridic acid ($H_2Ir\ Cl_6.aq$) by treatment with aqua regia.

Chloroiridic acid may be reduced with oxalic acid to give a range of iridium (III) salts of the type $M_3Ir\ Cl_6$ where M is ammonium or alkali metal, etc.

A blue hydrated oxide, $IrO_2.2H_2O$, may be prepared by treating chloroiridic acid with sodium carbonate.

Uses of Iridium salts

Chloroiridic acid provides the basis for the manufacture of supported and organoiridium homogeneous catalysts. Iridium trichloride in an organic solvent is used for coating electrodes and fused salt electroplating can be carried out with the complex cyanides.

Iridium electroplating is not widely practised.

Ruthenium, Ru, Atomic Weight 101.07

Ruthenium with osmium can be encountered as materials exhibiting many oxidation states. With ruthenium, oxidation states of +8, +7, +6, +3 are particularly common.

The volatile tetroxide, $Ru\ O_4$, is particularly useful although it is very reactive and somewhat hazardous. It can explode when the liquid contacts organic matter. The vapour may be absorbed in acids or alkalis to form various series of salts such as the perruthenate (+7), ruthenates (+6), chlorides (+3), nitrates (+3), etc.

The tetroxide is readily formed by oxidation of an alkaline ruthenate solution with hypochlorite, cerium (IV), permanganate, or by heating the metal in oxygen above 1000°C.

Ruthenium has a high affinity for nitric oxide and forms many nitrosyl compounds.

Ruthenium trichloride, obtained by distillation of the tetroxide into hydrochloric acid, followed by evaporation, is made in quantity and is used to form materials such as 'ruthenium red' (ammoniated ruthenium oxychloride) and ruthenium hexammine trichloride.

Ruthenium forms an oxide, the blue electrically conducting $Ru\ O_2$ (+4) which is obtained by roasting the metal in oxygen at 800°C. Another form of the oxide may be obtained by drying of the hydroxide precipitated from the trichloride or additions of base.

Uses of Ruthenium Salts

Ruthenium trichloride is the basis of many preparations for coating electrodes, particularly titanium, for use in corrosive environments. These form dimensionally stable anodes for world wide use in the chlorine industry, replacing the dimensionally unstable carbon anodes previously used.

Ruthenium nitrate species are of growing interest in nuclear waste disposal experiments where they are compatible with the glass made for storage and disposal of radioactive nucleides.

Ruthenium ammine complexes are used for nitrogen fixation and complexes such as $K_3$ or $(NH_4)_3$ $[Ru_2NCl_8(H_2O)_2]$ are used in electroplating.

Ruthenium red finds application as an electronmicroscope stain, and it is the standard stain for pectins in light microscopy. It stains cell surfaces but not the interior of intact cells and can be used in conjunction with osmium tetroxide.

Ruthenium (IV) oxides are used in thick film resistor inks for electronics applications.

Osmium, Os, Atomic Weight 190.2

Like ruthenium, osmium compounds can be commonly encountered in many oxidation states; however, the most important are +8 and +4.

The principal osmium compound is the tetroxide, $Os\ O_4$, which may be obtained by passing oxygen over the heated metal and condensing the product which forms almost colourless crystals. The material is water soluble and is sometimes referred to as "osmic acid' in this form.

The tetroxide can be slowly heated with hydrochloric acid and alcohol, to give chlorosmic acid, $H_2OsCl_6$, from which the usual series of ammonium and alkali salts can be made - the most useful being the relatively insoluble ammonium chlorosmate.

The tetroxide has a high vapour pressure at normal temperatures and the vapour particularly affects the eyes. The material is made and handled using special equipment and is normally supplied in sealed glass ampoules.

Uses of osmium salts

The tetroxide is widely used in solution or vapour form in electron microscopy as a fixative rather than a stain. Some structures are however stained by deposition of lower oxides or by organically bound but unreduced osmium.

Combinations of osmium tetroxide with sodium and zinc iodides are used for specific needs.

Osmium tetroxide is also used as a stereospecific reagent in (steroid) drug manufacture to give cis derivatives.

Thus olefines give cyclic osmate esters with osmium tetroxide in a suitable solvent which can then be hydrolysed to cis diols:-

$$\underset{\text{Olefine}}{>C{=}C<} + OsO_4 \longrightarrow \underset{\text{cyclic osmate ester}}{-C(-O-)-C(-O-)\,Os(=O)_2} \xrightarrow[Na_2SO_3]{\text{hydrolysis}} \underset{\text{cis diol}}{-C(OH)-C(OH)-}$$

Osmium tetroxide in aqueous or organic solvent can be used catalytically in hydroxylations used in conjunction with barium and silver chlorate or hydrogen peroxide.

# High Purity Chemicals

By L. P. Vergnano
JOHNSON MATTHEY CHEMICALS LTD., ORCHARD ROAD, ROYSTON,
HERTFORDSHIRE SG8 5HE, U.K.

## Introduction

This discussion includes the ultra purification of single elements and any combination of elements. One element may be combined with one or more others to give a vast number of products. Although this paper refers to inorganic materials it is not unusual to purify organic-metallic compounds as a route to an inorganic product.

Elements with short half lives and inert gases are excluded. These are shown in brackets in the text.

## Purity Levels

t.m.i. = Total metallic impurities which is the sum of the elements detected by optical emission spectroscopy expressed as parts per million (p.p.m.). Most of the materials under discussion fall in the range t.m.i. 1-10 p.p.m. At this stage it is necessary to appreciate how small is one part per million, i.e. 1 gram in a 1000 kilograms or 1 tonne.

## Historical

In general any chemical or physical separation must be considered a method of purification. The earliest applications of chemical processes were concerned with the extraction of metals. These arts were carried out without any theoretical background, but often with considerable skill, indicating long practice and a sound appreciation of materials.

Purification of materials can be traced back to the earliest civilisation. The first metal known was probably gold and was worked out of the river sands. Throughout the ages salt has been obtained by evaporation of sea water. This is most likely one of the earliest forms of purification by crystallisation.

There has been a progressive requirement for purer materials, and after the first World War there was a major effort to produce purer materials. In those days one would consider that the analytical grades of the materials were the purest available. These were very limited and would not include the wide range of individual elements and compounds available today.

For instance, the majority of rare earths were separated by means of fractional crystallisation and were certainly not pure by modern standards. Typically one produced salts of "didymium" which were a mixture of praseodymium and neodymium. The less abundant rare earths were not available; elements such as caesium, gallium, germanium, hafnium, lutetium, niobium, rhenium, rubidium, samarium, scandium, tantalum, yttrium and zirconium were only available in a very impure form in very small quantities.

Neither the techniques nor apparatus were available for carrying out the modern methods of purification.

Furthermore there were not the means of detecting low levels of impurities.

## The Chemists

To be successful in this sort of work the Chemists must have the following attributes: -

> A great enthusiasm for inorganic chemistry. They require to have a very complete knowledge of the chemistry of every element with which they are working. This is only acquired by experience in the field. They must be able to take into account all the reactions within a group and be able to devise for themselves the best method of preparation and purification.
>
> A good knowledge of physical methods is equally essential as is the ability to manipulate apparatus and to carry out more than one preparation at the same time. One should not be too frustrated when analysis indicates that the product is less pure than anticipated.

## Research and Development

Since one is dealing with a very wide variety of products it is essential to have a very liberal approach. The Chemists tend to specialise with a group of elements of which they have a particular knowledge and by pooling their knowledge a suitable process can frequently be devised.

Physical methods play a large part in modern purification techniques and the development of this type of apparatus can be a great asset. One is frequently working on a small to moderate scale and it is difficult to distinguish between research, development and production. Frequently the final product will come from the development area.

## Health and Safety

Since the range of compounds under discussion includes most elements there is a wide range of hazards. Known toxicity must be respected by taking the correct precautions.

Unfortunately the toxicological data for many chemicals and compounds are not available. Where this information is absent, it is essential to treat the material as toxic.

To this end preparation of unusual compounds should always be carried out in a well draughted area and conform with the Health and Safety at Work Act.

## Analysis

The techniques applied many years ago were very crude compared with todays sophisticated methods. The standard procedure which we adopt to monitor purification is optical emission spectrography. Facilities are such that within a few hours the material is examined for, say, seventy metallic elements on most inorganic substances. This is used as a standard in-process control.

Other facilities available are X-ray fluorescence, mass spectrometry, neutron activation and flame photometry, which are used to supplement data from emission spectrography. Other techniques used are ultra-violet atomic absorption and spectrophotometry which give a relatively quick answer for elements at very low levels. Needless to say, there are still occasions when it is necessary to resort to wet chemical methods, but in general terms one tends to use the instrumental techniques because they give quicker results.

Frequently it is found that none of the methods give a complete guide to the purity level. It is then necessary to use physical methods such as residual resistivity measurements, which are used in the case of ultra pure gallium, indium and other metals after the t.m.i. is $<$ 1 p.p.m. by emission spectrography.

It is absolutely essential that one has a reliable and rapid analytical service to produce High Purity Inorganic Chemicals, since the next purification step is governed by the existing impurity levels. Close liaison with the Analyst is essential so that guidance is given on the impurities likely to be present and the order of magnitude. In the electronic field it is sometimes necessary to fabricate a device to assess purity.

## Apparatus and Techniques

In the early days the tools to use and the techniques available were very limited. One generally used filtration, simple crystallisation, fractional crystallisation through given solvents and distillation. Since the 1939-1945 war there have been very many new techniques introduced; ion exchange is commonplace, various forms of chromotography are used, zone refining in particular has its applications. The environment has to be taken into account: cleaner working conditions are essential in the preparation of very high purity materials. Clean rooms, glove boxes and laminar flow cabinets are essential for this type of work. Probably one of the major advances is the present use of solvent extraction where it is quite possible to prepare materials of greater than 99.9999% purity.

The availability of materials of construction has changed so much over recent years that it is impossible for young Chemists fully to appreciate the difficulties of the earlier days. There were no plastics, so one must consider the difficulties in handling such materials as free hydrofluoric acid and fluorine. Temperature control in the region of + 1000$^{o}$C was very difficult, but these days modern sophisticated equipment is available and glass to glass seals are commonplace. In particular, the change from poor quality enamelled iron to glass lined steel vessels has been one of the greatest advances. Imagine the difficulties of making High Purity materials with the use of corks, rubber bungs and soda glass. Impurities were frequently present in the apparatus one was using. The availability of large scale industrial glassware has been one of the major advances. It is essential that large scale platinum and other specialised apparatus are freely available for preparing high purity materials.

Consideration must always be given to the possible

contamination from the walls of the reaction vessel or apparatus.

The availability of plastic materials has undoubtedly been the major contributor to purer chemicals. Care must be taken to ensure that the plastic does not contain leachable fillers, catalytic residues or metallic oils used in extrusion.

## Scale of Operation

Except for solvent extraction and ion-exchange it is unusual for this work to be carried out on a continuous scale. The scale of operation can vary from gram quantities to hundreds of kilograms.

## Preparation and Purification

The methods available are numerous and experience is generally the guiding factor. Over the years one acquires a knowledge similar to that of having "green fingers" in a garden. If the most abstruse combination of elements is required it is possible to predict the impurities most likely to cause trouble. Selection of raw materials must be made with great care. In the case of sodium chloride it would be very wrong to take the first raw material available since calcium, magnesium, potassium and rubidium are the most difficult elements to separate from sodium chloride. A material low in these elements must be obtained even if other impurities are exceptionally high. Similarly, in purifying hafnium compounds, one would look for a low zirconium content and vice versa. With rare elements one is frequently faced with only a single source of an inferior grade of raw material; hence the difficulties in purification can be considerable. Fortunately in Johnson Matthey Chemicals the platinum group metals, gold and silver are available and one can select the purest intermediate available. When looking for a raw material always question quality at the intermediate stage: it is sometimes purer than the end product. Unlike organic synthesis, it is unfortunate that there is no complete preparative book available that gives all the methods including the purification. In fact there are very few text books which cover the wide range of inorganic synthesis.

Where practical it is important to filter a solution after each purification step.

Scrupulously clean equipment must always be used. Leaching of impurities from a containing vessel is frequently necessary.

Equal care must be taken of the environment. The usual problems are silicon, iron, potassium and calcium.

Always check that the reagents used are of the quality compatible with the process.

Generally it is wiser to purify the individual element or one of its salts before attempting any combination. In general within a given group in the Periodic Table there are frequently one or more foreign elements which cause serious problems. Typical is the presence of magnesium in group la elements.

By examination of the individual groups in the Periodic Table it is possible to make some generalisations.

## Hydrogen, Lithium, Potassium, Sodium, Rubidium, Caesium and Ammonium Salts.

The chlorides are the most common salts available. In the case of sodium and potassium, precipitation with hydrochloric acid, in which the chlorides are insoluble, would give a product probably less than 50 p.p.m. in respect to its total metallic impurities. Higher purity can be obtained by re-precipitation. Potassium, calcium and rubidium are the most difficult elements to remove from other alkali metals. One may have to resort to zone refining, ion exchange, or even the preparation of mixed halides. Fractional crystallisation of mixed alkali halides e.g. $CsIBr_2$, $CsICl_2$ will separate sodium from caesium. To produce the base the most likely way is to form an insoluble pure oxalate and ignite:

$$(COOM)_2 \longrightarrow M_2CO_3 + CO\uparrow$$

(where M = alkali metal). The problem arises from the vessel in which to carry out this ignition since these strong bases attack almost all types of crucible material that are available; nevertheless we can obtain a product with total metallic impurities of less than 5 p.p.m.

Lithium carbonate can be purified by preparing the bicarbonate in a cold aqueous solution using carbon dioxide. The bicarbonate is decomposed by heating giving a pure carbonate, which has a low solubility at boiling point.

$$2Li_2CO_3 + H_2O + CO_2 \xrightarrow{0^oC} 2LiHCO_3 \xrightarrow{HEAT} 2Li_2CO_3 + CO_2\uparrow + H_2O$$

Fractional distillation of the metals, under the correct condition gives good results.

| Metal | Li | Na | K | Rb | Cs |
|---|---|---|---|---|---|
| B Pt $^{o}C$ | 1336 | 883 | 762 | 696 | 670 |

Heavy metals are invariably removed from aqueous solutions by filtration of the insoluble sulphides.

$$H_2S + MX_2 \longrightarrow MS\downarrow + 2HX$$

where M = heavy metal, X = cation.

Ammonium salts are relatively easy to prepare by direct reaction of pure ammonia gas and a suitably pure acid.

$$NH_3 + HX \longrightarrow NH_4X$$

Hydrogen can be purified by absorption on palladium and regeneration

Typical Compounds

Borates, Germanates, Halides, Hydroxides, Niobates, Nitrates, Oxyhalides, Phosphates, Sulphates, Tantalates.

## (Helium) Beryllium, Magnesium, Calcium, Barium and (Radium)

Beryllium must be treated separately since use can be made of its ability to produce basic salts. Crystallisation of the basic acetate $Be_4O(CH_3COO)_6$ and subsequent sublimation can give good results.

The acetate can be decomposed with acid to give other salts e.g. $8\,HCl + Be_4O(CH_3COO)_6 \xrightarrow{HEAT} 4\,BeCl_2 + 6CH_3COOH + H_2O$

## Vanadium, Niobium and Tantalum

The preparation of these high purity metals is difficult because they tend to form very stable occluded phases with non-metals, particularly with oxygen, nitrogen and carbon.

Good use can be made of the different states of oxidation. Vanadium oxytrichloride is prepared by reaction of the pentoxide with thionyl chloride.

$$V_2O_5 \; + \; SOCl_2 \longrightarrow 2VOCl_3 \; + \; 3SO_2$$

Subsequent hydrolysis followed by the formation of a vanadium alum with crystallisation can give a very pure product.

Niobium is invariably found in association with tantalum. Both metals can be easily freed from their contaminants by heating to red heat under vacuum since they both have very high melting points: Nb 2468$^{o}$C , Ta 3030$^{o}$C.

The pentachlorides are easily prepared by direct reaction of the metals with chlorine:

$$Ta \equiv Nb. \quad 2Nb \; + \; 5Cl_2 \longrightarrow 2Nb\,Cl_5$$

Separation can be effected by sublimation. The pentachloride can be hydrolysed to give the pentoxide.

$$2\,Nb\,Cl_5 + 5H_2O \longrightarrow Nb_2O_5 + 10\,HCl$$

$Ta \equiv Nb$

## Typical Compounds

Halides, Oxyhalides, Niobates, Tantalates, Vanadates.

## Chromium, Molybdenum and Tungsten

Chromium purification is carried out by crystallisation of chromium (VI) oxide in nitric acid. The oxide is less soluble in nitric acid than water and it is possible to wash out the impurities. Chromium (III) compounds may be prepared by the reduction of the chromium (VI) oxide with an alcohol in the presence of the respective acid e.g.

$$2\,CrO_3 + 6HCl + 3C_2H_5\,OH \xrightarrow{HEAT} 2CrCl_3 + 3CH_3\,CHO + 6H_2O$$

The so called ammonium molybdate of commerce has the approximate formula of $(NH_4)_6\ Mo_7\ O_{24}\cdot 4H_2O$. If this is dissolved in excess ammonia solution, ammonium (VI) molybdate $(NH_4)_2\ Mo\ O_4$ is formed. By crystallisation of this salt and precipitation of molybdic acid with subsequent washing, purification can be effected.

$$(NH_4)_2\ MoO_4 + 2HNO_3 \longrightarrow H_2MoO_4 \downarrow + 2NH_4\ NO_3$$

Sodium tungstate when treated with nitric acid in aqueous solution gives tungsten (Vl) oxide. This may be washed with water to assist purification.

$$Na_2\ WO_4 + 2HNO_3 \longrightarrow WO_3 \downarrow + H_2O + 2Na\ NO_3$$

### Typical Compounds

Alums, Ammonium Salts, Chromates, Halides, Oxides, Molybdates, Tungstates.

## Manganese, (Technetium), and Rhenium

Relatively pure manganese metal may be obtained by electrolysis of manganese sulphate in the presence of ammonium sulphate. The cathode is stainless steel and the anode is a lead sheet.

Palladium (II) chloride may be purified by removing Pt and Ir using this method. The filtrate is boiled with excess ammonia solution and acidified with hydrochloric acid to the insoluble dichlorodiammine palladium (II)

$$PdCl_2 + 2NH_3 \xrightarrow{HCl} Pd(NH_3)_2Cl_2 \downarrow$$

This may be washed free of impurities.

Osmium may be burnt in oxygen to give osmium (VIII) oxide which is purified by distillation.

$$Os + 2O_2 \longrightarrow OsO_4 \uparrow$$

Ruthenium (VIII) oxide is equally volatile and may be purified by distillation. It is prepared by passing chlorine through a solution of an alkaline ruthenate

$$K_2RuO_4 + Cl_2 \longrightarrow RuO_4 \uparrow + 2KCl$$

Rhodium (III) chloride is prepared by passing chlorine over rhodium powder heated to $400^{\circ}C$

$$2Rh + 3Cl_2 \longrightarrow 2RhCl_3$$

The chloride is heated with sodium chloride in the presence of chlorine to give sodium hexachlororhodate (III)

$$3NaCl + RhCl_3 \longrightarrow Na_3RhCl_6$$

This may be purified by recrystallisation.

Typical Compounds

Halides, Nitrates, Nitrites, Oxides, Oxyhalides, Sulphates, Sulphites.

Copper, Silver and Gold

All these metals may be purified by electrolysis. To avoid cross contamination, chemical processing is also necessary. Electrolytic copper may be dissolved in nitric acid to give the nitrate

$$Cu + 4HNO_3 \longrightarrow Cu(NO_3)_2 + 2NO_2\uparrow + 2H_2O$$

Recrystallisation assists purification.

Decomposition with sulphuric acid to the sulphate

$$Cu(NO_3)_2 + H_2SO_4 \xrightarrow{HEAT} CuSO_4 + 2HNO_3$$

followed by subsequent crystallisation gives a pure electrolyte.

Silver chloride is precipitated from silver nitrate with hydrochloric acid

$$AgNO_3 + HCl \longrightarrow AgCl\downarrow + HNO_3$$

The chloride is washed with water to remove impurities, subsequently dissolved in ammonia solution, filtering and taking care not to allow the solution to dry since this is an explosive. Addition of nitric acid will re-precipitate silver chloride which may be given further washing. The chloride is heated with dextrose and sodium hydroxide at $60^{o}C$ to give silver metal which may be washed free of sodium salts.

Chlorauric acid solution is prepared by dissolving the metal in aqua regia. This may be filtered to remove any insoluble silver chloride. Subsequent precipitation with sulphur dioxide will precipitate the gold which may be washed to purify.

$$Ga + 6\,HNO_3 \longrightarrow Ga(NO_3)_3 + 3\,NO_2\uparrow + 3H_2O$$

and may be recrystallised for further purification. The oxide is formed by heating the nitrate

$$4\,Ga(NO_3)_3 \xrightarrow{HEAT} 2\,Ga_2O_3 + 12NO_2\uparrow + 3O_2\uparrow$$

Gallium (III) halides can all be prepared by direct elemental combination and may be purified by distillation.

Indium metal may be purified by zone refining. Both the indium (I) (II) and (III) halides are formed by the direct combustion of the elements. Indium (III) halides tend to be more volatile than the indium (I) and (II) halides. This can assist in purification. Electrolyses of indium sulphate solutions is an aid to purification.

Thallium is usually contaminated with Pd, Ni, Cd, Zn and As. Purification can be carried out by reacting the metal with nitric acid to give thallium (I) nitrate

$$Tl + HNO_3 \longrightarrow Tl\,NO_3 + \tfrac{1}{2}\,H_2\uparrow$$

The enormous difference in solubility in water, 4g 100 $ml^{-1}$ at $0^oC$ and 593.9g $100ml^{-1}$ at $100^oC$, allows scope for purification.

This can be converted to the sparingly soluble thallium (I) sulphate by the addition of sulphuric acid

$$2Tl\,NO_3 + H_2SO_4 \longrightarrow Tl_2SO_4\downarrow + 2HNO_3$$

Electrolysis of the sulphate gives a very high purity metal.

Typical Compounds

Alums, Borates, Garnets, Halides, Nitrates, Oxides, Oxyhalides, Sulphates.

## Carbon, Silicon, Germanium, Tin and Lead

Carbon can be obtained from pure carbohydrates or hydrocarbons by burning in a deficiency of air.

Silicon (IV), germanium (IV) and tin (IV) chlorides are readily obtainable. Frequently they are cross contaminated. In addition the major contaminants are arsenic and iron. Fractional distillation over copper aids purification. Germanium (IV) oxide is easily formed by hydrolysis of the chloride.

$$GeCl_4 + 2H_2O \longrightarrow GeO_2\downarrow + 4HCl$$

The insoluble oxide can be washed to aid purification.

Silicon (IV) chloride is hydrolysed with water to form the bulky silicic acid which requires heating to form the silicon (IV) oxide.

$$SiCl_4 + 2\,H_2O \longrightarrow SiO_2\downarrow + 4HCl$$

This may equally be washed to aid purification.

Tin (IV) chloride does not hydrolyse easily with water. If it is added to an aqueous solution of ammonium chloride then ammonium hexachlorostannate is formed which can be purified by recrystallisation,

$$SnCl_4 + 2NH_4Cl \longrightarrow (NH_4)_2SnCl_6$$

which on heating in air will give pure tin (IV) oxide

$$(NH_4)_2SnCl_6 \xrightarrow{HEAT} SnO_2 + 2NH_3\downarrow + 2HCl + 2Cl_2\uparrow$$

Sulphur can be purified by sublimation. Alternatively, recrystallisation through carbon disulphide gives good results. The sulphur can be burnt in pure oxygen to give sulphur (IV) oxide

$$S + O_2 \longrightarrow SO_2\uparrow$$

Burning in excess oxygen over platinum will give sulphur (VI) oxide

$$2\ SO_2 + O_2 \xrightarrow{Pt} 2SO_3\uparrow$$

Elemental selenium may be burnt in oxygen which is passed through fuming nitric acid to give selenium (IV) oxide

$$Se + O_2 \longrightarrow Se\ O_2\uparrow$$

This may be resublimed for purification. Wet purification may be carried out by reduction of oxide to give selenium which is washed until neutral.

$$Se\ O_2 + 2H_2O + 2SO_2 \longrightarrow Se\downarrow + 2H_2SO_4$$

Tellurium powder is dissolved in nitric acid to form the basic nitrate

$$2\ Te + 9HNO_3 \longrightarrow Te_2O_3\ (OH)\ NO_3 + 8NO_2\uparrow + 4H_2O$$

The basic nitrate is purified by recrystallisation and ignited to give tellurium (IV) oxide. This may be dissolved in hydrochloric acid and reduced with hydrazine to give pure tellurium

$$TeO_2 + N_2H_4 \longrightarrow Te + N_2\uparrow + 2\ H_2O$$

<u>Typical Compounds</u>

Halides, Oxycompounds, Selenates, Selenides, Selenites, Tellurates, Tellurides, Tellurites.

## Fluorine, Chlorine, Bromine, Iodine and (Astatine)

Free fluorine is rarely required. Generally the source is high purity hydrofluoric acid 40% which may be purified by treating with lead carbonate which forms Pb ClF and removes the chloride. Treatment is also made with barium hydroxide to remove sulphate as insoluble Ba $SO_4$. Distillation is carried out in a platinum or pure silver apparatus.

Commercial chlorine may be purified by scrubbing through concentrated sulphuric acid. It may be condensed in a dry ice/acetone bath. The liquefied chlorine is repeatedly vapourised and condensed while non-condensable gases such as oxygen are removed by pump. Final purification is by high vacuum fractionation passing into receivers cooled by liquid nitrogen.

Bromine may be purified by fractional distillation over potassium bromide. This is to remove the other halogens. The bromine may be dried by distillation over phosphorus (V) oxide.

Iodine can be resublimed at room temperature in a glass casserole over a long period. It may also be purified by steam distillation over potassium iodide.

## Typical Compounds

Halides, Oxyhalides, Interhalogen Compounds.

Puratronic[R] products

Acoustic-optic devices

Acoustic surface wave filters

Bubble domain devices

Crystal growth

- fluxed melt and melt growth
- aqueous solution growth
- liquid encapsulation
- doping
- vapour/liquid phase epitaxy
- molecular beam epitaxy

Electroluminescent/photoconductive devices

Infra-red transmitting glasses

Lasers

Linear and non-linear optical materials

Microwave ferrites and garnets

Optical crystals

Phosphors

Piezoelectric and pyroelectric devices

Semiconductors

- compound preparation
- diffusion and ion-implantation doping

Sputtering

Vacuum and electron beam evaporation

Potential Users

Specpure(R) Products

| Industry | Examples of Specpure(R) Material used |
| --- | --- |
| Iron & Steel production plants | $Fe_2O_3$, $Cr_2O_3$ |
| Non-ferrous metal production plants | $Al_2O_3$, Cu |
| Chemical manufacturers - especially inorganic | All |
| Cement works | $CaCO_3$, $Al_2O_3$ |
| Large engineering companies - motor, rail transport, aircraft, civil engineering, shipbuilding | $Fe_2O_3$, $Al_2O_3$<br>For analysis of incoming raw materials. |

Government Research Establishments

Water and air pollution

Road building

Atomic Energy - probably the largest user industry for analysis

Defence/Army Scientific Research

Forensic

Agricultural

Explosives